Dineke Schokkin
A Grammar of Paluai

Pacific Linguistics

Managing editor
Alexander Adelaar

Editorial board members
Wayan Arka
Danielle Barth
Don Daniels
Nicholas Evans
Gwendolyn Hyslop
David Nash
Bruno Olsson
Bill Palmer
Andrew Pawley
Malcolm Ross
Dineke Schokkin
Jane Simpson

Volume 663

Dineke Schokkin
A Grammar of Paluai

The Language of Baluan Island, Papua New Guinea

DE GRUYTER
MOUTON

ISBN 978-3-11-077777-2
e-ISBN (PDF) 978-3-11-067517-7
e-ISBN (EPUB) 978-3-11-067522-1
ISSN 1448-8310

Library of Congress Control Number: 2019952200

Bibliographic information published by the Deutsche Nationalbibliothek
The Deutsche Nationalbibliothek lists this publication in the Deutsche Nationalbibliografie; detailed bibliographic data are available on the Internet at http://dnb.dnb.de.

© 2021 Walter de Gruyter GmbH, Berlin/Boston
This volume is text- and page-identical with the hardback published in 2020.
Photo credit: Dineke Schokkin
Typesetting: Integra Software Services Pvt. Ltd.
Printing and binding: CPI books GmbH, Leck

www.degruyter.com

To my Alup

Acknowledgements

This grammar is based on my PhD thesis, which was accepted at James Cook University, Cairns, Australia, in 2014. As such it has been a long time in the making, and consequently there are many people I am indebted to. I hardly know where to start with expressing my gratitude to the Baluan community. I cannot list all the people that have contributed to this work in one way or another, but I wish to mention the following by name. For providing texts for recordings, thanks are due to Alup Kaluwin, Bruno Komet, Carolyn Ambou, Cathy Posopat, Kalou Solok, Keket Maluan, Kireng Wari, Lalau Kanau, Lauan Memelam, Lorat Molean, Lynette Touain, Maiau Keket, Martin Salau, Ngat Poraken, Ngi Sokou, Ninou Kireng, Ninou Paromui, Ninou Pokat, Posopat Membup, Sakumai Yêp, Sapulai Papi and Wendy Lauan. Invaluable assistance with transcription, translation and elicitation was provided by Carolyn Ambou, Keket Maluan, Lorat Molean, Lynette Touain, Pulou Wari, Pwanou Silikara and Wendy Lauan. Lorat Molean has been of vital assistance with the collection of fish and plant names. Thanks are also due to Aêwai Salkimut and the late Soanin Kilangit, for introducing me to several people, and sharing stories. A special thanks goes to Cholai of Mouk village, for providing an opportunity to charge my equipment with his generator, and to Kilangit Bayang and his family, for hosting me and looking after me during my stays in Port Moresby. Last but not least, I want to thank Lorat Molean and Ninou Kireng for my Baluan names, Alup Sauka and Alup Komdal, which I will always carry with pride, and my Baluan family, Keket, Maiau, Maluan, Ngat and little Jema, for sharing their home, their food and their lives with me. *Wuro paran menengan.*

I thank all the people who made the logistical and practical side of fieldwork run smoothly. Thanks to Baluan community members Keket Maluan, Sapulai Papi and Lorat Molean, who provided the Manus Provincial Government with a letter of invitation, the research visa application proceeded swiftly. I thank the Manus Provincial Government for granting permission to carry out my fieldwork, and Jim Robbins and Georgia Kaipu of the National Research Institute of Papua New Guinea for assisting me with the visa application process. René van den Berg was very helpful in arranging my flights to and from Ukarumpa and other things related to my visit to SIL PNG in 2012.

I also thank my supervisors Sasha Aikhenvald, Bob Dixon and Ton Otto for their continuing support for my project both during and after my PhD candidature. My fieldwork was funded through Dixon and Aikhenvald's competitive grants, the Firebird Foundation for Anthropological Research, James Cook University (based on the JCU Minimum Standard of Resources policy) and a JCU Graduate Research Scheme grant. One field trip and the completion of the manuscript took place while I was employed as a Research Fellow at the Australian National University within the ARC Laureate Project 'The Wellsprings of Linguistic Diversity' (Grant No. FL130100111).

I am grateful to have met so many fellow academics during both my JCU and ANU time who have been very generous, inspiring and supportive in multitudes of ways, and I would particularly like to thank Grant Aiton, Angeliki Alvanoudi, Wayan Arka, Danielle and Wolfgang Barth, Juliane Böttger, Chiara Bresciani, Jessica Cleary-Kemp, Signe Dalsgaard, Steffen Dalsgaard, Don Daniels, Marie-France Duhamel, Mark Ellison, Beth Evans, Nick Evans, Christiane Falck, Murray Garde, Sonja Gipper, Tina Gregor, Valérie Guérin, Darja Hoenigman, Eri Kashima, Kate Lindsey, Alex Marley, Miriam Meyerhoff, Elena Mihas, Yankee Modi, Kate Naitoro, David Nash, Simon Overall, Bill Palmer, Chia-Jung Pan, Andy Pawley, Mark Post, Anders Rasmussen, Sonja Riesberg, Malcolm Ross, Mikko Salminen, Hannah Sarvasy, Anne Schwarz, Jeff Siegel, Jane Simpson, Ruth Singer, Hedvig Skirgård, Borut Telban, Charlotte van Tongeren, Daniela Vavrova, Lourens de Vries, Kasia Wojtylak, and Sihong Zhang. Further thanks are due to the two examiners of my thesis, Bernard Comrie and the late Frank Lichtenberk, and to an anonymous reviewer for Pacific Linguistics, for their valuable feedback which greatly improved the present work. Thanks to Elena Rhind and Amanda Parsonage (JCU) and Bianca Hennessy (ANU) for their great administrative support. Thanks to Bert Peeters for helping out with copyediting and formatting of the final manuscript.

Finally I would like to thank my friends and family, both near and far. Their continuous support is what keeps me going.

Contents

Acknowledgements —— VII

List of Figures —— XIX

List of Tables —— XXI

Conventions and abbreviations —— XXIII

1 The language and its context —— 1
  1.1 Introduction —— 1
  1.2 Sociolinguistic situation —— 2
  1.3 Sociocultural background —— 3
  1.4 Genetic affiliation —— 4
  1.5 Existing descriptions of neighbouring languages —— 5
  1.6 Data collection and methodology —— 5

2 Phonology —— 8
  2.1 Syllable structure —— 8
  2.2 Segmental phonology —— 9
    2.2.1 Overview of consonant phonemes —— 9
    2.2.2 Distribution and realisations of consonant phonemes —— 10
      2.2.2.1 Plosives —— 11
      2.2.2.2 Nasals —— 17
      2.2.2.3 Fricatives —— 20
      2.2.2.4 Liquids —— 21
    2.2.3 Vowel phonemes —— 21
      2.2.3.1 The vowel /i/ —— 22
      2.2.3.2 The vowel /e/ —— 22
      2.2.3.3 The vowel /ɛ/ —— 22
      2.2.3.4 The vowel /a/ —— 23
      2.2.3.5 The vowel /ɔ/ —— 23
      2.2.3.6 The vowel /o/ —— 23
      2.2.3.7 The vowel /u/ —— 23
      2.2.3.8 The approximants [j] and [w] —— 24
      2.2.3.9 V-V sequences and phonetic diphthongisation —— 26
      2.2.3.10 Merger of close-mid vowels —— 28

- 2.2.4 Minimal pairs — 30
  - 2.2.4.1 Minimal pairs for consonant phonemes — 30
  - 2.2.4.2 Minimal pairs for vowel phonemes — 31
- 2.2.5 Minor phonological processes — 31
  - 2.2.5.1 Vowel assimilation — 31
  - 2.2.5.2 Syllable reduction — 32
  - 2.2.5.3 Metathesis — 32
- 2.2.6 Phonological domains and sandhi rules — 32
  - 2.2.6.1 Phonological word — 33
  - 2.2.6.2 Phonological phrase — 34
- 2.2.7 Morphophonology — 34
  - 2.2.7.1 Assimilation and deletion of /n/ — 35
  - 2.2.7.2 Vowel alternation in directly possessed nouns — 36
- 2.2.8 Phonology of loans — 37
- 2.3 Prosody — 40
  - 2.3.1 Word stress — 40
    - 2.3.1.1 Contrastive stress — 41
    - 2.3.1.2 Stress assignment in disyllabic forms — 41
    - 2.3.1.3 Stress assignment in trisyllabic forms — 42
    - 2.3.1.4 Stress assignment in quadrisyllabic forms — 43
    - 2.3.1.5 Interaction of morphology and stress assignment — 43
    - 2.3.1.6 (Un)predictability of word stress — 47
  - 2.3.2 Intonation — 48
    - 2.3.2.1 Distinguishing speech acts — 49
    - 2.3.2.2 Signalling the boundaries of prosodic units — 50
    - 2.3.2.3 Signalling emphasis and contrast — 52
- 2.4 Orthography — 54

3 Word classes I: open classes — 57
- 3.1 Preliminaries — 57
- 3.2 The noun — 58
  - 3.2.1 Noun subclasses, type I: personal, local and common nouns — 59
  - 3.2.2 Noun subclasses, type II: direct and indirect possession — 60
    - 3.2.2.1 Kinship relations — 62

		3.2.2.2	Part-whole relations —— 64
		3.2.2.3	Spatial relations —— 68
		3.2.2.4	Attributes —— 71
		3.2.2.5	Fluidity between possessive constructions —— 75
	3.2.3	\multicolumn{2}{l}{Noun subclasses, type III: numeral and possessive classifiers —— 76}

- 3.2.3 Noun subclasses, type III: numeral and possessive classifiers —— 76
    - 3.2.3.1 Numeral classifiers —— 76
    - 3.2.3.2 Possessive classifier *ka-* —— 79
- 3.2.4 Noun morphology —— 80
    - 3.2.4.1 Compounding —— 80
    - 3.2.4.2 Reduplication —— 81
    - 3.2.4.3 Suffixation —— 83
    - 3.2.4.4 Derivation of other parts of speech from nouns —— 83
- 3.3 The verb —— 84
    - 3.3.1 Subclassification of verbs —— 85
        - 3.3.1.1 Transitive and active intransitive verbs —— 85
        - 3.3.1.2 Stative verbs —— 86
        - 3.3.1.3 Directionals —— 88
        - 3.3.1.4 TAM verbs or particles —— 91
        - 3.3.1.5 Existential verbs —— 91
        - 3.3.1.6 Light verb constructions —— 92
    - 3.3.2 Verb morphology —— 93
        - 3.3.2.1 Deriving another verb —— 94
        - 3.3.2.2 Word class-changing derivations —— 96
    - 3.3.3 Functions of nominalisations —— 101
- 3.4 Adjectives: overview and semantic classes —— 102
    - 3.4.1 Formal characteristics of adjectives —— 103
    - 3.4.2 Adjectives within the NP —— 103
    - 3.4.3 Predicative adjectives —— 104
    - 3.4.4 Adjectival morphology —— 104
        - 3.4.4.1 Reduplication —— 104
        - 3.4.4.2 Suffixation —— 105
- 3.5 Forms that appear in more than one word class —— 105
    - 3.5.1 Forms that appear as nouns and verbs —— 105
    - 3.5.2 Forms that appear as nouns and adjectives —— 106
- 3.6 Adverbs —— 107
    - 3.6.1 Types of adverbs —— 107
        - 3.6.1.1 Manner adverbs —— 107

		3.6.1.2	Adverbs of degree —— 109
		3.6.1.3	Temporal adverbs —— 111
		3.6.1.4	Spatial adverbs —— 112
		3.6.1.5	Modal/epistemic adverbs —— 113

    3.6.2 Derivational processes related to adverbs —— 115
        3.6.2.1 Reduplication —— 115
        3.6.2.2 Periphrastic derivation of adverbials with *la* —— 115

3.7 Overview —— 116

4 Word classes II: closed classes —— 118

4.1 Preliminaries —— 118
4.2 Pronouns —— 118
    4.2.1 The free pronoun paradigm —— 120
    4.2.2 The bound pronoun paradigm —— 121
        4.2.2.1 Subject forms —— 121
        4.2.2.2 Object forms —— 121
    4.2.3 Possessive forms —— 122
        4.2.3.1 The possessive/locative form *ta-* —— 122
        4.2.3.2 The possessive classifier *ka-* —— 123
4.3 Demonstratives —— 124
    4.3.1 Basic demonstrative forms —— 124
    4.3.2 Demonstratives with the formative *te-* —— 125
    4.3.3 Spatial demonstratives with *a=* —— 127
    4.3.4 Free demonstrative forms formed with *ta=* —— 128
    4.3.5 Overview of demonstrative forms —— 128
4.4 Adpositions —— 129
    4.4.1 The preposition *a=* —— 129
    4.4.2 The preposition *pari* —— 131
    4.4.3 The "preposition" *ta-* —— 132
4.5 Numerals —— 133
    4.5.1 The numerals one to ten —— 133
    4.5.2 Higher numerals —— 135
    4.5.3 Morphology and syntax of numerals —— 136
        4.5.3.1 Numerals as noun modifiers —— 136
        4.5.3.2 Numerals heading a predicate —— 138
    4.5.4 Counting money —— 138
4.6 Quantifiers —— 139
    4.6.1 Quantifiers referring to large quantities —— 140

            4.6.2   Quantifiers referring to small or indefinite
                    quantities —— 141
                    4.6.2.1   The forms *an-* and *nan=* with quantifiers —— 142
                    4.6.2.2   The form *nan=* with numerals —— 145
    4.7   Interrogative words —— 146
            4.7.1   Questioning identity: *sa, sap* and *sê* —— 147
            4.7.2   Questioning time and place —— 148
            4.7.3   Questioning relations with *samai-* —— 148
            4.7.4   Questioning purpose or reason —— 149
            4.7.5   Questioning manner —— 149
    4.8   Negation and mood markers —— 150
    4.9   Conjunctions and clause connectors —— 151
            4.9.1   Coordination —— 151
            4.9.2   Subordination —— 154
                    4.9.2.1   Temporal subordinate clause markers —— 154
                    4.9.2.2   Other types of subordinate clauses —— 155
    4.10 Interjections and formulaic words and phrases —— 156
    4.11 The formative *ta=* —— 157
            4.11.1   Use of *ta=* with nouns —— 158
            4.11.2   Use of *ta=* with adjectives —— 160
            4.11.3   Use of *ta=* with demonstratives —— 160
            4.11.4   Use of *ta=* with prepositions —— 161
            4.11.5   Use of *ta=* with numerals —— 161
            4.11.6   Use of *ta=* with quantifiers —— 162
            4.11.7   Use of *ta=*: summary —— 162

5   **The noun phrase —— 164**
    5.1   Structural features of the NP —— 164
    5.2   Determiner —— 165
            5.2.1   Personal pronoun as determiner —— 165
            5.2.2   Numeral 'one' as determiner —— 167
    5.3   Prenominal modifier —— 168
    5.4   Post-nominal modifier —— 168
    5.5   Possessor —— 169
            5.5.1   Form of direct possession —— 169
            5.5.2   Form of indirect possession —— 170
    5.6   Prepositional phrase —— 171
    5.7   Relative clause —— 172
    5.8   Demonstrative —— 172
    5.9   Coordination of NPs —— 173

6 Predicates I: verbal predicates —— 175
- 6.1 Indexing of S/A and O —— 176
- 6.2 Aspect —— 180
  - 6.2.1 Preverbal aspectual particles —— 181
    - 6.2.1.1 Core aspect —— 181
    - 6.2.1.2 Secondary aspect —— 189
  - 6.2.2 Postverbal aspectual particles —— 193
    - 6.2.2.1 Completive *nêm* —— 194
    - 6.2.2.2 Durative *wot* —— 195
- 6.3 Direction and associated motion —— 196
  - 6.3.1 *la* 'go to' —— 197
  - 6.3.2 *me* 'come' —— 198
  - 6.3.3 *wot* 'go horizontally, away from DC' —— 198
  - 6.3.4 *sot* 'go upwards' —— 199
  - 6.3.5 *sa* 'come upwards' —— 199
  - 6.3.6 *suwot* 'go downwards' —— 200
  - 6.3.7 *si* 'come downwards' —— 200
  - 6.3.8 *wen* 'move horizontally' —— 201
  - 6.3.9 *suwen* 'move downwards' —— 201
- 6.4 Reality status —— 201
  - 6.4.1 Forms of the irrealis marker —— 202
  - 6.4.2 Functions of the irrealis —— 203
    - 6.4.2.1 Unrealised event/state of affairs —— 203
    - 6.4.2.2 Immediate future —— 204
    - 6.4.2.3 Past habitual —— 204
    - 6.4.2.4 Marker of a dependent clause —— 206
  - 6.4.3 Dependencies between reality status and other verbal categories —— 206
- 6.5 Modality —— 209
  - 6.5.1 Desiderative/intentional *pwa* —— 210
  - 6.5.2 The particle *sa* —— 212
    - 6.5.2.1 Apprehensive *sa* —— 212
    - 6.5.2.2 Negated *sa* as deontic modal operator —— 215
    - 6.5.2.3 Negated *sa* as a marker of negative polarity in the future —— 216
  - 6.5.3 Irrealis and aspectual particles with modal overtones —— 217
    - 6.5.3.1 Irrealis and perfective *pe* —— 217
    - 6.5.3.2 Irrealis and imperfective *no* —— 218

6.5.3.3 Irrealis and continuative/habitual *to* —— 219
6.5.3.4 Irrealis and the directional *la* —— 220
6.6 Structural properties of the verb complex —— 221

7 Predicates II: non-verbal and copula predicates —— 225
7.1 Predicates with a noun as head —— 225
7.2 Predicates with an adjective as head —— 227
7.3 Predicates headed by a numeral —— 228
7.4 Predicates containing *ta-* or *ka-* —— 228
7.5 Predicates containing *pari* —— 230
7.6 Predicates headed by an interrogative word —— 231
7.7 Potential copula clauses —— 232
    7.7.1 Clauses with *to* 'be' —— 232
    7.7.2 Clauses with *la* 'go' —— 234
7.8 Comparative constructions —— 235

8 Grammatical relations and valency —— 238
8.1 Expression of core and peripheral arguments —— 238
    8.1.1 Core arguments S, A and O —— 238
    8.1.2 The E argument —— 239
        8.1.2.1 Extended intransitive verbs —— 240
        8.1.2.2 Extended transitive or ditransitive verbs —— 243
    8.1.3 Peripheral arguments —— 246
8.2 Valency —— 248
8.3 Valency-changing derivations —— 251
    8.3.1 Causative *pe-* —— 251
    8.3.2 Applicative *-(C)ek* —— 255
        8.3.2.1 The productive applicative —— 255
        8.3.2.2 Frozen applicatives with three core arguments —— 259
        8.3.2.3 Other frozen applicatives —— 259
        8.3.2.4 Applicatives: overview and conclusions —— 264
    8.3.3 Valency-reducing operations —— 265
        8.3.3.1 Reduplication —— 265
        8.3.3.2 Fossilised prefixes *ma-* and *ta-* —— 266
    8.3.4 Reflexive and reciprocal constructions —— 268
        8.3.4.1 Reflexives —— 268
        8.3.4.2 Reciprocals —— 269

9 Serial verb constructions — 272
9.1 Asymmetrical SVCs — 273
9.1.1 Cause-effect/resultative SVCs — 274
9.1.1.1 Cause-effect SVCs with intransitive V1 — 276
9.1.2 Adverbial SVCs — 276
9.1.3 Valency-increasing SVCs with *tou* or *lêp* — 278
9.1.4 SVCs with a posture verb — 279
9.1.5 SVCs with a directional — 281
9.1.5.1 Directionals with intransitive verbs — 281
9.1.5.2 Directionals with transitive verbs — 284
9.2 Symmetrical SVCs — 289
9.2.1 Sequences with *liliu* 'return; again' — 290
9.2.2 Lexicalisation of symmetrical SVCs — 291
9.3 SVCs: conclusions — 294
9.3.1 Formal properties of asymmetrical SVCs — 294
9.3.2 Combinations involving more than two verbs — 297
9.3.3 Grammaticalisation of SVC components — 298
9.3.4 SVCs as grammatical and phonological words — 300

10 Speech act distinctions and polarity — 302
10.1 Mood — 302
10.1.1 Interrogative mood — 302
10.1.1.1 Content questions — 302
10.1.1.2 Polar questions — 305
10.1.1.3 Indirect questions — 307
10.1.1.4 Tag questions — 307
10.1.2 Imperative mood — 308
10.1.2.1 Commands and requests — 309
10.1.2.2 Negative imperatives — 311
10.1.3 Dependencies between mood and verbal categories — 312
10.2 Polarity — 313
10.2.1 Answer to a polar question — 314
10.2.2 Predicative negation — 314
10.2.2.1 Negation of realis and verbless predicates — 314
10.2.2.2 Negation of irrealis predicates — 317
10.2.2.3 Negation of existential predicates — 317
10.2.2.4 Dependencies between negation and verbal categories — 319

		10.2.3	Non-predicative negation —— 320

- 10.2.3 Non-predicative negation —— 320
    - 10.2.3.1 The negative adverbial phrase *la pwên* —— 320
    - 10.2.3.2 Negated noun phrases —— 321
- 10.2.4 Intensification and moderation of negated clauses —— 321
- 10.2.5 Ellipsis of *ma=* —— 322

11 Clausal relations and clause combining —— 324
11.1 Dependent clauses —— 324
- 11.1.1 Relative clauses —— 325
    - 11.1.1.1 Possible syntactic functions in the RC of the common argument —— 326
    - 11.1.1.2 Some further types of relative clauses —— 335
    - 11.1.1.3 Relative clauses: conclusions —— 336
- 11.1.2 Complement clauses —— 337
    - 11.1.2.1 Verbs of Attention —— 338
    - 11.1.2.2 Verbs of Thinking/Speaking —— 340
    - 11.1.2.3 Verbs of Quotation —— 340
    - 11.1.2.4 Verbs of Liking —— 341
    - 11.1.2.5 Modal verbs —— 342
    - 11.1.2.6 Complement clause modifying adjective —— 344
    - 11.1.2.7 Complement clauses: conclusions —— 345
- 11.1.3 Adverbial subordinate clause —— 345
    - 11.1.3.1 Temporal subordinate clause —— 347
    - 11.1.3.2 Manner subordinate clause —— 349
    - 11.1.3.3 (Possible) consequence clause —— 350
    - 11.1.3.4 Concessive clause —— 353
    - 11.1.3.5 Conditional subordinate clause —— 354

11.2 Combining main clauses —— 358
- 11.2.1 Addition —— 358
    - 11.2.1.1 Temporal succession —— 358
    - 11.2.1.2 Same-event and new-event addition —— 359
    - 11.2.1.3 Elaboration —— 360
    - 11.2.1.4 Contrast —— 361
- 11.2.2 Alternatives —— 361

12 Pragmatics and discourse practices —— 362
12.1 Preliminaries —— 362
12.2 Terminology —— 363
- 12.2.1 Presupposition and assertion —— 363
- 12.2.2 Identifiability —— 364

        12.2.3   Activation — 364
        12.2.4   Topic — 365
        12.2.5   Focus — 365
    12.3 Information structure on the sentence level — 367
        12.3.1   Identifiability: definiteness and specificity — 367
            12.3.1.1   Definiteness — 367
            12.3.1.2   Specificity — 369
        12.3.2   Anaphors and cataphors — 370
            12.3.2.1   Pronouns as anaphors and cataphors — 370
            12.3.2.2   Demonstratives as anaphors and cataphors — 372
        12.3.3   Topicalisation — 376
            12.3.3.1   Topicalisation of S/A — 376
            12.3.3.2   Topicalisation of O — 377
            12.3.3.3   Topicalisation of an Oblique — 377
            12.3.3.4   Topicalisation of a Possessor — 378
            12.3.3.5   Discourse function(s) of topicalisation — 379
        12.3.4   Focus in Paluai — 381
            12.3.4.1   Focus and sentence accent — 381
            12.3.4.2   The particle *ya* — 384
            12.3.4.3   The particle *ma* — 386
    12.4 Discourse organisation at the paragraph level and beyond — 386
        12.4.1   The structure of narratives — 387
        12.4.2   Two examples — 387
            12.4.2.1   Example 1: The story of Parulabei and Komou — 387
            12.4.2.2   Example 2: Planting yams — 390

Appendix I Recordings metadata — 393

Appendix II Texts — 395

References — 423

Subject Index — 429

Author Index — 433

List of Figures

Figure 1	Language map of Manus Province —— 2	
Figure 2	Acoustic difference between [p] and [β] in the form /sɔpɔl/ —— 12	
Figure 3	Lenition of /k/ to [x] —— 15	
Figure 4	Comparison of /pʷ/ and /pu/ —— 16	
Figure 5	Comparison of /mʷ/ and /mu/ —— 19	
Figure 6	Paluai vowel inventory —— 21	
Figure 7	F1 and F2 values for the vowels of one female speaker —— 29	
Figure 8	Intensity contour showing primary stress on clitic /tɛ/ —— 46	
Figure 9	Pitch and intensity contours for the wordlist recording of /kalsɔn/ 'ankle decoration' —— 48	
Figure 10	Pitch contour of a content question —— 49	
Figure 11	Pitch contour of a polar question —— 50	
Figure 12	Pitch contour of a tag question —— 50	
Figure 13	Pitch contour of example (57) —— 51	
Figure 14	Pitch contour of example (58) —— 53	
Figure 15	Pitch contour of example (59) —— 53	

List of Tables

Table 1	The consonant phonemes of Paluai	10
Table 2	Phonological domains in Paluai	11
Table 3	Possible vowel sequences in Paluai	27
Table 4	Phonological domains in Paluai	33
Table 5	Vowel alternations in directly possessed nouns	37
Table 6	Orthographic symbols for consonant phonemes	54
Table 7	Orthographic symbols for vowel phonemes	54
Table 8	Overview of possible trisyllabic forms in Paluai	54
Table 9	Overview of possible quadrisyllabic forms in Paluai	55
Table 10	Terms for consanguineal kin	62
Table 11	Terms for affinal kin	63
Table 12	Tree and plant parts	67
Table 13	Spatial nouns	69
Table 14	Human characteristics	72
Table 15	A selection of nouns that can be both directly and indirectly possessed	74
Table 16	Semantic distinctions involving the numeral 'one'	77
Table 17	"Generic" nouns used for fractions and collections	78
Table 18	Directional paradigm	88
Table 19	Organisation of the directional paradigm along dimensions of absolute FoR and deixis	90
Table 20	TAM verbs and particles	91
Table 21	Examples of light verb constructions with *pe*	92
Table 22	Some lexicalised verb sequences	94
Table 23	Some nominalising reduplications	96
Table 24	Inherently reduplicated verbs	98
Table 25	Nominalisations formed by adding an -*a* or -*o* suffix	99
Table 26	Semantic classes of adjectives, with some examples	102
Table 27	Some manner adverbs	108
Table 28	Adverbs of degree	110
Table 29	Temporal adverbs	111
Table 30	Spatial adverbs	113
Table 31	Modal/epistemic adverbs	114
Table 32	Distinguishing criteria for word classes N, V, A and Adv	116
Table 33	Comparison of word classes with respect to distinguishing criteria	117
Table 34	Word class-changing derivational morphology	117
Table 35	The free pronoun paradigm	120
Table 36	The bound pronoun paradigm (subject forms)	121
Table 37	The bound pronoun paradigm (object forms)	121
Table 38	The possessive paradigm (with *ta*-)	122
Table 39	The possessive paradigm (with *ka*-)	123
Table 40	The demonstrative paradigm	129
Table 41	Prepositions and related forms in Paluai	129
Table 42	Numerals one to ten for animates	134

https://doi.org/10.1515/9783110675177-205

Table 43	Numerals one to ten for inanimates	134
Table 44	Multiplications of ten	135
Table 45	Multiplications of hundred	136
Table 46	Quantifiers referring to large quantities	139
Table 47	Quantifiers referring to small quantities	139
Table 48	Functions and distribution of *an-*	143
Table 49	Functions and distribution of *nan=*	144
Table 50	Interrogative forms	146
Table 51	Markers for the coordination of two main clauses	151
Table 52	Temporal subordinating conjunctions	154
Table 53	Other subordinating conjunctions	155
Table 54	Formulaic words and phrases	156
Table 55	Expressive/conative interjections	157
Table 56	Phatic/conative interjections	157
Table 57	Overview of uses of *ta=*	163
Table 58	Core aspect particles	181
Table 59	Secondary aspect particles	189
Table 60	Postverbal aspectual particles	194
Table 61	Forms of irrealis prefixes	202
Table 62	Overview of possible TAM combinations and meanings	223
Table 63	Question markers that can function as heads of a non-verbal predicate	231
Table 64	Verbs of emotion belonging to the extended intransitive subclass	240
Table 65	Other verbs belonging to the extended intransitive subclass	240
Table 66	Verbs belonging to the extended transitive subclass	244
Table 67	Intransitive frozen applicative verbs	260
Table 68	Some frozen applicative forms with *-ek*	261
Table 69	Some frozen applicative forms with *-sek*	262
Table 70	Some frozen applicative forms with *-tek*	263
Table 71	Frozen applicative forms with *-ngek*	263
Table 72	Frozen applicative forms with *-nek*	264
Table 73	Frozen applicative forms with *-lek*	264
Table 74	Participant role-marking functions of **akin[i]*	265
Table 75	Derived intransitive verbs with a *ma-* prefix	266
Table 76	Derived intransitive verbs with a *ta-* prefix	267
Table 77	Some cause-effect/resultative SVCs	275
Table 78	Some motion verbs frequently followed by directionals in SVCs	282
Table 79	Some symmetrical SVCs	289
Table 80	A selection of combinations with *tou*	292
Table 81	Potential historical stages of SVC lexicalisation	293
Table 82	Properties of asymmetrical Serial Verb Constructions	295
Table 83	Some lexicalised negative phrases	321
Table 84	Functions of CA in MC and RC	326
Table 85	Semantic subtypes of complement-taking verbs	338
Table 86	Types of adverbial subordinate clauses	346
Table 87	Semantic types of main clause linking	358
Table 88	Demonstrative paradigm	373

Conventions and abbreviations

Example sentences consist of four lines. The first line uses a practical orthography. The second line shows the underlying morphemes. The third line is a gloss, and the fourth line provides a free translation in English. In glosses, proper names are represented by initials. Constituents are in square brackets. Examples are numbered separately for each chapter. If elements are marked in other ways (e.g. with boldface), this is indicated where relevant. A source reference is given for each example; the reference code consists of the initials of the speaker (or 'Game' for the *Man and Tree* games), the date the recording was made, and the number of the utterance. An overview of recordings plus reference codes can be found in Appendix I.

The following abbreviations and symbols are used in glosses:

1	first person
2	second person
3	third person
ADJ	adjectiviser
ANIM	animate
APPL	applicative
CAUS	causative
CLF	classifier
CNTF	counterfactual
COMP	complementiser
COND	conditional
CONT	continuative
DEF	definiteness marker
DEM	demonstrative
DIM	diminutive
DIST	distal
du	dual
EMP	emphatic marker
EXCL	exclusive
FREE	free pronoun
FOC	focus marker
HAB	habitual
IMP	imperative
INANIM	inanimate
INCL	inclusive
INT	intermediate
INTJ	interjection
INTF	intensifier
IPFV	imperfective
IRR	irrealis
MOD	modal operator
NEG	negation marker
NOM	nominaliser
NS	non-singular
PART	partitive

pc	paucal
PERT	pertensive
PFV	perfective
pl	plural
POSS	possessive
PRF	perfect
PROG	progressive
PROX	proximate
RECIP	reciprocal
REDUP	reduplication
REL	relative clause marker
sg	singular
SPEC.COLL	specific collective
SUB	subordinate clause marker
TAG	tag question marker
ZERO	person/number reference with no overt realisation
-	affix boundary
=	clitic boundary
.	morpheme boundary

The following abbreviations are used as subscripts to constituents:

A	transitive subject
AdvCl	adverbial subordinate clause
AG	animate goal
AM	associated motion
Apo	apodosis
BEN	benefactive
CC	copula complement
Compl	complement clause
E	extended argument
LG	locative goal
LOC	locative
MAL	malefactive
MC	main clause
NVPRED	non-verbal predicate
O	transitive object
Obl	oblique argument
Pot	potential
Pro	protasis
Quot	quotative
RC	relative clause
S	intransitive subject
SVC	serial verb construction
Top	topicalised element

The following abbreviations are used in running text and in tables:

A	adjective
Adv	adverb
CA	common argument
CoreAsp	core aspect
DC	deictic centre
DIR	directional
FC	focal clause
FoR	frame of reference
intr	intransitive
LG	locative goal
Manip	implement argument of ditransitive AFFECT verb
Mod	modifier
N, NP	noun, noun phrase
n/a	not applicable
POc	Proto-Oceanic
ProObj	object bound pronoun
ProSubj	subject bound pronoun
PV	postverbal element
s.b.	somebody
SC	supporting clause
SecAsp	secondary aspect
s.t.	something
stat	stative
SVC	serial verb construction
TAM	Tense-Aspect-Modality
tr	transitive
V	verb
VC	verb complex

1 The language and its context

1.1 Introduction

Baluan is a small island located south-east of the main island of Manus Province in Papua New Guinea (PNG); see Figure 1. PNG has the highest rate of linguistic diversity in the world in terms of number of unrelated language phyla (Nettle 1999). Manus Province, located north-east of the PNG mainland, consists of the relatively large Manus Island (commonly referred to as the "big place" or "mainland" within the province) and a large number of surrounding islands. In the province, around 30 languages are spoken, all of which belong to the Oceanic subgroup of the Austronesian language family (Lynch, Ross, and Crowley 2002: 10).

Two languages are spoken on Baluan Island: Titan (ISO 639-3: ttv), which is also spoken in several locations on Manus Island and on a number of other islands in the province, and Paluai, more commonly known as Pam-Baluan (ISO 639-3: blq) (Simons and Fennig 2018). Although the latter name is more commonly known to the wider world, native speakers prefer to use the autodenomination Paluai to refer to their island, language and group identity. This practice will be followed throughout this work.[1] Figures from the 2011 Census show the total number of inhabitants of Balopa LLG (consisting of Baluan, Lou and Pam islands) to be 3,516 (National Statistical Office Papua New Guinea 2014). Including expatriate speakers, the number of Paluai speakers can be estimated to fall somewhere between 2,000 and 3,000.

The languages spoken in Manus Province belong to the Admiralties cluster, a higher-order subgroup of Oceanic (Lynch et al. 2002: 94). In contrast to the Oceanic subgroup as a whole, which is relatively well represented in the literature compared to various other language families, very little is known about the Admiralties languages: "the language situation [here] is complex and remains poorly understood" (Lynch et al. 2002: 123).

In this chapter a background to the grammatical description of Paluai will be given. First, general sociolinguistic information about the language will be provided, followed by a brief discussion of the sociocultural context; for a more exhaustive discussion, see Schokkin (2018). Next, an overview of

---

[1] Baluan is probably a derivation from the name Paluai, which originated in the colonial period. It is unclear how it came about, but outside of the island, the name Baluan rather than Paluai is used to refer to Paluai Island and its inhabitants.

https://doi.org/10.1515/9783110675177-001

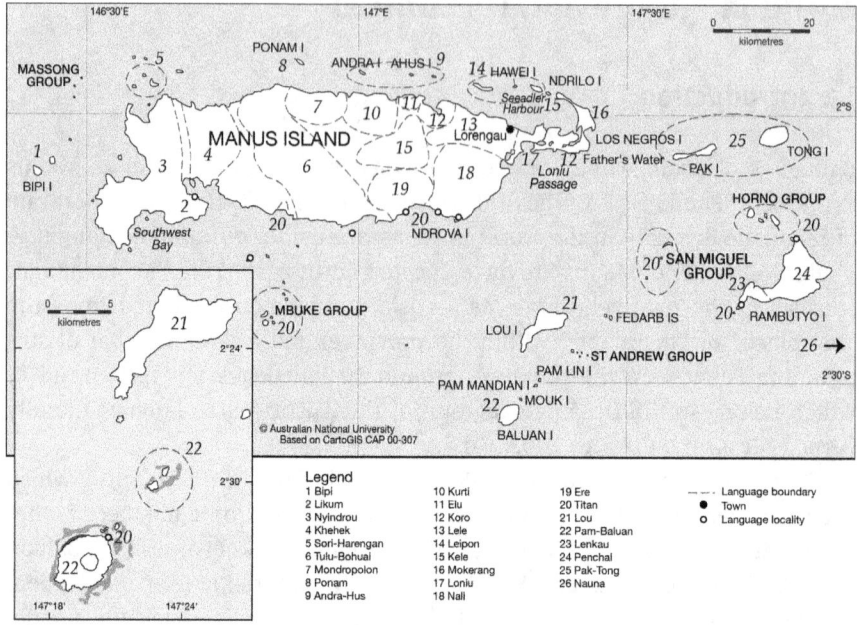

Figure 1: Language map of Manus Province.

the genetic affiliation of Paluai and existing sources on neighbouring languages is given. The final section of the chapter discusses data collection and methodology.

1.2 Sociolinguistic situation

The language described in this work is spoken in two locations: on Baluan Island and on nearby Pam Island (see Figure 1). There are minor lexical differences between the variety spoken on Baluan and the one spoken on Pam, but they are probably not sufficient to classify the varieties as different dialects. Native speakers insist that the varieties spoken on Baluan and Pam are the same. In addition, inhabitants of the two islands regard their customs as very much alike, and there is considerable intermarriage. The data on which the current description is based were collected exclusively on Baluan Island; I will therefore refer to the language variety as "Paluai" throughout the work, in accordance with the wishes of the Baluan speech community, as mentioned above. However, the reader should keep in mind that most of the description applies to the Pam variety as well.

On nearby Lou Island, a closely related variety is spoken that is commonly known as Lou (ISO 639-3: loj) (Simons and Fennig 2018). This variety is considered to have dialectical variation between villages, but the distinctions are minor and mainly of a phonological and lexical nature. In fact, distinctions between Lou and Paluai are minor as well, and probably also mainly phonological and lexical. Based on data in Stutzman (1997), it is estimated that the lexicons of the Baon dialect of Lou and of Paluai overlap for about 80–90%, and that there is very little grammatical difference. There are regular phonological correspondences between the two varieties; see Ross (1988) for more on sound correspondences between Admiralties languages. Paluai speakers consider Lou culture to be related to Paluai, but with some pronounced differences. Thus it can be said that Lou and Paluai are considered separate languages based primarily on cultural and political, rather than linguistic, grounds.

All members of the community on Baluan Island acquire Paluai from birth as their native language, with a few exceptions such as in-married women and the Titan speakers in Mouk village. In addition, often from a very early age, people acquire Tok Pisin, an English lexifier creole and an official language of PNG. English, although gaining ground as the language of mass media, government and education, is not commonly acquired in a naturalistic setting by people on the island, and people's first exposure to English occurs usually when they enter primary school around the age of eight. For expatriate Paluai, the situation is often very different. For them, because they are not part of a stable Paluai speech community, the pressure of Tok Pisin and English is much greater, in particular within mixed marriages. Children growing up away from Baluan Island often acquire only a passive command of the language, and have Tok Pisin (or sometimes English) as their first language. On Baluan Island, people of the older generations (about age fifty and upwards) often show at least a passive command of either Titan, Lou or both, and sometimes of some other Manus languages, due to long-standing contacts between various neighbouring speech communities. These forms of passive bilingualism are on their way out, since Tok Pisin is increasingly used as a lingua franca between language groups. See Schokkin (2017) for more on Paluai-Tok Pisin contact and its implications.

1.3 Sociocultural background

Linguistic research on the languages of Manus Province is relatively scarce, especially when compared to the considerable amount of anthropological work that has been done. Anthropologists have traditionally distinguished three main groups of people: the Usiai people, who predominantly live away from the coast on the main island; the Moanus (or Titan), who traditionally did not own land and live in stilt houses on the coast; and the Matankor people of the surrounding islands (Nevermann 1934; Bühler 1935). The distinction is emic and was given further credence by the early ethnographers; as a matter of fact, only the Titan can be regarded as a single ethnolinguistic unit. Margaret Mead (1930, 1934, 1956) wrote about child rearing and cultural change among the Titan people of Pere. Other classic texts are Fortune (1965) on religion, Schwartz (1963) on the Paliau Movement, areal culture, cultural totemism and cargo cults, and Carrier and Carrier (1989) on kinhip, exchange and trade. Valuable work on material culture in Manus has been done by Ohnemus (2002); other recent anthropological sources on Manus are Wanek (1996), Dalsgaard (2009) and, of particular importance to Baluan, the work carried out by Ton Otto: e.g. Otto (1991, 1992, 1997, 2002, 2008), Otto and Pedersen (2005), and Dalsgaard and Otto (2011).

The Baluan population is spread over six villages, most of them on the north coast, while a large group of people live elsewhere because they are working in urban areas (Otto 1992: 264).

1.4 Genetic affiliation

As mentioned, Paluai is a member of the Oceanic subgroup of the Austronesian language family. Austronesian is divided into at least ten primary subgroups, of which nine are located in Taiwan and the other one, Malayo-Polynesian, consists of the remainder of Austronesian languages (Blust 2013: 29). Due to the large number of first-order subgroups on Taiwan, the latter is now uncontroversially believed to be the homeland of the Austronesian family. Multiple waves of migration have led to the spread of the language family across the globe. Migration basically went from west to east, with the farthest of the Polynesian islands reached by about 1000 A.D. (Lynch 1998: 56). The Admiralties group was established as a primary subgroup of Oceanic based on shared innovations as compared to the protolanguage (Ross 1988; Lynch et al. 2002), following the conventional methodology of establishing subgroups in historical linguistics.

Intensive contact with non-Austronesian "Papuan" languages has led to major changes in some of the Oceanic languages of Melanesia. Although there are no non-Austronesian languages spoken in Manus Province at present, archeological evidence indicates that it was occupied at the time the Austronesian speakers arrived (Ambrose 2002), making it plausible that substrate phenomena still exist in the Oceanic languages spoken nowadays.

There are a number of publications about historical phonology and linguistics of the area by Blust (1981, 1998, 2007, 2008) and Ross (1988). Ross (1988: 316) divides the Admiralties cluster into two second-order groups:
1. the western Admiralties, containing Wuvulu and Aua, Seimat, and now extinct Kaniet;
2. the eastern Admiralties, divisible into two third-order subgroups:
    a). the south-eastern Admiralties, comprising Pak-Tong, Baluan-Pam, Lou, Lenkau, Penchal and Nauna;
    b). the Manus subgroup, which includes all languages of Manus Island and its remaining offshore islands.

1.5 Existing descriptions of neighbouring languages

A preliminary sociolinguistic overview of the language situation in Manus was given by Schooling and Schooling (1980). The only other Admiralties language with a published full-length grammar is Loniu (Hamel 1994). Mussau is a language spoken on the islands of the St. Matthias Group, located east of Manus in New Ireland, and is considered closely related to the languages of the Admiralties; a sketch grammar has been published by the Summer Institute of Linguistics (SIL) (Brownie and Brownie 2007). Various other members of SIL have worked in Manus Province, which has resulted in a sketch grammar of Seimat (Wozna and Wilson 2005) and two MA theses: Stutzman (1997) on the Lou verb phrase, and Hafford (1999) on Wuvulu. More limited data from SIL are available for Kurti, Nyindrou, Lele, Khehek, Nali and Bipi. For Titan, there is an extensive text collection by Meier (1907–1912) with German translations, which has been adapted, translated into English and provided with a sketch grammar by Bowern (2011). There is also a sketch grammar of Kele in Lynch et al. (2002). In addition to the thesis this work is based on, three further PhD theses on Admiralties languages were successfully completed in recent times: Hafford (2014) on Wuvulu, Boettger (2015) on Lele, and Cleary-Kemp (2015) on Koro.

1.6 Data collection and methodology

The linguistic data on which this work is based were collected during four field trips to Baluan Island, totalling about 11 months. In addition, a wordlist compiled by Ton Otto and several recordings made by him during field trips in the 1980s and 1990s were used.

The methodology used during field trips can be characterised as "immersion fieldwork", in which a researcher lives with a language community for an extended period of time, takes part in daily activities and actively learns to speak the language (Dixon 2010a). I got in touch with the Baluan community through Ton Otto and spent most of my time in Lipan village. I lived there with a local family, by whom I was adopted. I would partake in their daily life and accompany them to the garden, on fishing trips, to church and to customary activities.

The pilot trip and the first few weeks of the first major field trip were spent acquainting myself with the field situation and introducing myself and the project to various key members of the community, such as clan chiefs, ward councillors and the local court magistrate. There was a lot of community support for a language documentation project, as people felt a need to preserve their language and culture, which are both under pressure. Quite a few people were keen on participating in the project, and I managed to record people from villages all over the island. I also witnessed a large number of traditional ceremonies, which are part and parcel of daily life on Baluan, and recorded several. At a later stage, in particular during the second major field trip, when I was more fluent in the language, I ran several elicitation sessions and had a number of people play the *Man and Tree* game (Levinson et al. 1992). The games were recorded as well. Most recordings were made using a Zoom H4 digital recorder with either an external Rode NT1 cardioid condenser or Samsom SE50 head-mounted microphone. In situations where this setup was not practical, such as with public speeches, the in-built microphones of the Zoom H4 were used.

All recordings were transcribed in the field, and checked and translated with the assistance of a native speaker of Paluai. Although many people provided recordings and evidence in the form of field notes, I had a handful of consultants who all had good command of English and who assisted me on a regular basis with transcription and translation: three of them were women and three were men. For glossing, interlinearisation and database management, version 1.6.1 of the program Toolbox (SIL International 2017) was used. Furthermore, the analysis in this work is based on notes on spontaneous language production collected during the field stays, and on elicitation sessions that were carried out as much as possible in Paluai.

For grammatical analysis of the data, use has been made mostly of Dixon's Basic Linguistic Theory (Dixon 2010a, 2010b, 2012). In most cases, this framework turned out to be well-suited for the analysis of this particular language. Where the need for supplementary theory was felt, this was used in addition. Throughout the work, I have made an effort to clearly define the terminology I am using, in particular where the term is not one commonly encountered, or is used with several definitions in the existing literature, which is the case for a lot of linguistic terminology.

An effort has been made to collect speech samples from a wide variety of genres and speakers. Genres represented include: spontaneous conversation, semi-spontaneous conversation (elicited by the *Man and Tree* picture matching task), procedural texts, family histories, anecdotes, children's stories, traditional legends, public speeches and traditional chants.[2] Speakers come from a variety of age groups and from several villages around the island, and are fairly balanced between males and females. In Appendix I, an overview is given of texts collected, including information on length, genre and speaker; only spoken texts are involved, since I did not use any examples from chants for this work.

Speakers from the 10–20 year-old age cohort are absent from the data collection. People from this age group felt reluctant about being recorded. However, since this is not a sociolinguistic study of Paluai, this is not an insurmountable gap in the data. Moreover, observations in the field have allowed me to form an impression of the language use of younger people, and data from slightly older speakers (in particular 20–30 years old) can give an idea of how Paluai and Tok Pisin are in mixed use in the speech of the younger generations.

---

2 The chants were collected for a side project funded by the Firebird Foundation for Anthropological Research, titled *Collection of Traditional Song Genres on Baluan Island (PNG)*.

2 Phonology

2.1 Syllable structure

Generally, a reference grammar will discuss segmental phonology before syllable structure. However, the analysis of some phones and phonetic diphthongs depends on syllabicity, and therefore it is appropriate to first discuss which syllable structures are attested.

In Paluai, only a vowel can appear in the nucleus of a syllable, i.e. only vowels can be syllabic. Syllables do not have complex nuclei. Vowel sequences, therefore, are always analysed as consisting of either two syllable nuclei, or of a single syllable containing a vowel in the nucleus and a glide in the coda (see Section 2.2.3.9 below for a more elaborate discussion of diphthongs and vowel sequences). Syllables cannot have complex onsets or codas either, i.e., consonant clusters are not allowed. All consonants can appear in both onset and coda position, except for the alveolar fricative /s/. There seem to be no restrictions on CV or VC combinations. On the surface, allowed syllable structures are of the type (C)V(C), which means the following basic syllable structures are attested:

(1) V   /a/   'and'
     CV   /mɛ/   'come'
     VC   /ap/   second person dual, free pronoun
     CVC /kɛm/   'salt'

A vowel /i/ or /u/ may appear in the onset or in the coda of a syllable; it then loses its syllabicity and is realised as an approximant [j] or [w]. When a consonant (for instance a suffix) is added to the end of a word ending in [j] or [w], this increases syllable count, since extrasyllabic consonants are not allowed. The former approximant now becomes the nucleus of the added syllable. Because it gains syllabicity, the former approximant becomes a vowel (see Section 2.2.3.8 below for further discussion and exemplification). In underlying form, therefore, the following additional syllable structures are also allowed:

(2) V_oV_n
     V_nV_c
     $V_oV_nV_c$
     V_oV_nC
     CV_nV_c where V_o and V_c can only be /i/ or /u/

These always have (C)V(C) surface form. Examples are given below.

(3) /iɔ/ [jɔ] intermediate demonstrative
 /uɔ/ [wɔ] second person singular, free pronoun
 /ui/ [wuj] first person dual exclusive, free pronoun
 /au/ [aw] second person dual, free pronoun
 /iɔi/ [jɔj] 'stone'
 /uau/ [waw] 'move'
 /iɛp/ [jɛp] 'fire'
 /uak/ [wak] 'lizard sp.'
 /mui/ [muj] 'dog'
 /pɔu/ [pow] 'pig'

Monosyllabic forms are predominantly of the CVC type; V and VC syllables are not frequently attested as monosyllabic forms. In addition, words tend not to start with a V or VC syllable.[3] This means that the majority of words start with a consonant. Closed syllables seem to be much more common than in some other Oceanic languages (there are a fair number of Oceanic languages that do not allow (C)VC syllables at all). This is due to the fact that "[a]ll languages of the eastern Admiralties have lost not only original final consonants, but also the vowels that preceded them, and so allow many final consonants in contemporary word forms" (Blust 2013: 95). This historical change is probably the cause of the large amount of monosyllabic CVC forms in present-day Paluai, which may stem from earlier CV.CV forms that lost their final vowel.

2.2 Segmental phonology

2.2.1 Overview of consonant phonemes

The Paluai consonant inventory is shown in Table 1. Symbols between brackets indicate phones of which the phonemic status is marginal. There are four obstruent phonemes (three plosives and one fricative): /p/, /t/, /k/ and /s/; and four sonorants: /m/, /n/, /ŋ/ and /l/. In addition, Paluai has a labialised bilabial plosive /pʷ/ and a labialised bilabial nasal /mʷ/.

---

[3] An exception is the fairly large number of words that start with /a/, which may be due to fossilised morphology.

Table 1: The consonant phonemes of Paluai.

	Bilabial	Alveolar	Palatal	Velar	Glottal
Plosive	p	t		k	
Labialised plosive	pʷ			(kʷ)	
Nasal	m	n		ŋ	
Labialised nasal	mʷ				
Trill		(r)			
Tap or Flap		(ɾ)			
Fricative		s			(h)
Lateral approximant		l			

*marginal phonemes are indicated between brackets

The phonological status of the trill [r] and tap [ɾ] is unclear. They are in complementary distribution with [t] and occur in free variation with each other. Since [t] and the rhotics are contrastive in only one form, both rothics are analysed as allophones of /t/; this is discussed in Section 2.2.2.1.2. Almost all consonant phonemes appear in all positions: word-initially, word-medially and word-finally. The only fricative phoneme, /s/, is not attested syllable-finally. For the labialised stop /pʷ/ and nasal /mʷ/, contrast with /p/ and /m/ is neutralised word-finally. [kʷ] is not analysed as a separate phoneme; neither is [h], as discussed in Sections 2.2.2.1.1 and 2.2.2.3.2, respectively. Glides [j] and [w] can be analysed as non-syllabic realisations of the vowels /i/ and /u/, and are discussed separately in Section 2.2.3.8.

2.2.2 Distribution and realisations of consonant phonemes

In the sections below, consonant phonemes and their realisations are discussed. It is important to note that many phonological processes in Paluai operate in a domain either smaller or larger than the phonological word. Therefore, an overview of phonological domains, from largest to smallest, is given in Table 2, in order to assist the reader in understanding the following sections.

Table 2: Phonological domains in Paluai.

Phonological domain	Associated processes	Schematically
Major prosodic unit	Gradually declining pitch ending in a low boundary tone (L%)	
Minor prosodic unit	One pitch contour, ending in a high boundary tone (H%)	
Phonological phrase	Lenition of stops between vowels	/p/ → [β] / V_V /t/ → [r] / V_V /k/ → [x] / V_V
	Voicing of stops following a nasal	/p/ → [b] / N_ /t/ → [d] / N_ /k/ → [g] / N_
Phonological word	Primary stress assignment	
	Assimilation of /t/ to /l/	/t/ → [l] / l_
	Neutralisation of word-final /pʷ/ and /mʷ/	/pʷ/ → [p] / _# /mʷ/ → [m] / _#
Foot	Secondary stress assignment	
Syllable	Vowel-glide alternation between nucleus and coda	/u/ → [w] and /i/ → [j]
Segment		

2.2.2.1 Plosives

2.2.2.1.1 Bilabial stop /p/

/p/ occurs in all positions, as illustrated by (4).

(4) /paŋ/ 'rain'
    /sɔpɔl/ 'side; half'
    /tap/ first person plural inclusive, free pronoun

/p/ is realised as a voiceless unaspirated bilabial stop [p] phrase-initially or -finally, when following an obstruent, and in all positions in careful speech. It is often realised as a voiced bilabial stop [b] when it follows a lateral or nasal, and as a voiced bilabial fricative [β] between vowels. Phrase-finally, /p/ often has no audible release.

The fricative has only very brief friction, and can be distinguished mainly because it does not show the burst characteristic for stops in spectrograms, as illustrated in Figure 2.

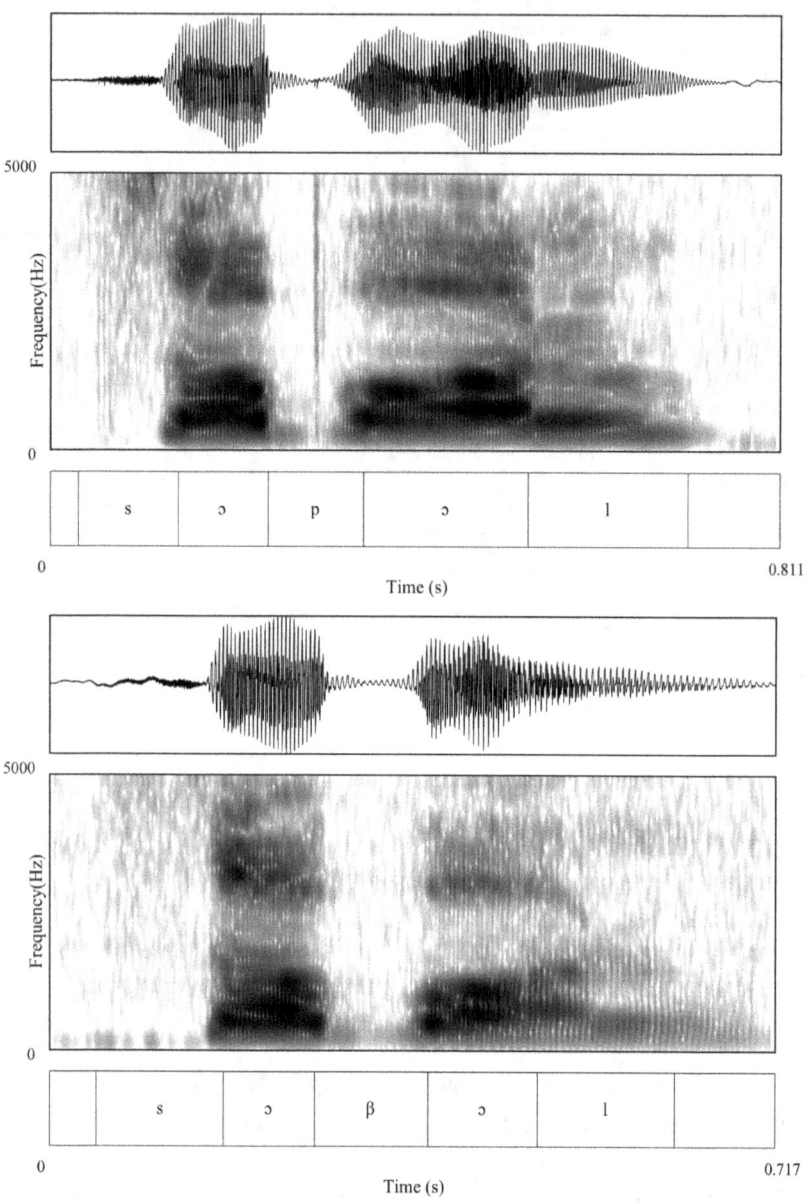

Figure 2: Acoustic difference between [p] and [β] in the form /sɔpɔl/.

2.2.2.1.2 Alveolar stop /t/
/t/ occurs in all positions, as shown in (5):

(5) /tum/ 'tidal wave'
    /patan/ 'on top of'
    /nɛt/ 'ocean'

/t/ is realised as a voiceless unaspirated alveolar stop [t] phrase-initially, when following an obstruent, and in all positions in careful speech. It is realised as a voiced alveolar stop [d] when it follows a nasal and as a voiced alveolar trill [r] or tap [ɾ] when it occurs between vowels. Within a phonological word boundary, it is realised as an alveolar lateral approximant [l] when following /l/. Phrase-finally, /t/ is sometimes aspirated.

The alternation between [t] and [ɾ] or [r] is a very regular process.[4] Almost all instances of [r] in connected speech can be shown to represent allophones of /t/, either because they are realised as [t] in other instances (e.g., in careful speech or when following an obstruent), or speakers would state that they could be realised as [t] when queried. Only when [r] occurs intervocalically in a multisyllabic morpheme, such as [mari] 'sleep', would speakers not accept replacing [r] with [t]. In some of these cases, morphological processes such as reduplication may give a decisive answer that [r] is indeed underlyingly /t/. Compare e.g. [kurun] 'small' with reduplicated [kutkurun] 'very small'. In the reduplicated CVC sequence, [r] no longer occurs intervocalically but is followed by a voiceless stop, and is now realised as [t]. We can therefore conclude that the form is underlyingly /kutun/, even though it never surfaces as such.

There is, however, one minimal pair in which [t] and [r] appear to be contrastive: [patan] 'on top of' and [paran] 'its stem'. The forms are shown in (6) below. They are both based on a root to which a 'pertensive' suffix indicating direct possession (cf. Section 3.2.2) is attached. /pata/ 'on top of' is an obligatorily possessed root and does not occur without a suffix. /pata/ 'stem' is not obligatorily possessed, and surfaces as [pat] when used without a possessive suffix. The difference in realisation between [pat] 'stem' and [paran] 'its stem' indicates that the [r] of [paran] is indeed an underlying /t/. /patan/ 'on top of'

---

4 There are no indications that there is predictable inter- or intra-speaker variation for [r] and [ɾ]. As these realisations appear to be in free variation, henceforth only [r] will be used to represent either.

is the only form in the data for which /t/ never changes to [r]; it is always realised as [patan].

(6) a. //pata// — 'top' → /pata-n/ [patan] 'on top of'
    b. //pata// [pat] 'stem' → /pata-n/ [paran] 'its stem'[5]

In addition, /t/ is realised as a lateral approximant [l] when following /l/ within a phonological word boundary. This alternation is seen for three grammatical morphemes that occur postnominally: /tɛ/, attested as dependent/subordinate clause marker and within the demonstrative system, the possessive classifier /ta/ and a formative /ta/ which has a range of functions (see Section 4.11). These forms change to [lɛ] and [la] following a noun ending in /l/, as shown below in (7). In addition, there are no [lt] sequences attested within monomorphemic forms.

(7) a. /samɛl=taŋ/   [samɛl laŋ]   'my knife'
    b. /mobail=tɛjɔ/ [mobeil leʲɔ] 'the mobile phone'

2.2.2.1.3 Velar stop /k/

/k/ occurs in all positions, as shown in (8).

(8) /kun/   'small basket'
    /kɔkɔn/ 'money'
    /mak/   'surgeonfish'

It is realised as a voiceless unaspirated velar stop [k] phrase-initially or-finally, when following an obstruent, and in all positions in careful speech. It can be realised as a voiced velar stop [g] following a nasal, and as an unvoiced velar fricative [x] between vowels. Phrase-finally, /k/ is sometimes unreleased.

Lenition of /k/ to [x] in the proper name *Lalau Kanau*, as part of the phrase *wong Lalau Kanau* 'I am Lalau Kanau', is illustrated by the spectrogram in Figure 3.

Whether the sequence [kw] should be analysed as a separate phoneme /kʷ/ or as an underlying /ku/ sequence is discussed in Section 2.2.2.1.1.

---

[5] The second vowel /a/ in the suffixed form was lost in the unsuffixed surface form of the root. This phenomenon has been reported for several other Admiralties languages as well.

2.2 Segmental phonology — 15

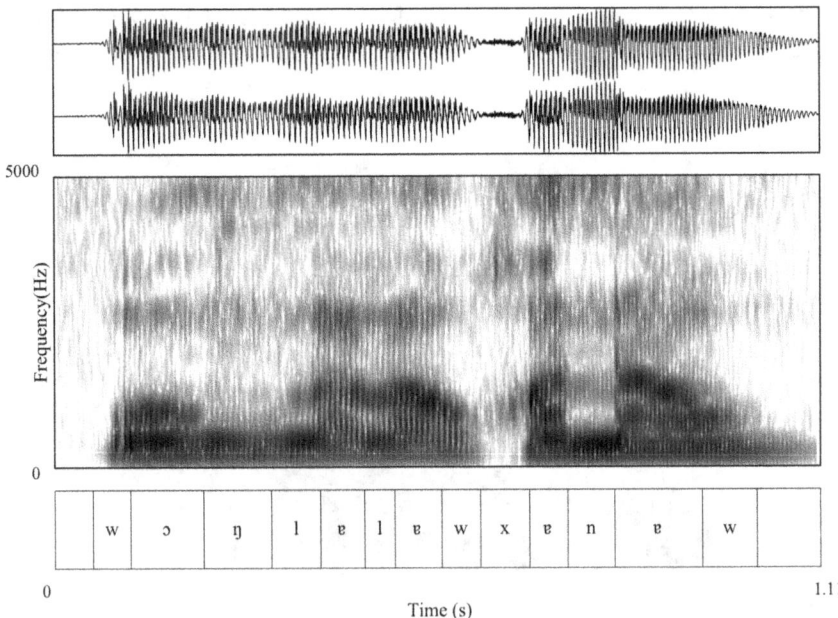

Figure 3: Lenition of /k/ to [x].

2.2.2.1.4 Labialised bilabial stop /pʷ/

Phrase-initially, and phrase-medially in careful speech, /pʷ/ is realised as a labialised voiceless bilabial stop [pʷ], with a complete closure of the vocal tract. In other registers, /pʷ/ is often realised as a voiced labial-velar approximant [w] between vowels or following a nasal, and there is no complete closure. Word-finally, /pʷ/ is realised as [p]. Below, /pʷ/ is shown in word-initial and -medial position.

(9) /pʷalɛi/   'ancestor spirit'
    /mʷalupʷɛ/ 'canoe type'

This phoneme occurs most often followed by /a/ and just a handful of times by other vowels. It is not attested before the close and close-mid back vowels /u/ and /o/. /pʷ/ can be distinguished from a [pu] CV sequence because the secondarily articulated vocoid is never syllabic.[6] This contrasts for instance with the noun /pul/ [pul] 'moon', with /u/ in the nucleus of the syllable, as illustrated in the

---

[6] Historically, however, this phoneme may have developed out of [p] followed by a rounded vowel, at least in some cases. See e.g. Blust (1981) and Lynch (2002) for discussion.

spectrograms in Figure 4. The /u/ in /pul/ is considerably longer and shows a steady state, whereas the [w] feature of the /pʷ/ in /kupʷɛn/ is just a short glide.

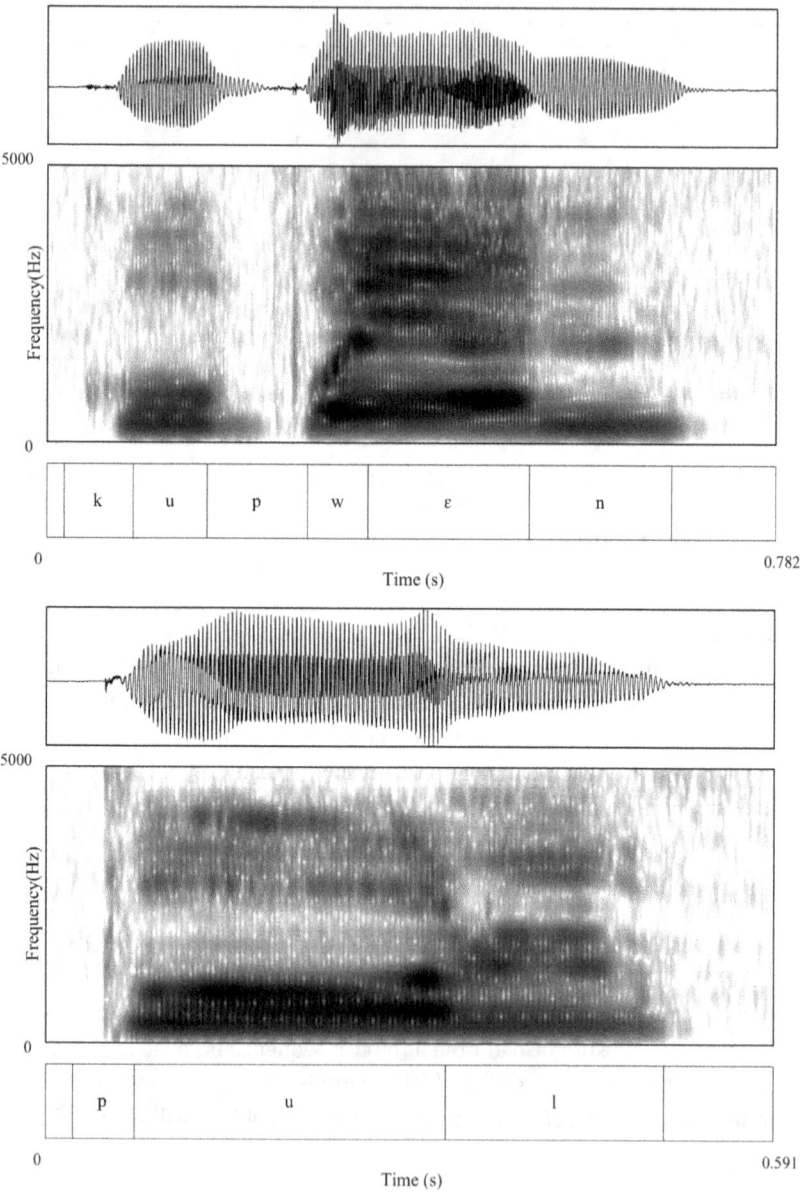

Figure 4: Comparison of /pʷ/ and /pu/.

For words ending in [p], a derived form must be available in the data to establish whether the underlying form is /p/ or /pʷ/, as is the case with nouns that have both an unpossessed root form and a suffixed directly possessed form (see Sections 3.2.2 and 5.5). This is illustrated in (10) for the form //kupʷɛ// 'branch'. The phoneme /pʷ/ retains the labial feature in the suffixed form, whereas it is realised as a plain stop [p] in the bare root form.

(10) a. //kupʷɛ// /kupʷɛ/ [kup] 'branch'
 b. //kupʷɛ// /kupʷɛ-n/ [kupʷɛn] 'its branch'

2.2.2.1.5 Is there a labialised velar stop phoneme /kʷ/?

The question whether Paluai has a labialised velar stop phoneme /kʷ/ is worth pursuing. Such a phoneme would result in a complete series of plain and labialised stops, and in some other Oceanic languages this sequence can indeed be analysed as a phoneme. There are a small number of forms that could be analysed as having /kʷ/, such as /kuam/ 'fifth-born son' which is often pronounced [kʷam]. However, in the forms that start with a [ku] ~ [kʷ] sequence, including /kuam/, the [u] can also be analysed as a full vowel, forming the nucleus of the first syllable. Thus, there are no forms for which a [kʷ] sequence is unambiguously a single segment. Because of the small number of forms, and the possibility of an alternative analysis, [kʷ] is not regarded as a separate phoneme.

2.2.2.2 Nasals

2.2.2.2.1 Bilabial nasal /m/

/m/ occurs in all positions, as is shown below. It has only one realisation: bilabial nasal [m].

(11) /man/ [man] 'seagull'
 /kamɛi/ [kamej] 'rainbow runner (fish species)'
 /kɛm/ [kɛm] 'salt'

2.2.2.2.2 Alveolar nasal /n/

/n/ occurs in all positions. It is usually realised as the alveolar nasal [n].

(12) /nɛi/ [nej] 'rat'
 /kanɛi/ [kanej] 'tree species with edible nut'
 /kun/ [kun] 'small basket'

Assimilation and elision of /n/ occurs frequently, but seems to be limited to a handful of frequent grammatical morphemes (particles and suffixes) ending in /n/; it is not attested in items from open word classes and is therefore analysed as a morpho-phonological process (see Section 2.2.7.1).

2.2.2.2.3 Velar nasal /ŋ/
/ŋ/ occurs in all positions. It has only one realisation: the velar nasal [ŋ].

(13) /ŋan/ [ŋan] 'white ant, termite'
/sɔŋap/ [sɔŋap] 'refugee'
/niŋ/ [nɪŋ] 'see'

2.2.2.2.4 Labialised velar nasal /mʷ/
Word-initially and -medially, /mʷ/ is realised as a labialised bilabial nasal [mʷ]. Word-finally, it is always realised as [m]. Similarly to the labialised stop /pʷ/, this phoneme appears almost exclusively followed by /a/ and only a few times followed by /ɛ/, /e/ or /ɔ/. It is not attested before the close and close-mid back vowels /u/ and /o/. /mʷ/ can be distinguished from a [mu] CV sequence because the secondarily articulated vocoid is never syllabic.[7] Contrasting with this is, for instance, the noun /mui/ [muj] 'dog', with /u/ in the nucleus of the syllable. As can be seen from the spectrograms in Figure 5, the [u] in /mui/ is considerably longer and has a steady state, whereas the [w] feature of /mʷ/ in /kɔmʷɛt/ is only a short glide.

(14) /mʷɛn/ [mʷɛn] 'man, person'
/kɔmʷɛt/ [kɔmʷɛt] 'grasshopper'

For forms ending in [m], a derived form must be available in the data to establish whether the underlying form is /m/ or /mʷ/. In (15), this is illustrated for the form //wumʷa// 'house', which either occurs as a bare root /wumʷ/, or suffixed for direct possession, as /wumʷa-n/.

(15) a. //wumwʷa// /wumʷ/ [wum] ~ [um] 'house'
b. //wumwʷa// /wumʷa-n/ [wumʷan] ~ [umʷan] 'his/her house'

---

[7] Historically, however, this phoneme may have developed out of [m] followed by a rounded vowel, at least in some cases. See e.g. Blust (1981) and Lynch (2002) for discussion.

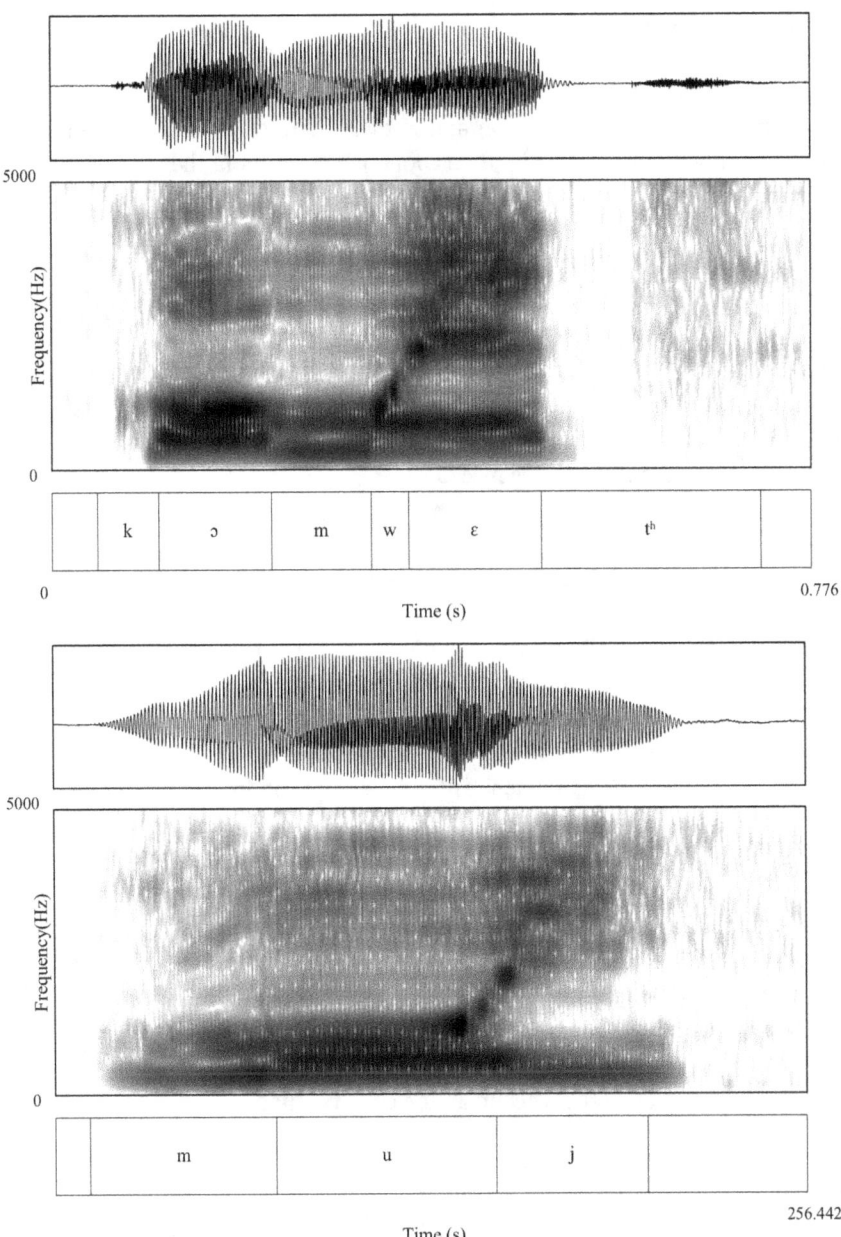

Figure 5: Comparison of /mʷ/ and /mu/.

2.2.2.3 Fricatives

2.2.2.3.1 Alveolar fricative /s/

/s/ cannot occur in coda position and is therefore attested only word-initially and -medially, not word-finally. It has only one realisation: the voiceless alveolar fricative [s].

(16) /sakɔ/ 'bait'
/lisat/ 'fish species'

The fact that /s/ is not attested in coda position is probably due to a historical process of fortition, in which it underwent a change to /t/ in that position. This is evident from several bare stems and derived forms based on the same root. Compare for instance the forms in (17a-d).

(17) a. /nut/   [nut]   'breast'
     /nusu-n/   [nu.sun]   'her breast'
   b. /lolot/   [lọ.lọt]   'be cool (V)'
     /lolosi-n/   [lọ.lọ.sm]   'cool (derived A)'
   c. /kulut/   [ku.lut]   'rubbish'
     /kulusu-n/   [ku.lu.sun]   'its rubbish'
   d. /masia-n/   [ma.si.ʲan]   'appreciation'
     /matmasia-n/ [mat.ma.si.ʲan] 'great appreciation'

Examples (a) to (c) show a contrast between bare stems and suffixed forms. In the bare stems, an /s/ in (word-final) coda position changed to /t/. In the corresponding suffixed forms, however, the /s/ syllabifies with the following syllable, which further consists of an underlying vowel (lost in the bare forms) and the suffix consonant. Since in these forms, /s/ was not in coda position, it did not change to /t/. Example (d) shows the same process, but here there is an alternation between a bare form, where /s/ occurs in onset position, and a CVC-reduplicated form, where /t/ occurs in the coda position.

2.2.2.3.2 Glottal fricative [h]

The glottal fricative [h] is discernible a number of times in the data, but it is not considered a phoneme in the present analysis. It is never contrastive and there is much variation among speakers: in the case of some speakers, an occasional [h] is heard, but there is no such sound in the speech of many others. [h] sometimes seems to occur following an open syllable with /ɛ/ or /ɔ/ in the nucleus, or it is inserted between two vowels in hiatus. The interjection /main/ 'I don't know!', for instance, is realised variably as [ma.ɪn], with a hiatus, [ma.ʲɪn], with an epenthetic glide, or [ma.ʰɪn]. The variation with regard to [h] seems not to be sociolinguistically conditioned and/or register-specific, but a more elaborate study would be needed to conclude this with certainty.

2.2.2.4 Liquids

2.2.2.4.1 Alveolar approximant /l/
The phoneme /l/ is realised as the alveolar lateral approximant [l] in all positions.

(18) /lan/   'south, south wind'
     /kulut/ 'rubbish'
     /kal/   'taro'

2.2.3 Vowel phonemes

Seven vowel phonemes are distinguished based on two dimensions: vowel height, and articulation in the front versus the back of the oral cavity. The vowel inventory is schematically represented in the chart below (Figure 6). This chart is based on mean F1 and F2 values of 100 vowel tokens from one female speaker; note that the Paluai vowel system is basically triangular and not quadrilateral in its organisation.[8] Vowel length is not contrastive. Front vowels are always unrounded and back vowels are always rounded. Only vowels can be syllabic, i.e. they can form a syllable nucleus (see Section 2.1 for more on syllable structure). Thus, there are no

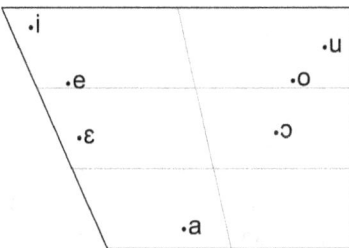

Figure 6: Paluai vowel inventory.

---

8 The chart was generated using this website: http://adambaker.org/formant-chart/formant-chart.html (accessed 1 November 2018).

syllabic consonants. Glides [j] and [w] can basically be analysed as non-syllabic vowels /i/ and /u/, and are discussed separately in Section 2.2.3.8.

2.2.3.1 The vowel /i/
The vowel /i/ is realised as a close unrounded front vowel [i] and appears in VC, CV and CVC syllables. In syllables checked by a nasal, /i/ is realised as [ɪ].

(19) /ip/ third person plural, free pronoun
/pi/ 'mosquito'
/nik/ 'fish'

2.2.3.2 The vowel /e/
Most older speakers realise the vowel /e/ as an unrounded front vowel [e̝] that is perceptibly higher than the cardinal vowel [e]. Many younger speakers, however, realise it closer to cardinal [e], or may even lower it slightly to [e̞]. /e/ may be in a process of merging with /ɛ/ for a number of younger speakers; see also Section 2.2.3.10. /e/ occurs mostly in CVC syllables; it is attested only rarely in CV and not at all in VC syllables.

(20) /lep/ 'take'
/se/ 'who'

2.2.3.3 The vowel /ɛ/
The vowel /ɛ/ is realised as an open-mid unrounded front vowel [ɛ]. When occurring in a phonetic diphthong with /i/, /ɛ/ is usually raised to [e], and by some speakers even to [ɪ].

(21) /ɛp/ first person plural exclusive, free pronoun
/mɛ/ 'come'
/lɛm/ 'hole'

The fact that /ɛ/ is raised in phonetic diphthongs means that for [e͡i] sequences, it is sometimes hard to establish whether the first vowel is actually phonemically /ɛ/ or /e/. This may partly have given rise to the merger we now see occurring. Phonetic diphthongs only occur across syllable boundaries, and since /e/ is rarely attested in open syllables otherwise, throughout this work [e͡i] sequences are analysed as underlyingly /ɛi/; also see Section 2.2.3.9 below for more on vowel sequences.

2.2.3.4 The vowel /a/

The phoneme /a/ is realised as a central open vowel [a] and has only one realisation. It appears in a range of environments: by itself and in VC, CV and CVC syllables.

(22) /a/   'and'
     /ap/   second person plural, free pronoun
     /ma/   'and, but'
     /mat/   'die'

2.2.3.5 The vowel /ɔ/

This vowel is realised as an open-mid rounded back vowel [ɔ]. /ɔ/ is not attested in VC syllables. When followed by /u/ and forming a phonetic diphthong, /ɔ/ is raised to [o].

(23) /lɔ/   'covering sheath of coconut'
     /sɔk/   'burn'

The fact that /ɔ/ is raised in phonetic diphthongs means that for [o͡u] sequences, it is sometimes hard to establish whether the first vowel is actually phonemically /ɔ/ or /o/. Phonetic diphthongs only occur across syllable boundaries, and since /o/ is rarely attested in open syllables otherwise, throughout this work [o͡u] sequences have been analysed as underlyingly /ɔu/; also see Section 2.2.3.9 below for more on vowel sequences.

2.2.3.6 The vowel /o/

This vowel is realised as a rounded back vowel [o̝], higher than the cardinal vowel [o]. It occurs in CV and CVC syllables, but not in VC syllables. For younger speakers, the contrast between this phoneme and the back vowel /u/ may be disappearing. This is discussed further in Section 2.2.3.10.

(24) /lolot/   'be cool'
     /pon/   'sea turtle'

2.2.3.7 The vowel /u/

The vowel /u/ is realised as a close rounded back vowel [u] and has only one realisation. It appears by itself and in CV and CVC syllables, but not in VC syllables.

(25) /u/   third person dual, free pronoun
/su/ 'comb'
/kun/ 'small basket'

2.2.3.8 The approximants [j] and [w]

The approximants [j] and [w] are analysed as non-syllabic allophones of /i/ and /u/ respectively. These approximants (or glides or semivowels) are phonetically distinguishable from their full vowel counterparts because they involve a greater restriction of the vocal tract as compared to full vowels (Ladefoged and Maddieson 1996: 323). [j] represents a palatal approximant, whereas [w] represents a voiced labialised velar approximant.

Thus, a [wV] or [jV] sequence in the onset of a syllable is underlyingly a /uV/ or /iV/ sequence, and a [Vw] or [Vj] sequence in the rhyme is underlyingly a /Vu/ or /Vi/ sequence.[9] There are three main arguments in favour of the present analysis. The first one is that phonological processes that only occur intervocalically, such as lenition of /k/ to [x], are attested when there is a flanking [j] or [w]; compare Figure 3 above. Secondly, there are systematic alternations observable due to morphological processes. For instance, a word-final [w] or [j] becomes syllabic when a suffix is attached, as exemplified below.

(26) /kamɔu/ ['ka.mow] 'speech' → /kamɔu-n/ [ˌka.mo.'un] 'its speech'
/sui/   [suj]   'soup'   → /sui-n/   ['su.ɪn]   'its soup'

As the above examples show, /u/ or /i/ in coda position is realised as a glide, resulting in a [Vw] or [Vj] sequence. However, when a possessive suffix /n/ is attached, increasing the syllable count by one, the underlying /u/ or /i/ now forms the nucleus of a new VC syllable. Because it has become syllabic, the /u/ or /i/ is pronounced as a full vowel and in some cases even stressed. In fast speech, a certain degree of phonetic diphthongisation is often discernible.

This process is most evident at the end of phonological words. Example (27) shows how the process works word-medially.

---

9 From a historical linguistic viewpoint, an alternative analysis is possible for at least some of the [Vj] and [jV] sequences. There are regular sound correspondences between [j] in Paluai and [r] in several other Admiralties languages: cf. e.g. Lou *ramat* ~ Paluai *yamat* 'person', Lele *por* ~ Paluai *poy* 'residue after boiling coconut oil'. Thus, a number of instances of [j] may in fact historically represent a separate phoneme, and not an underlying /i/. However, this study aims at a synchronic analysis of Paluai phonology, and thus this line of enquiry will not be further pursued here.

(27) /iuep/ ['ju.ʷẹp] 'two' → /ta=iuep/ [taj.'wẹp] 'SPEC.COLL=two'

In the base form of the numeral 'two', the [u] occurs in the nucleus of the first syllable and is thus syllabic. There is also often an epenthetic [w] between this [u] and the next VC syllable. When the proclitic /ta/ (see Section 4.11) is attached to this base form, the syllable count increases from two to three. However, in connected speech, the syllable count is usually again reduced to two, with the /i/ in the coda of the first syllable realised as a glide [j]. The /u/ that was in the nucleus of the first syllable of /iuep/ now forms the onset of the second syllable and is realised as a glide [w]. In addition, because stress can only fall on a full vowel in the nucleus of a syllable, stress has shifted from the /u/, which became nonsyllabic, to the /e/ vowel in the nucleus of the second syllable.

Word-initially, the process can be observed in vowel-initial verbs with the proclitic bound subject pronoun /i/ attached to them. However, there is a fair bit of inter-speaker variability discernable here, since some speakers syllabify the /i/ with the first syllable of the verb, realising it as [j], whereas others keep it as a separate syllable with /i/. Below, the two possibilities are shown.

(28) /i=apui/ [i.'ʲa.βuj] ~ ['ja.βuj] '(s)he cooked (it)'

This contrasts with the situation for the third singular irrealis prefix /ki/, which has CV syllable structure. In this form, the /i/ segment is never realised as [j] since it never syllabifies with the verb, so there is no comparable variation observed for this inflected form:

(29) /ki-apui/ [ki.'ʲa.βuj] '(s)he will cook (it)'

There is considerable coarticulation of a glide and a preceding vowel. For /ɛ/ and /ɔ/ this means that a following glide in a similar place of articulation causes some raising, resulting in realisation as [e] or [o]. Thus, /pɛi/ 'stingray' is realised as [pej] (not *[pɛj]), and /pɔu/ 'pig' is realised as [pow] (not *[pɔw]). However, when the glide is articulated "across" from the vowel (front glide with back vowel, or back glide with front vowel), or when the glide precedes the full vowel, this effect is not observed.

The third argument for the current analysis is based on distributional patterns of [w] and [j]. Firstly, [uw] sequences are not attested syllable-finally, indicating that there are no syllables whose rhyme contains an underlying /uu/ sequence. Syllable-final [ij] sequences do occur, but they can be analysed as underlying disyllabic /ii/ sequences (see also note 5). Secondly, [w] and [j] occurring syllable-initially, i.e. in the syllable onset, are frequently dropped before full vowels with similar place of articulation. [w] is usually dropped before the close back vowel /u/, sometimes before /ɔ/ and once before /a/. [j] is dropped before the close front vowel /i/. There thus seems to be a dispreference for the same syllable to both contain a full vowel /i/ or /u/ and the corresponding approximant.

Summing up, [j] and [w] are treated as vowels in phonological processes, they are often optional, never contrastive, and [j] is in complementary distribution with /i/ and [w] with /u/. This provides adequate evidence to analyse them as allophones of /u/ and /i/, conditioned by a distributional rule: these allophones only occur in the non-syllabic onset and coda positions.

2.2.3.9 V-V sequences and phonetic diphthongisation

There are no phonological diphthongs in Paluai; a syllable nucleus can only contain one vowel. Thus, although Paluai contains many V-V sequences and many vowel-glide or glide-vowel sequences, these are not analysed as diphthongs. The glide-vowel and vowel-glide sequences are analysed as consisting of /u/ or /i/ in a syllable onset or coda, which is then non-syllabic and realised as [w] or [j] (see Section 2.2.3.8 above). Other V-V sequences consist of two syllable nuclei. In fast speech, there is diphthongisation, but this is a phonetic rather than a phonological process. Table 3 gives an overview of the possible V-V sequences across syllables in multisyllabic morphemes. Note that instances of /i/ and /u/ listed here are syllabic ones, and should be distinguished from the glide realisations discussed above.

What is immediately evident from Table 3 is that /e/ and /o/ are hardly attested in vowel sequences. This is in line with the above-mentioned fact that they predominantly occur is in CVC syllables. The second inference from the data in the table is that opening V-V sequences (i.e. from a close vowel to an open vowel) do not diphthongise. There is often an epenthetic glide: for instance, /tiok/ is realised [tiʲɔk] and /lɛut/ is realised [lɛʷut]. However, such a glide is usually not obviously distinguishable. Sequences that can be realised as a phonetic diphthong are all closing (i.e. from an open vowel to a close vowel) and are 1) those that start with /a/ and end in a close to an open-mid vowel: [ai], [ae], [aɛ] and [au]; 2) those that start with open-mid back /ɔ/ and end in close-mid front [e] – which are very rare – and 3) those that start with open-mid back /ɔ/ and end in close back /u/, or start with open-mid front /ɛ/ and end in close front /i/. Still, all those sequences can also be realised with a slight epenthetic glide, or with a very slight intervocalic [h].[10]

---

[10] These observations are in line with the tendency for closing diphthongs to be falling (i.e. losing prominence) and for opening diphthongs to be rising (i.e. gaining prominence). In most

Table 3: Possible vowel sequences in Paluai.

Sequence	Example	Phonetic diphthong possible
/ii/	[piːŋ] 'extinguish'	n/a
/ie/	[nɛsiʲe̯t] 'anger'	no
/iɛ/	[ariʲɛk] 'clear the garden'	no
/ia/	[liʲan] 'anchor'	no
/iɔ/	[tiʲɔk] 'betel pepper vine'	no
/io/	not attested	n/a
/iu/	not attested	n/a
/ee/	not attested	n/a
/ei/	not attested	n/a
/eɛ/	not attested	n/a
/ea/	not attested	n/a
/eo/	not attested	n/a
/eo/	not attested	n/a
/eu/	not attested	n/a
/ɛɛ/	not attested	n/a
/ɛi/	[pɛʲin] ~ [pe͡in] 'woman'	yes
/ɛe/	not attested	n/a
/ɛa/	[lɛʲam] 'greedy'	no
/ɛo/	[malɛʲow] 'plant sp.'	no
/ɛo/	not attested	n/a
/ɛu/	[lɛʷut] 'weed'	no
/aa/	[mʷaʔaj] 'plant sp.'	n/a
/ai/	[pʷaʲit] ~ [pʷa͡it] 'sea anemone'	yes
/ae/	[naʲe̯t] ~ [ne͡e̯t] 'canoe part'	yes
/aɛ/	not attested	n/a
/aɔ/	[nanaɔp] 'tree sp.'	no
/ao/	not attested	n/a
/au/	[maʷut] ~ [me͡ut] 'sink'	yes
/ɔɔ/	[tɛpʷɔʔɔm] ~ [tɛpʷɔʷɔm] 'right now'	n/a
/ɔi/	not attested	n/a
/ɔe/	[lɔʲe̯ŋ] ~ [lɔ͡e̯ŋ] 'roast'	yes
/ɔɛ/	[alɔʷɛn] 'long, tall'	no
/ɔa/	[alɔʷaj] 'daylight'	no
/ɔo/	not attested	n/a
/ɔu/	[jɔʷun] ~ [jo͡un] 'its tail'	yes

of the V-V sequences the first vowel (i.e. syllable) is stressed and thus the most prominent. This prominence pattern coincides with that of closing diphthongs, but not with that of opening diphthongs, in which the second V is most prominent. Thus, closing phonetic diphthongs are attested whereas opening phonetic diphthongs are not.

Table 3 (continued)

Sequence	Example	Phonetic diphthong possible
/oo/	not attested	n/a
/oi/	not attested	n/a
/oe/	not attested	n/a
/oɛ/	not attested	n/a
/oa/	not attested	n/a
/oɔ/	not attested	n/a
/ou/	not attested	n/a
/uu/	[paruʷul] ~ [paruʔul] 'coral sp.'	n/a
/ui/	[muʲin] 'green coconut'	no
/ue/	[kuʷel̥] 'blue-spotted parrotfish'	no
/uɛ/	[nuʷɛn] 'wild tuber'	no
/ua/	[kuʷaŋ] 'clamshell sp.'	no
/uɔ/	[tuʷɔp] 'chew betelnut'	no
/uo/	not attested	n/a

A third important observation is that V-V sequences consisting of identical vowels are relatively rare, and realised in different ways. /ii/ sequences are realised as long vowels (e.g. [piːng] 'extinguish'). /aa/ sequences are realised with an epenthetic glottal stop [ʔ]. /uu/ and /ɔɔ/ sequences are realised either with a glottal stop or an epenthetic glide [w]. In these instances, the inter-speaker variation seems not to be conditioned by register-specific or sociolinguistic factors. Sequences of /ee/, /ɛɛ/ and /oo/ are not attested.

2.2.3.10 Merger of close-mid vowels

As mentioned, there appears to be an ongoing vowel merger for the close-mid vowels /e/ and /o/. Younger speakers in particular appear not to perceive /e/ and /ɛ/ as contrastive, and the same applies to /o/ and /u/. These speakers would maintain that e.g. /jek/ 'feel' and /jɛk/ 'hit', or /pol/ 'coconut' and /pul/ 'moon', sound the same. Others, particularly those over sixty, maintain there is a contrast between the sounds. As we have seen, distribution of the close-mid vowels is already quite limited: they are predominantly attested in CVC syllables, and absolute figures of occurrence compared to the other vowels are not high. Only 145 out of 2232 lexical items in the current Paluai dictionary contain one or more instances of /e/ (6.5%), and only 79 have /o/ (3.5%).

In order to further the investigation, F1 and F2 values were measured at the mid-point of 100 vowel tokens from one speaker, a woman of Lipan

2.2 Segmental phonology

village in her late thirties. These tokens were extracted from a wordlist recording made in 2015, measured using Praat (Boersma and Weenink 2018), and plotted using the R package phonR (McCloy 2016; R Core Team 2017). F1 and F2 values of each individual token are plotted in Figure 7, in addition to the mean values for each vowel (represented by the place of the IPA symbol) and ellipses, which are calculated based on the covariance of the tokens and a supplied confidence level of 95%, using Hotelling's T^2 distribution (a multivariate analog of the *t* distribution) to account for uncertainty when the number of tokens is low.

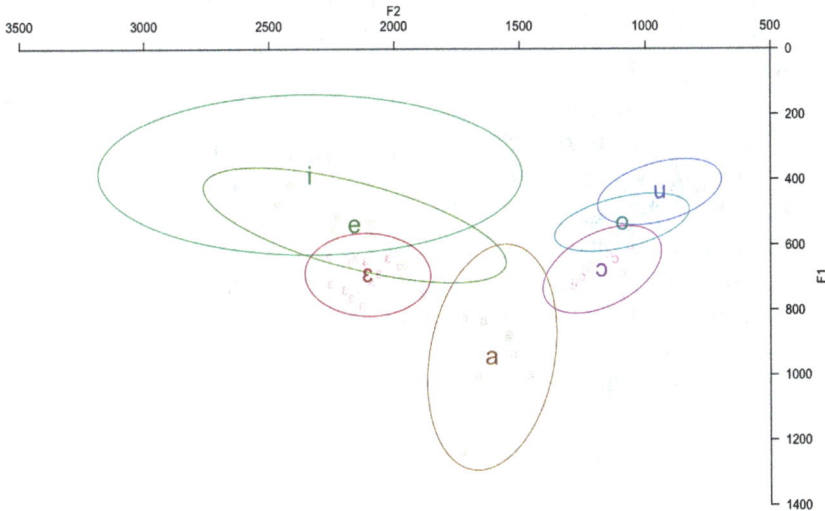

Figure 7: F1 and F2 values for the vowels of one female speaker.

While ellipses for the other vowels are mostly separate from each other, the ellipses for /e/ and /o/ almost entirely overlap those of other vowels, indicating that they may indeed be in the process of merging for this particular speaker.

If this indication, based on a small number of tokens from one speaker, is borne out by systematic analysis of a larger dataset, it would mean that Paluai is in the process of changing from a seven-vowel system to a five-vowel one. See also Hamel (1994) for discussion of a potentially similar process in the related language Loniu. A sociophonetic study investigating vowel systems across generations on Baluan Island would be informative, but is beyond the scope of this work.

2.2.4 Minimal pairs

Phonological oppositions involving phonemes with similar places or manners of articulation can be illustrated by means of minimal pairs.

2.2.4.1 Minimal pairs for consonant phonemes
Contrasts between the nasals /m n ŋ/ are shown in (30) for all positions.

(30) a. Onset
/mui/ [muj] 'dog'   /min/ [mɪn] 'hand'
/nui/ [nuj] 'squid sp.' /nin/ [nɪn] 'fight'
/ɲui/ [ɲuj] 'snore'   /ɲin/ [ɲɪn] 'scratch'
b. Onset (intervocalic)
/mamat/ [ma.mat] 'be awake'
/manak/ [ma.nak] 'elder'
/maɲat/ [ma.ɲat] 'work'
c. Coda
/uɔm/ [wɔm] 'chop'   /mɔm/ [mɔm] 'fish sp.'
/uɔn/ [wɔn] 'ladle, scoop'   /mɔn/ [mɔn] 'pandanus'
/uɔŋ/ [wɔŋ] first person sg   /mɔŋ/ [mɔŋ] 'dry (tree)'
           (free pronoun)
/iɛm/ [jɛm] 'lime stick'
/iɛn/ [jɛn] 'lie'
/iɛŋ/ [jɛŋ] 'slice'

Contrasts between the plain and labialised bilabials /m mʷ/ and /p pʷ/ are shown in (31). As mentioned before, these are only contrastive in onset position, the distinction being neutralised in the coda.

(31)   /mat/   [mat]   'die'   /mak/   [mak]   'emerge'
        /mʷat/ [mʷat] 'bandicoot'   /mʷak/ [mʷak] 'armband'
        /pet/   [pet]   'clean fish'   /pɔn/   [pɔn]   'compensate'
        /pʷet/ [pʷet] 'peel fruit'   /pʷɔn/ [pʷɔn] 'cover'

Contrasts between the alveolar plosive /t/ and fricative /s/ are shown in (32), for both word-initial and intervocalic onset positions. /t/ is realised as [r] or [ɾ] intervocalically, and /s/ is not attested in coda position (see above).

(32) a. Onset
/san/ [san] 'cut' /suk/ [suk] 'deceive'
/tan/ [tan] 'dig out (eggs)' /tuk/ [tuk] 'beat'
b. Onset (intervocalic)
/nusu-n/ [nu.sun] 'her breast' /musin/ [mu.sɪn] 'soft (betelnut)'
/nutu-n/ [nu.run] 'her belongings' /mutin/ [mu.rɪn] 'broken'

2.2.4.2 Minimal pairs for vowel phonemes

Contrasts between the front vowels /i/, /e/ and /ɛ/, and between the back vowels /u/, /o/ and /ɔ/, are shown in (33) and (34), respectively:

(33) /tik/ [tik] 'creep' /iik/ [jik] 'search for'
/tek/ [tek̬] 'cut' /iek/ [jek̬] 'feel'
/tɛk/ [tɛk] 'put' /iɛk/ [jɛk] 'hit'
/pil/ [pil] 'ladle' /iit/ [jit] 'chip off'
/pel/ [pel̬] 'roll' /iet/ [jet̬] 'separate'
/pɛl/ [pɛl] 'tree sp.' /iɛt/ [jɛt] 'frog'

(34) /pul/ [pul] 'moon' /suk/ [suk] 'shore'
/pol/ [pol̬] 'coconut' /sok/ [sok̬] 'tail of hair'
/pɔl/ [pɔl] 'bird sp.' /sɔk/ [sɔk] 'burn'
/pun/ [pun] 'origin' /sui/ [suj] 'soup'
/pon/ [pon̬] 'sea turtle' /soi/ [soj̬] 'demolish'
/pɔn/ [pɔn] 'compensate' /sɔi/ [sɔj] 'be upset (of stomach)'

2.2.5 Minor phonological processes

2.2.5.1 Vowel assimilation

There is some vowel assimilation in Paluai. For instance, the imperfective particle /nɔ/ is sometimes realised as [nɛ] when the following verb form has a front vowel in its first syllable, for example in the phrase /nɔ iɛn jɛt/ [nɛ jɛ jɛt] 'keep on stirring'. In addition, when one syllable of a phonological word contains /o/, vowels in other syllables tend to assimilate to that vowel. Thus, the first person singular versus third person singular possessive forms of the body part /min/ 'hand' are [minɔŋ] and [minan], respectively; see Section 2.2.7.2 for discussion of this morphophonological vowel alternation. However, for the body part noun

/potposo/ 'chest', the final vowel is [o̜] for each inflected form: cf. [po̜tpo̜so̜n] 'my chest' and [po̜tpo̜so̜n] 'his chest'. This suggests the existence of some form of assimilation process that overrides the typically occurring vowel alternation.

However, these are only minor patterns, and no systematic vowel harmony processes similar to those observed in the directly possessed noun paradigms and prefixed verbs forms of other Manus languages are found in Paluai (cf. Hamel (1994) for Loniu; Lynch et al. (2002) and Blust (2013) for other Admiralties languages).

2.2.5.2 Syllable reduction

Vowels are sometimes dropped in fast speech, for instance to facilitate production. Examples of vowel deletion (resulting in a reduced syllable count) are shown in (35):

(35) /pa.lɔ.si/ → [pal.si] 'in the past, long ago'
/ta.tɛ.iɔ/ → [ta.rɛ] intermediate demonstrative, free form

Another illustration of vowel deletion is the relatively high incidence of haplology in Paluai, a process whereby one of two consecutive identical or similar syllables is deleted. Examples of haplology are given in (36):

(36) /tɛ.mɛ.nin/ → [tɛ.mɪn] 'like, as follows'
/li.liu/ → [liw] 'again'
/iu.iu.et/ → [ju.e̯t] 'question'

2.2.5.3 Metathesis

Two irregular directly possessed nouns and their related unpossessed forms show traces of a historical process of metathesis, which is not productive in present-day Paluai. The following cases are attested; in the second one, there is also fortition of the /s/ in coda position:

(37) [ja.mat] 'person' → [jam.tan] 'its owner'
[ma.ŋat] 'work' → [maŋ.san] 'its work'

2.2.6 Phonological domains and sandhi rules

Different phonological processes operate in different phonological domains. This was shown in Table 2, which is repeated below as Table 4 for convenience.

Table 4: Phonological domains in Paluai.

Phonological domain	Associated processes	Schematically
Major prosodic unit	Gradually declining pitch ending in a low boundary tone (L%)	
Minor prosodic unit	One pitch contour, ending in a high boundary tone (H%)	
Phonological phrase	Lenition of stops between vowels	/p/ → [β] / V_V /t/ → [r] / V_V /k/ → [x] / V_V
	Voicing of stops following a nasal	/p/ → [b] / N_ /t/ → [d] / N_ /k/ → [g] / N_
Phonological word	Primary stress assignment	
	Assimilation of /t/ to /l/	/t/ → [l] / l_
	Neutralisation of word-final /pʷ/ and /mʷ/	/pʷ/ → [p] / _# /mʷ/ → [m] / _#
Foot	Secondary stress assignment	
Syllable	Vowel-glide alternation between nucleus and coda	/u/ → [w] and /i/ → [j]
Segment		

This section discusses the distinction between the phonological word and phrase level based on segmental phonological processes. Suprasegmental features such as intonation and stress assignment are discussed in Section 2.3.

2.2.6.1 Phonological word

In Paluai, the following processes operate at the phonological word level.
1. Primary stress is assigned on a word-per-word basis, with additional secondary stress if the word is three or more syllables long;
2. Word-final contrast between labialised and non-labialised consonants is neutralised;
3. Assimilation of /t/ to [l] only occurs within word boundaries;
4. Pauses are possible between words, but not within words.

The assimilation of /t/ to [l] can be regarded as a case of internal sandhi. It occurs when a host word ending in /l/ is followed by an enclitic starting with /t/, but not when there is a boundary between two phonological words, e.g. a lexical verb followed by a direct object NP, two verbs in a serial verb construction, or a noun followed by an adjective or numeral functioning as a modifier. The neutralisation of labialised consonants is an example of external sandhi, since it only occurs at word boundaries.

2.2.6.2 Phonological phrase

Other phonological processes operate both within and across phonological word boundaries, and thus cannot be analysed as either internal or external sandhi. Most notably, lenition of the stops /p/ and /k/ to fricatives [β] and [x], and lenition of the stop /t/ to [r] or [ɾ], can occur when these phonemes are flanked by vowels, either within or across phonological word boundaries. These processes are observed not only across the boundary between a clitic and its host, or within monomorphemic forms, but also between e.g. a verb and its following direct object NP, or a noun and a following modifier. The same is the case with voicing of stops following a nasal.

As they operate both within and across word boundaries, these processes are analysed as operating at the level of the phonological phrase, rather than that of the phonological word. Importantly, these operations are optional, and dependent on register: in careful speech they are often not applied. As such, they contrast with the sandhi rules that operate at the phonological word level: these have been found to apply without exception.

The phonological phrase often seems to coincide with a minor prosodic unit. No evidence has been found of phonological phrases that consist of more than one minor prosodic unit. Nevertheless, it is possible that a minor prosodic unit sometimes consists of more than one phonological phrase. Because all processes operating at the level of the phonological phrase are optional, there are no clear criteria to determine this with certainty.

2.2.7 Morphophonology

This section discusses a number of phonological processes with limited distribution that interact with Paluai morphology, and can thus be analysed as morphophonological phenomena.

2.2.7.1 Assimilation and deletion of /n/

When /n/ is the final segment of a number of grammatical morphemes, it is the subject of assimilation and elision processes that do not occur elsewhere. The processes are optional. They are attested for the following morphemes: the nominalising suffix *-(n)an* (discussed in Section 3.3.2.2.3.2), the default pertensive suffix *-n* on nouns (see Section 3.2.4.3 and cross-references there), the perfect aspect particle *an* (Section 6.2.1.1.3) and the progressive aspect particle *yen* (Section 6.2.1.2.3). /n/ optionally assimilates to [m] before /p/ or to [ŋ] before /k/, resulting in a homorganic consonant cluster, but only when it is the final segment in these morphemes. Likewise, /n/ is optionally elided when it precedes a sonorant (i.e., the glide [j], another nasal, or the lateral /l/).

Due to their restricted distribution, these processes are analysed as morphophonological alternations (Trubetzkoy 1969). Their optionality puts them on a par with the alternations operating at the level of the phonological phrase, rather than those that operate at the phonological word level (cf. Section 2.2.6 above). However, in contrast to other alternations at the phrase level, which occur regardless of word class, assimilation and elision do not occur when /n/ is the final segment of a content word rather than a function word. For instance, the final /n/ of the lexical source of progressive *yen*, the verb 'lie', does not assimilate or elide under the influence of the following segment, and neither does the process operate on the boundary between a lexical verb ending in /n/ and a following element, such as an object NP.

In addition, the processes found for /n/ are the only instances of anticipatory assimilation (i.e., assimilation to a following segment), whereas other assimilation processes can be analysed as progressive (i.e., triggered by a preceding segment). They are therefore regarded as qualitatively different from other, purely phonological, processes in the language. Below, /n/ assilimation and elision, respectively, are illustrated for each of the morphemes in which they are attested.

(38) Nominalising suffix *-an*
    a. Assimilation
        /tun-an pau/   [tunam baw]   'boiling oil'
        (gloss: boil-NOM oil)
    b. Deletion
        /pɛ-mat-an iamat/ [pɛmara jamat] 'murder'
        (gloss: CAUS-die-NOM person)

(39) Pertensive suffix -n
   a. Assimilation
      /kɔlɔ-n pɛin/   [kɔlɔm bei͡n]   'the woman's voice'
      (gloss: voice-PERT woman)
      /para-n kɛi/   [paraŋ gej]   'tree trunk'
      (gloss: stem-PERT tree)
   b. Deletion
      /mata-n iamat/   [mara jamat]   'human eye'
      (gloss: eye-PERT person)
      /kɔlɔ-n lau/   [kɔlɔ law]   'opening of a fishing net'
      (gloss: mouth-PERT net)
      /lipa-n mui/   [lipa muj]   'dog's teeth'
      (gloss: tooth-PERT dog)

(40) Perfect aspect particle *an*
   a. Assimilation
      /ŋa=an pɛ/   [ŋam bɛ]   'I have done (it)'
      (gloss: 1sg=PFT do)
   b. Deletion
      /ip=an iɛi=i/   [ipa jeʲi]   'they have shaved him'
      (gloss: 3sg=PFT shave=3sg)

(41) Progressive aspect particle *yen*
   a. Assimilation
      /wɔ=jɛn pe jɛp/ [wɔjɛm bɛ jɛp] 'you keep making fire'
      (gloss: 2sg=PROG make fire)
   b. Deletion
      /wɔ=jɛn jɛt/   [wɔjɛ jɛt]   'you keep stirring'
      (gloss: 2sg=PROG stir)

2.2.7.2 Vowel alternation in directly possessed nouns

Nouns that are obligatorily directly possessed (i.e., that receive an obligatory pertensive suffix), such as most kinship terms, have a CV or CV.CV root; examples are //tama// 'father' or //tina// 'mother'. The final vowel of this bound root shows regular alternations when suffixed with a first or second person pertensive suffix, as opposed to a third person suffix. This is illustrated in Table 5 below. Since these alternations are dependent on inflectional morphology, they can be regarded as morphophonological alternations, operating

Table 5: Vowel alternations in directly possessed nouns.

Final vowel of root form	Final vowel of 1sg and 2sg suffixed forms	Example of root and 1sg, 2sg and 3sg suffixed forms
/a/	[ɔ]	//tama// 'father': [tamɔŋ], [tamɔm], [taman]
	[a]	//tina// 'mother': [tinaŋ], [tinam],[11] [tinan]
/u/	[u]	//nupu// 'bottom': [nupuŋ], [nupum], [nupun]
/ɛ/	[e]	//kɛ// 'leg': [keŋ], [kem], [kɛn]
/ɔ/	[o]	//kasɔ// 'near': [kasoŋ], [kasom], [kasɔn]
/i/	[ɪ]	//mapi// 'fontanel': [mapɪŋ], [mapɪm], [mapin]

separately from regular phonological alternations. For /u/, however, no alternation is attested. In all likelihood, a historical version of the 1sg and 2sg pertensive suffixes contained a vowel that caused alternations to the final vowel of the root. It is beyond the scope of this work to reconstruct these historical suffixes.

2.2.8 Phonology of loans

The last decades have seen an increasing presence of Tok Pisin and in particular English loan words, which contain segments that are not native to Paluai and therefore do not adhere to Paluai phonotactics. Tok Pisin contains many lexical items originating in English that have already been adjusted to the phonological template of most Oceanic languages, with, among others, shifts from the voiceless labiodental fricative [f] to a bilabial stop [p] and from dental fricatives to alveolar stops, replacement of schwa with a full vowel, reduction of consonant clusters and diphthongs, etc. Compare the examples in (42):

---

[11] There is some variation involving the form [tinam]: [tinɔm] is found as well.

(42) English orthography   Tok Pisin orthography   Paluai realisation
    finger   pinga   [pɪŋga]
    friend   pren   [prɛn]
    three   tri   [tri]
    community   komuniti   [kɔmuniti]
    cucumber   kukamba   [kukamba][12]

Most people will adhere to Tok Pisin phonology also when using English loans. One of the main exceptions is the labiodental fricative [f], which most speakers are able to produce with ease. This means Tok Pisin and English realisations are currently used side by side. There is a sociolinguistic factor to this as well: the younger and more educated the speaker, the more likely it is that he or she will adhere to English rather than Tok Pisin phonology. Interestingly, when using an English loan word in Paluai speech, speakers will tend to use a realisation adhering to Tok Pisin phonotactics. In contrast, when using the same word in Tok Pisin speech, they will more often retain the English pronunciation. In other words, it seems as if they are more eager to adhere to the phonological template of Paluai and thus quite consciously to adjust a word's realisation if it is used in a Paluai environment.

Other phonological adjustments are attested as well. Firstly, the affricates [tʃ] and [dʒ] are problematic for Paluai speakers. This has been resolved by changing the affricate into a [si] sequence, thereby adding an extra syllable to the word. In fast speech, reduction of this extra syllable results in palatalisation of the [s] fricative. The second affricate of *church*, on the other hand, is deleted entirely.

(43) *Japan*   /si.a.pan/   [sʲapan]
    John   /si.on/   [sʲɔn]
    church   /si.os/   [sʲɔs]

Secondly, Tok Pisin words that start with the glottal fricative [h] are sometimes pronounced without the initial [h] sound. [hamamas] coexists with [amamas]

---

[12] It is often neither possible nor useful to decide whether a certain word was borrowed into Paluai from English via Tok Pisin, or whether it was borrowed directly from English. Tok Pisin has always been leaning heavily on English with regard to its lexicon, and continues to do so. With the introduction of mass media and in particular the internet to remote areas of PNG, one can probably safely say that many words from English will be introduced simultaneously into Tok Pisin and the local vernaculars.

'happy', [haŭs] with [aŭs] 'house', and [has] with [as] 'bottom'. In the latter case, [h] is clearly overgeneralised. This seems to be a further indication that [h] is not perceived as a phoneme by native speakers of Paluai, although it is sometimes phonetically realised.

Thirdly, an epenthetic vowel may at times be inserted to (partly) reduce consonant clusters. Usually, the inserted vowel is a copy of the vowel in the syllable nucleus. A few examples are given in (44). The process is not very consistent, however, and observation in the field suggests that many people are clearly capable of pronouncing consonant clusters such as [str] or [spr]. It seems that insertion of vowels occurs mainly in an attempt to adhere to Paluai phonotactics in a more conscious way.

(44) stret [stɛ.rɛt] 'straight, right'
 slip [si.lip] 'sleep'
 plawa [pa.la.wa] 'flower'

Lastly, the Tok Pisin and English voiced stops [b] and [g] are sometimes devoiced in careful speech, even though the devoiced variants are not allophones of /p/ or /k/, as is normally the case in Paluai.[13] In similar fashion, voiced bilabial or labiodental fricatives (represented in Tok Pisin orthography with the letter *v*) are sometimes realised as [p] or [b] in careful speech, in analogy with the Paluai alternation between stops and fricatives. Examples of these processes are given below. Interestingly, no examples were found of voiced stop /d/ being realised as [t], so /d/ may be the only loan phoneme in Paluai.

(45) gat [kat] 'have'
 hangamap [haŋkamap] 'hang (up)'
 gavman [kabman] 'government'
 vanuatu [panuatu] 'Vanuatu'
 balus [palus] 'airplane'
 bal [pʷal] 'ball'[14]

---

[13] An interesting exception is the proper name *Baluan*, which to the best of my knowledge is never pronounced [paluan], possibly to preserve a clear distinction with the original name of the island, *Paluai*.

[14] This loan is probably not realised as [pal] because this sequence also means 'penis'.

2.3 Prosody

Prosody refers to suprasegmental features of speech, as opposed to segmental phonology. Relevant features for Paluai include word stress and intonation.

2.3.1 Word stress

In words that contain more than one syllable, some syllables are more prominent than others; this prominence is referred to with the term "(word) stress". Acoustically, word stress predominantly correlates with intensity: stressed syllables have slightly higher intensity than unstressed syllables. Pitch appears to play no role in stress assignment at the word level, but functions to distinguish speech acts, to signal the end of a prosodic unit and, together with intensity, to apply contrastive focus (see Section 2.3.2 below). Duration does not appear to be a factor in stress assignment either. Paluai does not have contrastive vowel length and, on auditory or visual inspection, stressed vowels do not appear to be significantly longer than unstressed vowels; however, to be certain, measurements of duration would need to be taken and compared. On the other hand, there is a prosodic feature that is often used for emphasis and involves considerable vowel lengthening. The lengthened vowel can even be prolonged for several seconds and is also realised considerably louder. An example is [um paran mɛnɛŋaːːn] 'a vèèèry big house'.[15]

The findings in this section are based on a 2015 wordlist recording provided by a female speaker of Lipan village aged in her late thirties. The speaker was asked to repeat each form three times; the second iteration (in which pitch was usually kept level) was used to determine intensity through visual inspection in Praat. For some disyllabic forms, intensity of the two syllables is virtually identical, making it hard to determine which one the stress falls on. This phenomenon is seen when a vowel sequence in a CV.VC form is realised as a phonetic diphthong and the syllables are no longer kept separate (phonetically speaking).

In each word, there is one syllable that receives primary stress. In words that contain three or more syllables, there will usually be another syllable with secondary stress. This is usually not a syllable adjacent to the syllable that receives

---

[15] It is possible that there are other acoustic correlates of stress besides intensity, pitch and duration, as is the case in other languages. A careful phonetic study could resolve some of the unpredictability currently seen in Paluai, but is not within the scope of the current work.

primary stress. Stress assignment is fairly straightforward for disyllabic forms, but becomes increasingly unpredictable for three- and quadrisyllabic forms.

2.3.1.1 Contrastive stress

For two minimal pairs in the data, stress appears to be contrastive: the two forms in each pair potentially differ in meaning only as a result of stress assignment. Details are shown in (46) below.

(46) /arɛi/ ['a.rej] 'bite' vs. /arɛi/ [a.'rej] 'speak'
/pɔrɔk/ ['pɔ.rɔk] 'be painful' vs. /pɔrɔk/ [pɔ.'rɔk] 'strength'

In addition, there appears to be a systematic difference in stress of disyllabic items formed by derivational reduplication of a monosyllabic verb form, depending on whether the process derives a nominalistion or an intransitive verb; this will be discussed in more detail below in Section 2.3.1.5.2. Overall, however, it is concluded that, in Paluai, contrastive stress is a marginal phenomenon at best, because of the paucity of forms and the fact that the language shows rampant homophony elsewhere.

2.3.1.2 Stress assignment in disyllabic forms

Stress assignment of CV.CV and CV.VC forms is straightforward in most cases: these forms are typically stressed on the penultimate syllable. There are few CV.CV forms in present-day Paluai, probably due to historical changes. These mainly involve word-final vowel loss and the loss of articles, which in most cases fused with the noun (Lynch et al. 2002: 38). Stress assignment, however, may nevertheless fundamentally be based on a trochaic system, with an alternation of stressed-unstressed trochaic feet. The data in (47) are CV.CV forms (on the left) and CV.VC forms (on the right). An exception to the trochaic pattern is shown by the form /namʷi/ 'small'.

(47) /mari/ ['ma.ri] 'sleep (V)' /lɛut/ ['lɛ.ʷut] 'weeds'
/panu/ ['pa.nu] 'place, village' /maut/ ['ma.ut] 'sink'
/namʷi/ [na.'mʷi] 'small (A)'

CV.CVC disyllabic forms make up a third category. Nouns and adjectives with this syllable pattern predominantly show stress on the penultimate syllable, with only a handful of exceptions. Verbs and other word classes show much more variation in stress patterns and are thus less predictable than nouns.

Below, examples are given for nouns, adjectives, verbs and other forms, respectively.

(48) Nouns
/kulut/ ['ku.lut] 'rubbish' /puron/ [pu.'rɔn] 'activity'
/maɲat/ ['ma.ɲat] 'work' /pusɔk/ [pu.'sɔk] 'island'
/punat/ ['pu.nat] 'soil'

(49) Adjectives
/kɔnun/ ['kɔ.nun] 'heavy'
/kajan/ ['ka.jan] 'dark' (from V /kai/)
/kokin/ ['kọ.kɪn] 'hot' (from V /kok/)

(50) Verbs
/wɔrup/ ['wɔ.rup] 'climb down' /sirak [si.'rak] 'wake up'
/tapui/ ['ta.puj] 'shoot' /pilɛl/ [pi.'lɛl] 'laugh'

(51) Other word classes
/pamɔu/ ['pa.mow] 'four' /mɛnɔt/ [mɛ.'nɔt] 'many'
/sokom/ [sọ.'kọm] 'some'

Forms with CVC.CVC and CVC.CV syllable structure also show the abovementioned pattern: mostly trochaic feet in nouns, more variation in other word classes. Some examples of CVC.CVC and CVC.CV forms are given in (52).

(52) ['kal.sɔn] 'ankle decoration' [suk.'pɛk] 'corner'
['kɔm.tal] 'morning star' [luk.'suŋ] 'green'
['nuk.nan] 'his uncle'
['kɪn.ti] 'sleep'
['sal.pi] 'lightning'

2.3.1.3 Stress assignment in trisyllabic forms

For words with more than two syllables it is increasingly difficult to find uncontroversial forms that consist of a single morpheme. Many of them are derived forms, either historically or synchronically, and may therefore show effects of (earlier) productive morphological processes on stress assignment. There are many possible syllable combinations and stress patterns for trisyllabic forms, examples of which are shown in Table 8 on page 54 at the end of this chapter.

Forms of three or more syllables usually, but not always, have one syllable that carries primary stress and a second syllable that carries secondary stress.

2.3.1.4 Stress assignment in quadrisyllabic forms

There are few forms of more than three syllables in Paluai; some of the forms attested in the data are shown in Table 9 on page 55 at the end of this chapter. All of them show a certain degree of morphological complexity; it is therefore quite safe to say that there are no morphemes of more than three syllables in Paluai. In most cases, stress falls on either the final or the penultimate syllable

2.3.1.5 Interaction of morphology and stress assignment

As mentioned before, there are morphological processes that may play a role in the assignment of stress: suffixation, reduplication and cliticisation. These are discussed below.

2.3.1.5.1 Suffixation

There is one suffix that attracts stress and therefore has a major effect on stress assignment. The applicative suffix -*ek*, which is realised as [εk] and is a reflex of POc **akin[i]*, is used to productively derive applicative forms from transitive verbs (for a full account, see Section 8.3.2). In addition, it is also attested as a fossilised suffix in a large number of verbal and adverbial forms. In both cases, it receives primary stress. There are no known exceptions. Since this suffix is historically a disyllabic form, this makes sense: it formed a trochaic foot, but in Proto Eastern Admiralties lost its final syllable.[16] All forms ending in [εk], regardless of syllable count, will have primary stress on the final syllable. Forms of three and four syllables often also carry secondary stress on the first syllable.

2.3.1.5.2 Reduplication

Although reduplication is of limited productivity in the modern language, there are many inherently reduplicated forms, in which earlier reduplication processes may have been retained in "frozen" form. Reduplication in Paluai is based on a CVC-, a VC-, or a CV-template (Marantz 1982) and is always

---

16 The status of **akin[i]* as a free or bound form in POc is unclear (cf. Evans 2003). It may have been a preposition or a free verbal form serialised with the main verb, that was later reanalysed as a suffix. For stress-assignment purposes, however, there is no difference between the two analyses.

prefixing: the first CVC-, VC- or CV- sequence of the base is taken and then prefixed to the stem. This means that the maximal unit of reduplication is one syllable; there is no full reduplication of entire stems if they are two or more syllables long. The process often doesn't coincide with syllable boundaries, however. With CVC- or VC-reduplication, the coda of the reduplicant can be recruited from the onset of the second syllable of the base: compare e.g. /ku.tun/ 'small' with /kut.ku.tun/ 'very small'. This may reflect historical syllable boundaries that shifted due to loss of final consonants and/or vowels. See Sections 3.2.4.2 and 3.3.2.1.2 for a fuller account of the semantic aspects of reduplication for nouns and verbs respectively.

As mentioned above, there is a productive pattern of reduplication for transitive verbs. This process derives either a nominalised form or an intransitive verb. The former carries stress on the penultimate syllable, the latter on the final syllable. This is illustrated in (53) below.

(53) /ŋan/ 'eat (tr.)' → [ˈŋan.ŋan] 'food' vs. [ŋan.ˈŋan] 'eat (intr.)'
/tuk/ 'beat' (tr.) → [ˈtuk.tuk] 'beat (N)' vs. [tuk.ˈtuk] 'beat (intr.)'

For reduplicated verbs, this particular stress assignment often also seems to hold when further derivational processes take place. This is shown in the trisyllabic forms below, which have primary stress on the penultimate syllable. These forms stem from a fully reduplicated verb form carrying stress on the final syllable, which is suffixed to derive an adjective. This stress pattern is often, but not always, retained when the form receives its suffix. This is the case even when a non-reduplicated counterpart of the verb is no longer attested (i.e., when the reduplication process has ceased to be productive), as is exemplified in (55).

(54) Verb   [puj.ˈpuj] 'be soft'
    Adjective [puj.ˈpu.jin] 'soft'

(55) Verb   *[lɛŋ.ˈlɛŋ] 'shine?' (unattested, but cf. N *pulêng* 'dawn, sunset' and *kanan puleng* 'tomorrow')
    Adjective [lɛŋ.ˈlɛ.ŋɪn] 'shiny, dazzling, brilliant'

An alternative analysis may be that only one suffix attracts stress (the *-ek* suffix discussed in Section 2.3.1.5.1 above), and that other suffixes are extrametric and thus don't affect stress assignment. This does not seem to be the case, however, as some derived adjectives, e.g. *siksikan* 'sour' (from the stative verb *sik(sik)* 'be sour'), carry stress on the final syllable.

Another way in which reduplication plays a role in stress assignment is seen in the stress pattern of inherently reduplicated trisyllabic nouns that may historically be nominalisations, but for which no verbal counterpart is currently attested. As mentioned above, fully reduplicated nouns, derived by a productive process, usually carry stress on the first syllable of what is usually a disyllabic form. This stress pattern seems to be retained even when the form contains an additional syllable, which is in these cases not analysable as a suffix. Interestingly, secondary stress is not clearly present on the final syllable either, as would be expected, but rather on the penultimate syllable.

(56) [ˈkul.ˌku.lu] 'fight'
 [ˈpɔl.ˌpɔ.lɔt] 'traditional song type'

One possible explanation for the emergence of these forms is that Paluai used to also have a process of full reduplication of disyllabic forms, yielding e.g. forms such as [ˈku.lu.ˌku.lu] or [ˈpɔ.lɔt.ˌpɔ.lɔt], with trochaic stress patterns. At a later stage, the unstressed second syllable was dropped, yielding a trisyllabic form with primary stress retained on the first syllable and secondary stress on the now adjacent penultimate syllable.

2.3.1.5.3 Cliticisation

There are a number of forms following nouns that are analysed as clitics, based on the fact that they form one phonological word with their host unit: they include /tɛ/, attested as a dependent/subordinate clause marker and a formative within the demonstrative system, the possessive classifier /ta/ and a formative /ta/, the latter of which has a range of functions (see Section 4.11). While the last two are homophones, they are not considered related, as they show different behaviour and functions, and probably derive from different historical sources. Relevant processes for their analysis as clitics are discussed above in Sections 2.2.2.1.2 and 2.2.6.

When these elements attach to a host, primary stress is assigned only once to the resulting form, another indication that they are indeed clitics. Interestingly, primary stress falls in some cases on the demonstrative or possessive classifier; in this regard, the clitics show cross-linguistically unusual behaviour. This does not extend to other forms analysed as clitics, such as the subject and object bound pronouns, which never receive stress. Moreover, when /tɛ/ functions as a dependent clause marker rather than being part of a demonstrative form, it does not attract stress; likewise, the formative /ta/ is never attested as receiving stress in its own right.

In Figure 8, intensity and pitch contours are shown for the phrase *ipengan poron nik teyo* (gloss: 3sg=PFV eat bone-PERT fish LIG-DEM.INT) 'he ate the fish bone'. *poron nik teyo* forms one phonological word, which is part of the larger phonological phrase (and prosodic unit) *ipe ngan poron nik teyo*. The plot was extracted using Praat; the intonation contour is represented by a solid line, whereas the intensity contour is shown by a dashed line. Clearly, the highest intensity peak is on the first syllable of *teyo*. While *teyo* in this example is the final word in a major prosodic unit (thus showing a low boundary tone), the same phenomenon can be observed when this is not the case, and *teyo* occurs IU-medially.

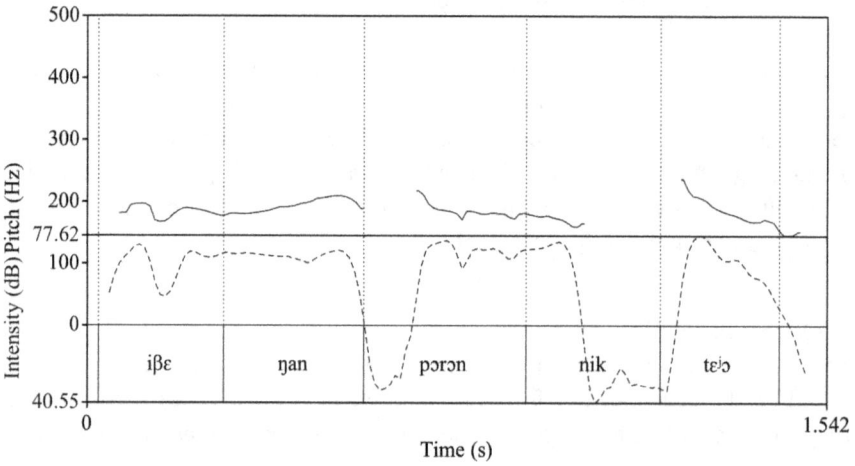

Figure 8: Intensity contour showing primary stress on clitic /tɛ/.

As is discussed in more detail in Chapter 12, complex forms with *te* not only function as situational deictic demonstratives, but play an important role as discourse deictics. It is plausible that primary stress on /tɛ/ and /ta/ is implemented for purposes of information structure, such as e.g. topicalisation or (contrastive) focus. In the example plotted in Figure 8, the referent of the phrase *poron nik teyo* was introduced into the discourse a little earlier, and the demonstrative is added since it now has definite reference. Also, it is an important prop in the wider story arc, as eating the fish bone led to a crucial event around which the entire story pivots: the dog vomiting into the canoe and eventually the canoe capsizing because of this. All these factors could play a role in *teyo* receiving primary stress in this case. However, to draw any more detailed and generalisable conclusions, a more systematic study of the relationship between Paluai prosody and information structure would need to be carried out.

2.3.1.6 (Un)predictability of word stress

As pointed out before, there is considerable unpredictability and variability in the area of word stress. It is very likely that phenomena at levels higher in the phonological hierarchy interact with lexical stress patterns. This may apply to the Admiralties languages more so than is generally the case cross-linguistically. For instance, Hamel (1994: 24–25) remarks about Loniu:

> Stress seems to play no role at the lexical level, since it may occur on different syllables depending on the structure of the word, phrase or clause, and there is apparently no phonological motivation at the lexical level for the choice of which syllable will receive the stress, whether it be primary or secondary. It is possible that stress is predictable only at the phrase or sentence level. It may be that stress assignment is a matter of rhythm, and that the overall contour of an utterance requires only that primary stress be penultimate or final within the utterance – whether the utterance is a single word, a phrase, or a clause.

One factor that appears to play a role in Paluai is that, as shown above, primary word stress can shift to a demonstrative or possessive classifier enclitic, possibly due to information structure demands.

There are also factors related to cross-linguistic differences in acoustic correlates to stress. In Paluai, word stress mainly correlates with intensity, but the actual differences in intensity between syllables are often very small, potentially making them hard to pick up auditorily for a fieldworker whose native language may exhibit quite different acoustic correlates to stress, and who is thus sensitive to different cues compared to native speakers. See e.g. Cutler (2012) for a discussion of the influence of L1 prosody on L2 perception.

In the wordlist recording that was used to determine the acoustic correlates of stress, the speaker was asked to repeat each word three times. These three repetitions each clearly form minor prosodic units, which together form one major one. For each sequence of words, there is rising pitch on the first iteration, level pitch on the second one, and falling pitch on the third, i.e. a low boundary tone to signal the end of a major prosodic unit that comprises one sequence. While the intensity peak is always on the same syllable of the word, regardless of where it occurs in the major prosodic unit, there is a difference in relative intensity between iterations. The final word in a major prosodic unit has lower intensity overall, and the intensity differences between syllables are "flattened out" even more. It thus gets even harder to detect stress auditorily for utterance-final word forms. This process is illustrated in the plot in Figure 9. The figure shows that with the falling pitch on the third

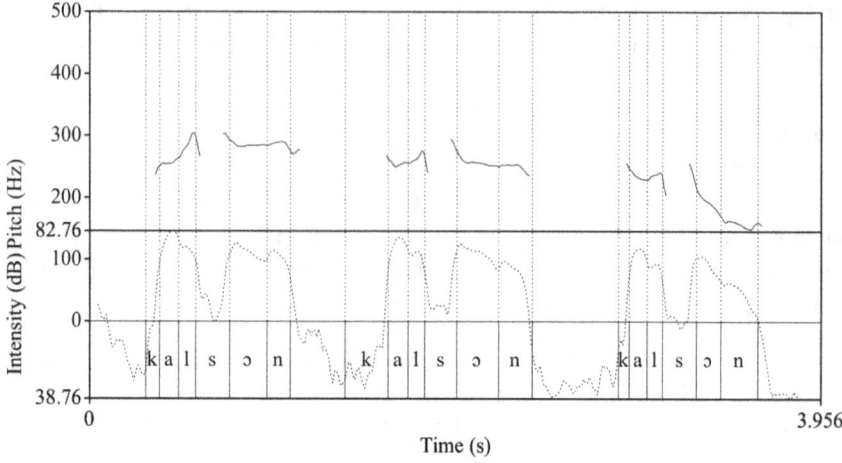

Figure 9: Pitch and intensity contours for the wordlist recording of /kalsɔn/ 'ankle decoration'.

iteration, intensity diminishes and flattens out considerably. Auditorily, for the author at least, this has the effect of an apparent stress shift to the final syllable of a major prosodic unit.

These observations bring up two important points. Firstly, the acoustic data corroborate the observations made in the field that stress is likely to be a psycholinguistic reality on the lexical level for Paluai native speakers. I was often corrected when misplacing stress on words uttered in isolation, but it was hard for me to figure out what exactly I had done wrong and why, most likely because of interference from my native language. Secondly, what this shows is that acoustic analysis can be an important tool towards a better understanding not only of the segmental phonology of a previously undescribed language, but also of its suprasegmental processes.

2.3.2 Intonation

Speech act types, in particular statements and questions, are distinguished by pitch variation, i.e. different intonation patterns. Further types of speech acts, such as commands, do not seem to differ significantly from statements with respect to intonation. Intonation, probably in addition to other prosodic features, is also used to signal the end of minor and major prosodic units, and for emphasis and contrast.

2.3.2.1 Distinguishing speech acts

Visual inspection of pitch contours in Praat has brought to light the following differences between statements and questions:
- Statements seem to have slightly lower mean pitch than questions;
- Questions have greater fluctuation in pitch than statements, with maximum pitch, but maybe also minimum pitch, generally exceeding that of statements;
- Statements start with a relatively high pitch that gradually declines, with a fall at the end of a major prosodic unit;
- Questions show peaking intonation (a sharp rise and fall in pitch) on the final element of the prosodic unit;
- The question word in content questions is realised with a peak in intensity;
- The pitch peak in questions lies around 160–200 Hz for male speakers, and 250–300 Hz for females;
- Tag questions show dipping intonation (a sharp fall and rise in pitch) on the tag element /ɛ/ (usually realised [ɛh]), which is the final element of the prosodic unit.

Figure 10 shows the pitch contour of the content question *Wokum sa ro kulum?* 'What do you have in your mouth?'. There is a clear pitch peak on the final element, *kulum*, and a peak in intensity on the question word *sa* 'what'. An example of a pitch contour for a polar question is shown in Figure 11, for the question *A kel sun tao?* 'And your canoe?'. Again, there is a peak in pitch on the final element, *tao*. Figure 12 shows the pitch contour of a tag question, *Wopwa wo la pe*

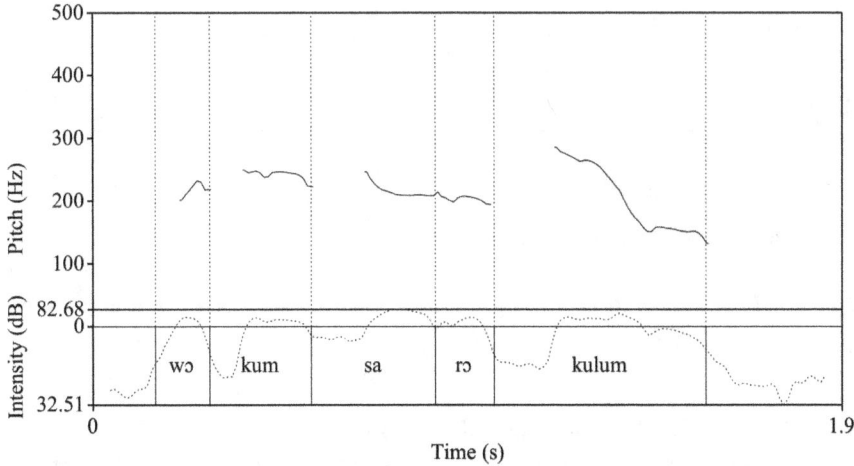

Figure 10: Pitch contour of a content question.

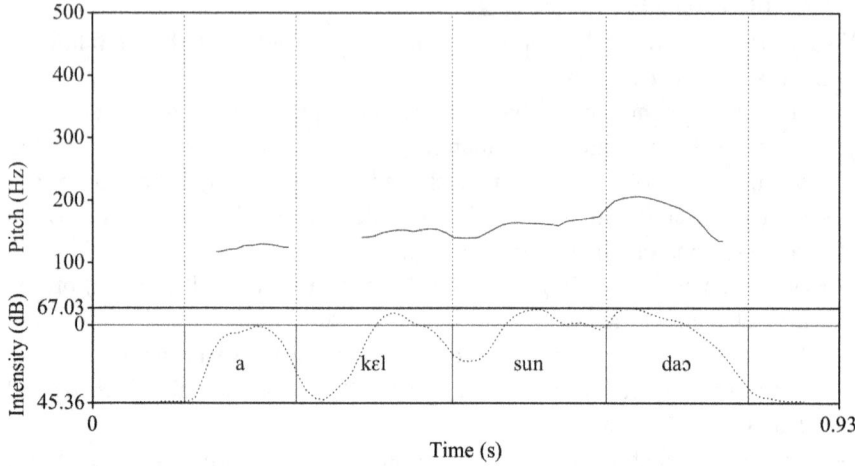

Figure 11: Pitch contour of a polar question.

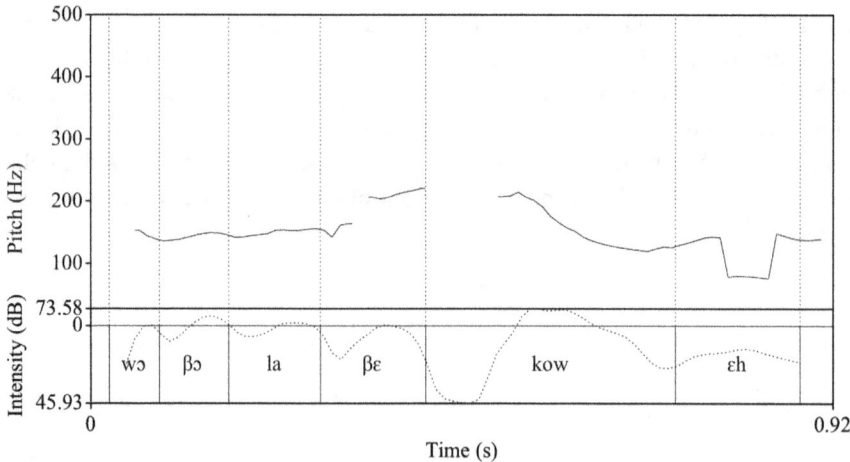

Figure 12: Pitch contour of a tag question.

kou, e? 'You wanted to go fishing, right?'. Here we see a sharp fall and rise in pitch on the element e.

2.3.2.2 Signalling the boundaries of prosodic units
An utterance consists of a major prosodic unit, which can contain several minor prosodic units. Minor prosodic units end in a high boundary tone,

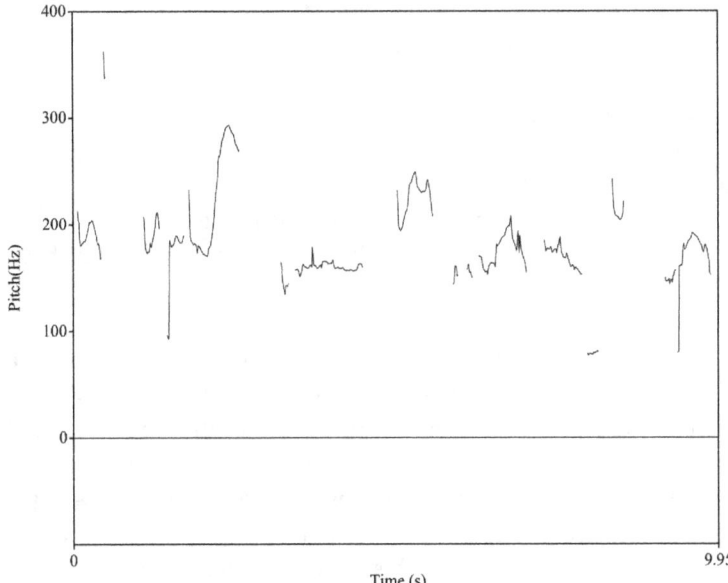

Figure 13: Pitch contour of example (57).

meaning pitch stays level or rises slightly. This indicates that the major prosodic unit they form part of is not finished. The end of the major prosodic unit is signalled by a low boundary tone, i.e. a fall in pitch. Throughout a major prosodic unit, pitch gradually declines, and at the end the speaker has generally reached the bottom of their pitch range. Major prosodic units can be very long, especially in narratives; sometimes they include multiple coordinated main clauses. An example is given in (57):[17]

(57) [i‿↘jɔ ‖ i‿rɛ pʷapʷa sẹ‿ran ↗muj | a‿pʷapʷa‿ran ↗'ŋɔjaj | aj‿'pɔtnan dɛ‿u rɔ rɔk la‿bian pal↗'si | a i‿↘nẹm ‖]
    i   yo ↘
    3sg DEM.INT
    i   te-yo   pwapwa sê   ta-n   mui ↗
    3sg LIG-DEM.INT story   DIM   CLF.POSS-PERT dog

---

17 An undertie (‿) indicates absence of a break; the single (|) and double (‖) pipes are used for minor and major prosodic boundaries, respectively. A rise in pitch is indicated by a rising arrow (↗) and a fall in pitch with a falling arrow (↘).

a	pwapwa	ta-n		ngoyai ↗		
and	story	CLF.POSS-PERT		cuscus		
a=i	potnan	te	u=to tok	la	pian	palosi ↗
OBL=3sg	time	REL	3du=CONT stay	go.to	good	past
a	i=nêm ↘					
and	3sg=be.finished					

'That's it. That is the little story of the dog, and of the cuscus, about the time when they were getting on well in the past; and it's finished.' (LL010711_0092)

Figure 13 shows the pitch contour of example (57). For clarity's sake, the entire sequence of about ten seconds long is represented as a whole and thus not transcribed within the graph, in order to give the reader a better appreciation of the integration of several minor prosodic units into one major one. As can be seen, minor prosodic units are realised with rising pitch resulting in a peak, after which pitch resets. The peaks of the minor units gradually decline throughout the major prosodic unit, and there is a fall in pitch at the end of the major prosodic boundary.

2.3.2.3 Signalling emphasis and contrast

Prosodic features can also signal emphasis, or else a contrast between two elements (see Section 12.3.4.1 for a discussion of contrastive focus). The first of the two elements is realised with additional intensity, and there is usually a peak in pitch on the emphasised or contrasted element(s). Example (58) below shows an utterance with contrastive focus on *nganngan* 'food' and *muyou* 'snake'.

(58) [ŋa‿ma‿akẹp ŋan'↗ŋan pʷen | ma‿ŋa‿no akẹp 'mu↘jow]
 *ngamaakêp **nganngan** pwên, ma ngano akêp **muyou**.*
 'I didn't pick up the **food**, but I picked up the **snake** instead.'
 (Game1_021012_0510)

Pitch and intensity of the utterance are plotted in Figure 14. There is a small peak in pitch on the second syllable of *nganngan* and the first syllable of *muyou*. In addition, intensity is highest on the negator *pwên* and on the contrasted element *muyou*.

There is evidence to suggest that the pitch and intensity range with contrastive focus are also dependent on genre, and perhaps show inter-speaker variation. Example (59) below was taken from a public speech:

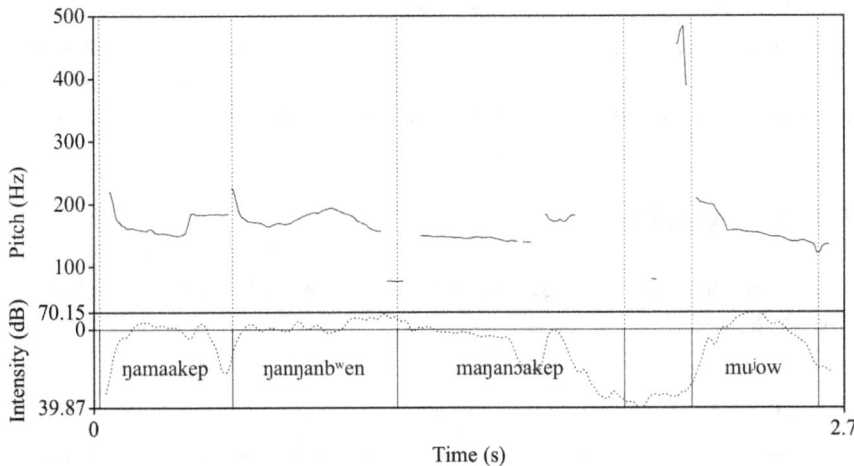

Figure 14: Pitch contour of example (58).

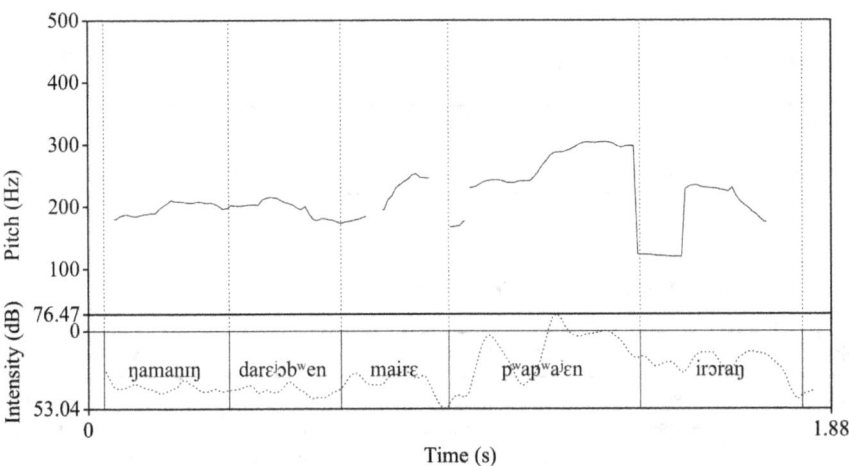

Figure 15: Pitch contour of example (59).

(59) [ŋa‿ma‿nɪŋ tarɛɔ‿pʷẹn | ma‿i‿rɛ pʷapʷaʲ↗'ɛn i‿rɔ ↘raŋ]
ngamaning tareo pwên, ma i re **pwapwaen** iro rang.
'I didn't see all that, but the **story** is mine.' (YK290411_1_0030)

Figure 15 shows pitch and intensity for example (59). It shows that the peak on the last syllable of the emphasised element is about 300 Hz, whereas pitch lies around 200 Hz elsewhere. In addition, the difference in intensity between the emphasised element and the rest of the utterance is greater than for the other examples shown in this section. This is an indication that the pitch and intensity range speakers have available to them may be greater in the case of public speeches.

2.4 Orthography

In the remainder of this work, data will be represented in a practical orthography, as shown in Table 6 for consonants and Table 7 for vowels.

Table 6: Orthographic symbols for consonant phonemes.

Phoneme	/p/	/t/	/k/	/s/	/m/	/n/	/ŋ/	/mʷ/	/pʷ/	/l/	[j]	[w]
Character	⟨p⟩	⟨t⟩ ⟨r⟩	⟨k⟩	⟨s⟩	⟨m⟩	⟨n⟩	⟨ng⟩	⟨mw⟩	⟨pw⟩	⟨l⟩	⟨y⟩	⟨w⟩

Table 7: Orthographic symbols for vowel phonemes.

Phoneme	/a/	/ɛ/	/e/	/i/	/ɔ/	/o/	/u/
Character	⟨a⟩	⟨e⟩	⟨ê⟩	⟨i⟩	⟨o⟩	⟨ô⟩	⟨u⟩

Table 8: Overview of possible trisyllabic forms in Paluai.

Structure	CV.CV.CV	CV.CV.VC	CV.CV.CVC	CVC.CV.CVC	CV.CV.V	V.CV.CV
Nouns	[ˈka.la.ˌsi] 'headdress' [ˈmʷa.lu.ˌpʷɛ] 'canoe type' [ˈkɔ.ŋu.ʲa] 'snoring'*	[ˌna.na.ˈɔp] 'tree sp.' [ˈpa.ru.ˌʷul] [ˌna.lu.ˈʷaj]	[ˌsa.ri.ˈkow] 'bridge' [kɔ.ˈlɔ.raj] 'song type'	[kuj.ˈku.jej] 'goosebumps' [ˈpaŋ.ˌpa.ŋaj] 'thought' [ˌpot.ˈpo.son] 'his chest'	[ˈmi.mi.a] 'urine'* [ˌsi.si.ˈa] 'broom'*	[ˈa.kɛ.kɛ] 'plant sp.'
Verbs	not attested	[ˌpo.ˈna.ʷut] 'be unskilled'	[ˈnu.ru.ˌβuj] 'mature' [ˌpu.lu.ˈtow] 'stick'	[kul.ˈpɔ.pɔt] 'whistle'	not attested	[ˈi.li.ˌli] 'stand up'

Table 8 (continued)

Structure	CV.CV.CV	CV.CV.VC	CV.CV.CVC	CVC.CV.CVC	CV.CV.V	V.CV.CV
Adjectives	not attested	[ˌko.ro.'an] 'violent' [ˈŋa.peˌan] 'bitter'	[ˈmɛ.nɛˌŋan] 'big' [ˌpi.li.'pil] 'yellow' [lu.'ka.jaw] 'light green'	[nak.'na.kan] 'bright'* [ˈkɔlˌpɛ.pɛn] 'sluggish' [ˌsik.si.'kan] 'sour'*	not attested	not attested
Other	not attested	[ˈsa.maˌin] 'what of' [ˈsɛ.lɛˌuk] 'one part of s.t. long'	[ˌmi.si.'min] 'five hundred' [mɔ.'nɔ.kin] 'afterwards'*	not attested	not attested	not attested

*indicates a derived form

Table 9: Overview of possible quadrisyllabic forms in Paluai.

Structure	CV.CV.CV.CV	CV.CV.CV.CVC	CV.CV.CV.VC	CVC.CV.CV.CVC
Nouns	[kuˌru.ru.'pa] 'making mounds in garden'	[ˌku.na.wa.'jɛn] 'his breath' [ˌpa.ra.'jo.jɔp] 'power' [ˌsa.ne.'ru.an] 'younger brother of deceased'*	[ˌpe.re.li.'an] 'his middle'	not attested
Verbs	not attested	[ˌku.na.wa.'jut] 'rest, regain breath'* [teˌre.pe.'lɛk] 'run'*	not attested	not attested
Adjectives	not attested	[ˌti.na.wa.'jɛn] 'enormous' [ˌnu.ru.'pu.jan] 'mature (of woman)'	not attested	[ˌsup.su.pu.'rɔn] 'rough'
Other	not attested	[ˌpa.nu.ra.'sip] 'first'*	[na.'re.ŋiˌaj] 'many' [ˌpe.ru.ru.'ɛk] 'ignorant' [ˌsa.pu.lu.'ɛk] 'round'	not attested

*indicates a derived form

The phoneme /t/ is always represented with ⟨t⟩ when it is known that a surface realisation [r] or [ɾ] is an underlying /t/. If this is not certain, it is represented with ⟨r⟩. Thus, for this phoneme the orthography is potentially overspecified. There are three digraphs in the system: ⟨ng⟩, ⟨mw⟩ and ⟨pw⟩. This will not give cause for ambiguity. Since /g/ is not a phoneme in Paluai, there are no [ng] sequences. /mu/ or /pu/ sequences that could be potential variants of the labialised consonants are represented as ⟨mu⟩ or ⟨pu⟩ and can thus be distinguished from them.

In line with native speaker preferences, non-syllabic /i/ and /u/ are written differently at the beginning and end of a word. Word-initially, they are written ⟨y⟩ and ⟨w⟩ respectively, whereas word-finally they are written ⟨i⟩ and ⟨u⟩, even though they are realised as [j] and [w] in both cases.

3 Word classes I: open classes

3.1 Preliminaries

In Paluai, the two major word classes are nouns and verbs. These form truly open classes to which new items are added constantly, for example through borrowing or through derivational mechanisms such as compounding. There is also a class of adjectives, but most and possibly all of its members are derived. Even so, there are criteria allowing to distinguish members of this class from both nouns and verbs. Distinguishing adverbial forms is slightly more problematic, but there are grounds on which a separate class of adverbs can be distinguished. To these two classes new members can be added as well, although this happens on a smaller scale than in the case of nouns and verbs.

This chapter discusses the four open word classes: noun, verb, adjective and adverb. These classes contain lexical items, or content morphemes, that carry lexical meaning and can be contrasted with functional or grammatical morphemes, which will be discussed in the next chapter and which generally form closed classes (but see Section 4.9 for discussion on borrowed subordinating conjunctions). In any case, the distinction between open and closed classes is a gradient rather than an absolute one. There is a correlation between the size, "lexicality" and "openness" of a word class. Nouns form the biggest class, which shows the most time-stable lexical meanings and the largest number of borrowings and neologisms. This is followed by verbs, adjectives and adverbs, respectively.

In this chapter, the criteria for distinguishing the various word classes will be discussed, together with morphological and periphrastic word-class-changing derivations. Open word classes can be subdivided into several subclasses, based on formal and/or semantic criteria. Where relevant, distinct subclasses will be discussed, together with the criteria on which they are based. An overview of modifiers that are relevant for a specific word class will be provided here as well, although such modifiers, strictly speaking, are generally part of a closed class (an example are TAM particles, which modify verbs). How modifiers work inside the noun phrase and verb complex will not be discussed in great detail here, but in Chapters 5 and 6, respectively. At the end of the current chapter, Tables 32 and 33 give an overview of word classes and the criteria to distinguish them, and Table 34 gives an overview of word-class-changing morphological processes.

3.2 The noun

Nouns form an open class consisting of several thousand members. Distributionally, nouns can be distinguished because they typically function as the head of a NP that, as a constituent, forms an argument of a verb. They can also be the head of a nominal predicate in a verbless clause. Nouns cannot take bound subject pronouns,[18] or TAM particles and prefixes. This distinguishes them from verbs. In addition, nouns cannot take an adverbial modifier of degree, which distinguishes them from adjectives. In contrast to members of other word classes, nouns can be part of a possessive construction and be modified by demonstratives, numerals, certain quantifiers as well as other nouns and adjectives. In addition to this, they can be replaced by pronouns and certain demonstratives.

Number marking is not obligatory. However, there are a number of strategies to indicate number on nouns (see Section 5.2), where animate nouns seem to have a greater tendency to be marked for number, in line with observations for other Oceanic languages (Lynch et al. 2002: 39). There is also no flagging of grammatical relations by bound morphology within the NP. Contrary to the majority of Oceanic languages, but in line with developments in the Admiralties subgroup (Blust 2013), Paluai nouns are not preceded by articles.

Nouns can be subcategorised in a number of ways, based on semantic grounds and concomitant grammatical properties. One distinction is based on the possessive construction(s) a noun can enter into. This is the subject of Section 3.2.2. Another distinction that can be made is a common one for Oceanic languages, between personal, local and common nouns (Section 3.2.1). A third distinction for nouns is disappearing fast. The semantics of a noun determines which numeral classifier is to be used with it, so that nouns can be grouped into subclasses based on the form of the numeral they take. For the forms of numerals, see Section 4.5; for the various numeral classes, see Section 3.2.3.

Nouns are often derived from verbs, but verbs derived from nouns seem to be very uncommon. There are several nominalising operations, such as reduplication, suffixation and zero derivation, which will be discussed in Section 3.3, where the focus is on verbs. There are minor processes that derive adjectives and adverbs from nouns. Common morphological operations on nouns that do not result in a change of word class are reduplication, compounding and suffixation. There are many irregularities in noun morphology. Nouns do not have prefixes, infixes or circumfixes.

---

18 The pronominal system is discussed in Section 4.2. There is a formal difference between free and bound pronouns only for the 1sg subject form.

3.2.1 Noun subclasses, type I: personal, local and common nouns

As is common for Oceanic languages (Lynch et al. 2002: 37), Paluai nouns can be grouped into three different subclasses: personal, local and common nouns.

Personal nouns include proper names of people. These generally do not appear in a possessive construction. An exception is formed by the so-called birth order terms (cf. Schokkin 2018). These terms sometimes appear in an indirect possession construction, e.g. *Alup ta-ng* 'my first-born daughter'. They are best analysed as "hybrid" forms used both as personal names and as referential devices.

Local nouns include place names and familiar places that need no further specification, such as 'shore', 'bush', etc. A subclass of these are "spatial nouns": directly possessed nouns referring to a location relative to an object and often used where other languages would use a preposition instead. For more extensive discussion, see Section 3.2.2.3. The remainder of nouns are common nouns.

In many Oceanic languages, nouns can be grouped based on the article they take: personal and common nouns often take a different article. Since Paluai, like most other Admiralties languages, has lost its articles, subclasses of nouns are primarily distinguished on semantic grounds. Local nouns can also be distinguished based on a syntactic criterion: when forming a locative constituent, they do not need to be introduced by a preposition *a=*. All common nouns, on the other hand, must be introduced by this preposition, which is added as a proclitic to a 3sg pronoun. In example (1), the local noun (place name) *Lou* is used; the local noun *suk* 'shore' appears in example (2). Example (3) shows that *kanum* 'garden' is a common and not a local noun, since it has to be accompanied by the preposition *a=*.

(1) *eppwa kala yik ngoyai Lou*
 ep=pwa   ka-la   yik   ngoyai   Lou
 1pl.EXCL=want.to IRR.3sg-go.to search.for cuscus   Lou
 'We were about to go and hunt cuscus on Lou Island.' (NP210511_2_0004)

(2) *urêsuwen suk*
 wurê=suwen   suk
 1pc.EXCL=move.down   shore
 'We went down to the shore.' (MK060211_0006)

(3) *ipat mun to ai kanum teo*
 i=pat   mun   to   a=i   kanum   te-yo
 3sg=plant banana   be   OBL=3sg   garden LIG-DEM.INT
 'He planted bananas in the garden.' (WL020611_0029)

3.2.2 Noun subclasses, type II: direct and indirect possession

A second way to group nouns is based on how they are used in possessive constructions. Unless otherwise indicated, the terms "possession" and "possessive" are used here to refer to a purely grammatical construction that usually points to a relation of close association characterised by an entity, the Possessor (Po), "having", "owning" or "controlling" another entity, the Possessee (Pe). This close association encompasses a number of semantic relations with different characteristics; possession or ownership in the strict sense is only one of them. Because this range of relations is expressed by a limited number of grammatical constructions, these are often grouped together under the header "possessive constructions".

Formally, many Oceanic languages distinguish two types of possessive construction: direct and indirect (cf. Lichtenberk 2009a). The terms "alienable" and "inalienable", which are also often used with regard to possessive constructions, will be used exclusively to refer to the semantic sense of a possessive relation. The two pairs of terms, "direct" and "indirect" on the one hand and "inalienable" and "alienable" on the other, are kept separate because it is not at all certain whether we can assume a one-to-one correspondence between a formally direct and semantically inalienable type of possession, on the one hand, and a formally indirect and semantically alienable type of possession, on the other hand. This issue will be discussed in more detail below.

In the case of direct possession, a suffix cross-referencing person and number of the Po, referred to as a "pertensive" suffix, is added directly to the noun. With indirect possession, a pertensive suffix is added to a possessive classifier either preceding or following the noun. Nouns that are only found in direct possessive constructions are bound roots, i.e. they are obligatorily suffixed, and end in a vowel. A large number of nouns have two forms: a bound root ending in a vowel, which is used in a direct possessive construction (a long form) and an independent form without the vowel, which is used by itself or in an indirect construction (a short form).[19] Many other nouns only appear in indirect possessive constructions. In Paluai the Po follows the Pe in direct and indirect possessive constructions alike, with the exception of constructions involving the alimentary classifier *ka-*, where Po precedes Pe (see Sections 3.2.3.2 and 5.5.2). The formal characteristiscs of direct and indirect possessive constructions are

---

19 Historically word-final consonants and the vowels that precede them were dropped in Paluai and other Eastern Admiralties languages (Ross 1988: 330; Blust 2013: 95). This explains why the final vowel was retained in the suffixed forms, but was lost from the forms used independently.

discussed in more detail in Section 5.5; the paradigms of pronominal possessive suffixes are the topic of Section 4.2.3.

There are four formal subclasses of nouns with regard to possession. A noun:
1. Must appear in a direct possessive construction (as a bound root);
2. May optionally appear in a direct possessive construction (as a long form) or unpossessed (as a short form), but never appears in an indirect possession construction;
3. May appear either in a direct possessive construction (as a long form) or (as short forms) in an indirect possessive construction or unpossessed;
4. May appear in an indirect possessive construction or unpossessed.
5. Must appear in an indirect possessive construction, and never appears unpossessed.

The rules for subclasses 1 and 5 are the strictest, as these nouns must always be used in a possessive construction. The rules for subclasses 2 to 4 are somewhat looser, since these nouns are not obligatorily possessed. To the first subclass belong most nouns that refer to kinship relations, some nouns that refer to human and non-human attributes or characteristics, and spatial nouns that function prepositionally. The second subclass consists primarily of body part terms. The third subclass mainly contains nouns referring to attributes; it is discussed in Section 3.2.2.4.3. To the fourth subclass belong all nouns, including a number of kin terms, that have not been encountered in a direct possessive construction. The fifth subclass appears to have only one member: the kin term *pên* 'daughter'.

As stated previously, many more semantic distinctions than 'ownership' in the strict sense can be linguistically expressed with a possessive construction. The following semantic distinctions will be discussed here:
– Kinship relations, both consanguineal and affinal
– Part-whole relations, including:
    – Human and animal body parts
    – Tree and plant parts
    – Non-animate object parts
– Spatial relations, extended to temporal relations
– Human characteristics and propensities
– Attributes and characteristics, usually of non-humans

Spatial relations are often metaphorically extended from part-whole relations. In addition, a construction can be ambiguous between two interpretations. The various possibilities are discussed one by one in the following subsections.

3.2.2.1 Kinship relations

Kinship terms are mentioned in a range of works as the most likely candidates for obligatory direct possession: see e.g. Lévy-Bruhl (1914); Chappell and McGregor (1989, 1996); McGregor (2009); Aikhenvald and Dixon (2013); and, specifically on Oceanic languages, Lichtenberk (2009a). Generally, the explanation provided is that in a great number of cultures, kinship is perceived as inalienable. The relation between Po and Pe is thought of as something that cannot be undone: blood ties are for life. Paluai kinship terms form a closed subclass of twenty-five nouns, twenty-one of which obligatorily occur in a direct construction. These terms function both referentially (in which case they refer to the relative as a third person when addressing somebody else in the second) and vocatively (when the relative is addressed in person). An overview of the terms is given in Table 10 and Table 11.

Table 10: Terms for consanguineal kin.

Generation	Paluai term	English equivalent	Used for[20]
+4	*makapua-*	great-great-grandparent	all relatives in this generation
+3	*apua-*	great-grandparent	all relatives in this generation
+2	*tupu-*	grandparent	all relatives in this generation
+1	*tama-*	father	F, FB, FFBS, FFBS etc.
	tina-	mother	M, MZ, MFBD etc.
	nukna-	uncle	MB, MFBS, MFZS etc.
	sae-	aunt	FZ, FFBD, FFZD etc.
0	*tou-*	older same-sex sibling	FeS, FeBS, MeZS etc. (ego m); FeD, FeBD, MeZD etc. (ego f); WeZH, HeBW (WeZ, HeB)
	nae- / *sanei-*[21]	younger same-sex sibling	FyS, FyBS, MyZS etc. (ego m); FyD, FyBD, MyZD (ego f); WyZH, HyBW (WyZ, HyB)
	patne-	sister (ego m)	FD, FBD, MZD etc.
	mwane-	brother (ego f)	FS, FBS, MZS etc.
	pwai	(cross-)cousin	FZC, FFBDC, FFZDC etc.; MBC, MFBSC, MFZSC etc.

---

20 F = father, M = mother, B = brother, Z = sister, S = son, D = daughter, H = husband, W = wife, C = child, e = older, y = younger

21 *nae-* is predominantly used for "real" siblings, whereas *sanei-* can be used for both "real" and classificatory siblings.

Table 10 (continued)

Generation	Paluai term	English equivalent	Used for[20]
−1	naru-	son	S, BS, FZSS, MBSS, FBSS, MZSS (ego m); ZS, FZDS, MBDS, FBDS, MZDS (ego f)
	pên ta-	daughter	D, BD, FZSD, MBSD, FBSD, MZSD (ego m); ZD, FZDD, MBDD, FBDD, MZDD (ego f)
	wolia-	niece/nephew	ZC, MBDC, MZDC, FBDC, FZDC etc. (ego m); BC, MBSC, MZSC, FBSC, FZSC etc. (ego f)
−2	maêwe-	grandchild	all relatives in this generation
−3	apua-	great-grandchild	all relatives in this generation
−4	makapua-	great-great-grandchild	all relatives in this generation

Table 11: Terms for affinal kin.

Generation	Paluai term	English equivalent	Used for
all	nope-	in-law	term indicating all kinds of affines of all generations e.g. SWF/M, DHF/M, BWF/M, ZHF/M, SWB/Z, DHB/Z
+1	nana-	father/mother-in-law	HF, WF; HM, WM
0	asoa-	husband	H
	paria-	wife	W
	ipa-	sister-in-law (ego f)	patne- of H, W of mwane-
	polam	in-law (general)	affines of ego's generation
	kauwat	in-law; friend; tradespartner	affines mainly of ego's generation
−1	kat naru-	daughter-in-law	SW

Paluai has an elaborate kinship system, discussed in more detail in Schokkin and Otto (2017). Suffice it to say here that many kinship terms are classificatory: for example, father's brother's children and mother's sister's children are grouped with real siblings, and so on.

There are four exceptions to the rule that kinship terms should be directly possessed: these are *pên* 'daughter', *polam* 'in-law', *kauwat* 'in-law' and *pwai* 'cousin'. Cousins and in-laws have a "special" status in Paluai culture, and both statuses can be said to be diametrically opposed. With in-laws, one is in a taboo relationship, which means they have to be paid respect. *Polam* is preferably used as a respectful term to address and greet the person.

Cross-cousins, on the other hand, are people with whom one can freely joke. This is especially the case with second or third cousins, when people are *pet pwai kesin*. With *pet pwai laleusip*, the children of your MB and FZ, the joking relationship is still quite civilised. Joking relations between cousins are very widespread in Melanesia and typically are rather bawdy with a lot of sexually tinted jokes. Since there is a cross-sibling link in the ascending generation, cousins on one's father's side do not belong to his (and therefore one's own) lineage. Cousins on one's mother's side do belong to her lineage, but not to one's own. Traditionally, cousins negotiate in finding marriage partners for each other, but these partners have to have a sufficiently distant relationship.

Thus, *polam*, *kauwat* and *pwai* may stand for somewhat "looser" relationships, albeit for very different reasons. This looser relationship may be reflected in the linguistic form with which it is referred to. Interestingly, *kauwat*, *pwai* and *polam* seem to be the only symmetrical kinship terms (i.e., if one person calls the other *pwai*, the reverse is also true).

The fact that *pên* 'daughter' is not directly possessed is harder to explain. Since marriage patterns are exogamous, daughters will marry out of their own clan and their children will belong to the daughter's husband's clan. This may be reflected in the kinship term being indirectly possessed. Interestingly, *pên* always occurs in an indirect possession construction and never as an unpossessed form, also when used as an address term. This is not the case with *polam* and *pwai*. Thus, it can still be considered as obligatorily possessed (like other kinship terms); the possessive construction just takes a different form.

3.2.2.2 Part-whole relations

3.2.2.2.1 Human and animal body parts

Like kinship terms, body part terms are likely candidates for obligatory direct possession. Since the part cannot, in general, be severed from the whole without serious harm, this part-whole relation is seen as inalienable.

For Paluai a distinction can be made between primary and secondary body parts. Primary body parts form an (in theory) closed subclass of single morphemes referring to various parts of the body. The larger open subclass of secondary body parts consists of compounds, of which the first term (mostly bearing a -*n* pertensive suffix) refers to a part of the second term, which is chosen from the closed class of primary body parts. Some examples are given below: primary body parts in (4) and (5), secondary body parts in (6) and (7).

(4) *minong*
    mina-ng
    hand-1sg.PERT
    'my hand'

(5) *nupum*
    nupu-m
    bottom-2sg.PERT
    'your bottom'

(6) *numun parung*
    numu-n   patu-ng
    hair-PERT head-1sg.PERT
    'my head hair'

(7) *kapun kêm*
    kapu-n   ke-m
    bladder-PERT leg-2sg.PERT
    'your calf'

Sometimes the first terms of these secondary body parts are not used by themselves. For instance, when a speaker wants to say 'my hair', she always needs to specify the body part that she is referring to. The class of secondary body parts is in principle an open class, because for many items any primary body part could theoretically be placed in the second slot.

Body parts, both external and internal, usually appear in a direct possessive construction. Some bodily substances, like *apirô-* 'sweat', *ngingia-* 'saliva' and *kunawaye-* 'breath', usually do as well. However, in certain contexts the unsuffixed stem can be used. This is possible with most, perhaps all, body part terms. For instance, for blood the suffixed root *noye-* is used when referring to blood inside the body, but the unsuffixed stem *noy* when referring to bloodstains or blood in a bag for transfusion, for example. In addition, body part terms can be used as bare stems when they are used in generic contexts. The final vowel is absent in these cases; a few examples are provided in (8) and (9).

(8) *ipmape ai min pwên, te i repwo ippe ai masin*
ip=ma=pe   a=i   min   pwên
3pl=NEG₁=make OBL=3sg hand   NEG₂
te   i   te-pwo   ip=pe   a=i   masin
SUB 3sg LIG-DEM.PROX 3pl=make OBL=3sg machine
'It is not made by hand, but by machines.' (Game1_021012_0165)

(9) *nupan mat*
nup-an   mata
wash-NOM eye
lit. 'washing of the eye' [name of traditional ceremony]

An interesting point about body part terms is that when they are used as bare stems, they do not appear in indirect possession constructions, but always as unpossessed stems. Thus, when a speaker wants to express a possessive relation that includes a body part, this can only be done by means of a direct possessive construction. Bodily substances that can be regarded as waste products, such as *ne* 'faeces' and *mimia* 'urine', are always indirectly possessed.

3.2.2.2.2 Tree and plant parts

Tree and plant parts are usually suffixed with *-n* and are thus formally very similar to human and animal body parts.[22] The problem, however, is that there are no forms showing a first or second person suffix, which makes it hard to establish whether the relation should be characterised as possession or association.[23] For a more elaborate discussion of this problem, see Section 3.2.2.4.3. Table 12 shows some frequently encountered forms.

A number of the terms in the table (*sanga-*, *koropu-* and *noye-*) are also used to refer to human body parts. Many of the forms end in *-n*. Whether this final *-n* should be considered a suffix or part of a "frozen" form is a matter for discussion. Although terms for plant parts are generally used in combination with e.g. a name for a tree species, such as in *yêbin tiok* 'piper betle leaf', they are also often used by themselves, as in the example below.

---

[22] As is set out in Section 5.5, *-n* is the pertensive suffix for the third person singular and all non-singular persons. The first person singular pertensive suffix is *-ng*, and the pertensive suffix for second person singular is *-m*.
[23] Of course, one could try to elicit this by asking if it would be possible for a tree (if trees could talk) to use e.g. the directly possessed form *yêpi-ng* 'my leaf', but this would not be readily comparable because the tree would in this case assume an anthropomorphic character.

3.2 The noun — 67

Table 12: Tree and plant parts.

Part	Meaning	Term with human reference
akon	root	
koropun	base of tree trunk	koropu- 'abdomen'
kumun	sprout	
kupwen	branch	
kuêsamen	string with which a coconut is attached to the tree	
lo	covering sheath of coconut	
mangun	dry coconut with meat	
muin	green coconut with juice and little meat	
noyen	sap	noye- 'blood'
paraken	stalk of coconut frond	
paran	stem, trunk	
puan	fruit, seed	
payan	soft meat inside a coconut	
pukien / pwakien	unripe coconut (only water, sour taste)	
pungun	coconut husk	
(pun) kuen	stalk with which fruit is attached to the tree	
pusuan	coconut shell	
saêrin	coconut palm frond	
sangan	area between the roots of a tree	sanga- 'groin'
sangarun	area between the "fork" of the branches of a tree	
sapon	top of a plant, crown of a tree	
silin	sucker (of banana, pineapple)	
(su)puron	node on bamboo stalk	
yapit	"side leaves" of coconut frond	
yêpin	leaf	
yon	tree leaf (bigger sized, such as banana leaf)	

(10) *yep iret ai yêpin teo*
 yep i=tet a=i yêpin te-yo
 fire 3sg=spread OBL=3sg leaf LIG-DEM.INT
 'The fire spread through the leaves.' (KM190211_0020)

The forms ending in *-n* are never used as bare stems (i.e., without the *-n*, as is the case for e.g. body parts), indicating that speakers do not perceive *-n* as a suffix and may regard it as part of the stem. Supporting evidence for this claim is that the forms ending in *-n* are used as citation forms, in contrast to nouns referring to body parts, where the short form is the citation form. Finally, in cases like the example above, it could be said that *yêpin* is referring to 'its leaves' in general, without specifying the Po, or that the Po is elided. However, the story from which this example was taken is about clearing the bush in order to make space for a garden, so that, in fact, *yêpin* refers to the leaves of many trees. Nevertheless, the plural is not indicated on the pertensive form, which gives rise to the assumption that *yêpin* and other names of plant parts ending in *-n* refer to the part in a generic sense, just as the bare stem does in the case of body parts.

3.2.2.2.3 Non-animate object parts

As with other part-whole relations, parts of non-animate objects are generally expressed through a direct possessive (or associative) construction (on the latter, see Section 3.2.2.4.3 below) that can usually be analysed as a compound. Some of these are metaphorical extensions of human body parts, such as *pole-* 'lower arm'. Some examples are given in (11) and (12); compounding is discussed in more detail in Section 3.2.4.1.

(11) *kerein kel*
 kerei-n kele
 edge-PERT canoe
 'canoe edge'

(12) *polen pil*
 pole-n pil
 lower.arm-PERT ladle
 'handle of a soup ladle'

3.2.2.3 Spatial relations

Spatial relations are generally expressed by directly possessed nouns of a closed subclass called "spatial nouns". Although there seems to be a small class of genuine prepositions (most notably *a=*; see Section 4.4.1), these have very generic semantics, and when more specification is needed a spatial noun is used. Some spatial nouns are clear metaphorical extensions of body parts; others are more opaque. Table 13 gives an overview.

As is clear from the table, many spatial nouns are also used as body parts or modifiers of secondary body parts. Evidence that, synchronically, these forms are genuine directly possessed nouns, rather than frozen forms, comes from the fact that direct possession is productive for many terms. For instance *kaso-*, *naêmwa-* and *pwalinga-* are attested with first and second person pertensive suffixes. This

Table 13: Spatial nouns.

Form	Spatial meaning	Related term
asilo-	(at the) side of	
kaso-	near to	
kô-	(on) top of	kô- '(on) top of the body'
lalo-	inside (of)	
mata-	(in) front of	mata- 'eye'
monok / monoki-	(at the) back, behind (also used for temporal relations)	
moyen / moyenga-	between, among, in the middle of	moyen lipa- 'front teeth'
naêmwa-	(at the) back of	naêmwa- 'backside'
nau-	(on the) surface of	naun mina- 'back of hand'
naso-	(at the) end of (s.t. elongated)	
nisie-	(at the) middle of (s.t. elongated)	nisie- 'middle of the body'
panu / panua-	(at the) front, in front of (also used for temporal relations)	
parayo-	(in) front of (a human being)	
pata-	(on) top of	
pururua-	(in the) middle of	pururuan patu- 'middle of skull'
pwalinga-	(together) with (comitative)	
pwapwa-	(at the) side of	pwapwa- 'cheek'
pwoyo-	under(neath)	paye 'down below'
soyo-	(at the) pointed front of s.t.	e.g. soyon puyusu- 'tip of nose'
sue-	(at the) side of	suen mata- 'side of face'
ya-	(in the) top of	yan sipe- 'top of shoulder'

is not the case for e.g. plant part terms. Examples (13) to (15) show spatial nouns with first and second person possessors.

(13) *ip lau karo au kasông*
   ip   lau   ka-to   wau   kaso-ng
   3pl people IRR.NS-HAB   move   near-1sg.PERT
   'The people will walk near me.' (LK25011_0075)

(14) *ngaro naêmwom*
   nga=to   naêmwa-m
   1sg=be   backside-2sg.PERT
   'I am behind you.' (field notes 20/08/2012)

(15) naêng ta Pwanou iro pwalingong
　　　naê-ng　　　　　　　　　ta=Pwanou　i=to　　pwalinga-ng
　　　younger.sibling-1SG.PERT　SPEC.COLL=P.　3SG=be　with-1SG.PERT
　　　'My younger brother Pwanou stayed with me.' (KM060111_0015)

An example of *parayo-* and *naêmwa-* with the general *-n* suffix is given in (16). The terms here refer more generally to the front and the back of a scene, rather than to the front and back of specific objects.

(16) lôk teo, iro parayon le iro naêmwan?
　　　lôk　　　te-yo　　　　i=to　　parayo-n　　le　i=to　　　naêmwa-n
　　　vegetable　LIG-DEM.INT　3SG=be　front-PERT　or　3SG=be　backside-PERT
　　　'The vegetable, is it at the front or at the back?' (Game1_021012_0281)

The terms *monok* and *panua* in particular are used to refer to temporal notions of 'before' and 'after', in addition to their primary meanings 'front' and 'back'. Some examples are given in (17) to (19); here, too, constructions can be ambiguous between a spatial and a temporal interpretation.

(17) kope sopol pari panu a kope sopol pari monok
　　　ko-pe　　　　sopola　　　　　pari　　　　　　panua　a
　　　IRR.1SG-make　one.half.round　belonging.to　front　and
　　　ko-pe　　　　sopola　　　　　pari　　　　　　monok
　　　IRR.1SG-make　one.half.round　belonging.to　behind
　　　'I will make the front side and the back side [of a grass skirt].' (AK160411_1_0012)

(18) yamat te iro au panuan teo
　　　yamat　　te　　i=to　　　wau　　panua-n　　te-yo
　　　person　REL　3SG=CONT　walk　front-PERT　LIG-DEM.INT
　　　'The person who went before her / in front of her...' (KW290611_0012)

(19) monokin pêng tangunan
　　　monoki-n　　　pêng　ta=ngunan
　　　behind-PERT　day　SPEC.COLL=five
　　　'After five days, ...' (LM260511_1_0046)

Sometimes, other body part terms are found in temporal expressions:

(20) perelia-n   panu
     middle-PERT place
     'midday, noon'[24]

3.2.2.4 Attributes

3.2.2.4.1 Human characteristics and propensities

Interestingly, the direct possessive construction is not limited in its use to concrete parts of the body. Other human traits that may be seen as inalienable can, and often must, also be directly possessed. Table 14 gives an overview of these. Some of the terms, such as *arona-*, *ngaya-*, *nuruna-* and *pwaku-*, can also refer to characteristics of non-human entities. This is the case in examples (21) and (22), whereas in (23) and (24) they refer to human characteristics or belongings.

(21) *aronan suian yapi re ep Paluai ro pe*
     arona-n   sui-an   yapi te   ep   Paluai to   pe
     procedure-PERT   fry-NOM sago REL 1pl.EXCL   P.   HAB do
     'The procedure of frying sago, such as we Paluai are used to doing it.' (CA120211_1_0015)

(22) *kope puk mapian la rai*
     ko-pe   puk   mapia-n   la   ta-i
     IRR.1SG-PFV   open knowledge-PERT   go.to POSS.CLF-3sg.PERT
     'I will inform her about it.' [*lit.* 'I will open the knowledge of it to her'] (KW290311_0008)

(23) *ngamalêp nansê masiom pwên*
     nga=ma=lêp   nan=sê   masia-m   pwên
     1sg=NEG₁=take   PART=DIM   appreciation-2sg.PERT NEG₂
     'I don't appreciate what you did at all.' [*lit.* 'I do not take a bit of your appreciation'] (LL010711_0062)

---

24 *Panu* has many senses, among which 'the visible world'. Here it probably refers to 'day' (as opposed to 'night').

Table 14: Human characteristics.

Form	Meaning/paraphrase
arona-	character; way s.t. should be done, procedure; sense, foreknowledge (prescriptive/moral dimension)
ayo-	way
kasu-	bad behaviour
longoa-	(temporary) habit, characteristic
lopwa-	(designated) place, location; position in society
maloa-	shadow; reflection in water; spirit, soul (also: picture, photograph)
mamarou / mamarou-°	way of doing s.t., style, characteristic behaviour (moral dimension; overtones of "good behaviour")
mapia-	knowledge
masia-	appreciation
napu-	(food) taboo
ngai / ngaya-°	name
ngola-	language
nia-	pregnancy (lit. 'stomach')
nonou / nonoua-°	promise; a sign of s.t. that cannot be changed
nurun / nuruna-°	possessions; everything that typically belongs to s.t. or s.b.; (more abstract) business
perina-	bad behaviour, bad attitude
pepi	kindness, generosity (not directly possessed)
pelingara-	namesake
pwaku-	style, type (e.g. of dancing)
sayoa-	"ordinariness"[25]
tangoa-	(temporary) habit, characteristic; effort, attempt

° Direct possession is not obligatory. The first form represents the short variant, which occurs unpossessed or in indirect possession constructions; the second form shows the directly possessed stem.

(24) *Ngat, i repwo malopwom pwên*
 Ngat i te-pwo ma=lopwa-m pwên
 N. 3sg LIG-DEM.PROX NEG$_1$=place-2sg.PERT NEG$_2$
 'Ngat, this here is not your place.' [said to somebody who died on Lou Island] (NP210511_2_0055)

Some items, such as *nurun(a-)* and *mamarou*, can appear either in a direct or an indirect possessive construction, which is in line with what Lichtenberk

---

25 This term is used in the phrase *wuk sayoa-* which means 'mock' (*lit.* 'declare ordinary').

(2009a) refers to as "fluidity" in Oceanic possessive systems. It is not easy to pinpoint the decisive factors as to which construction is chosen when. One speaker mentioned that *nuruna-* used with direct possession is more all-encompassing, referring to each and every thing associated with a person or clan (both concrete items and abstract notions), whereas with indirect possession it refers to more commonplace and concrete things like someone's clothes or personal belongings carried on the body. Another speaker mentioned that the indirect possessive construction is sometimes used (with additional loudness and higher pitch) in order to emphasise a contrast between the speaker's ways and someone else's:

(25) *mamarou **rang** i remenin teo*
 mamarou ta-ng i temenin te-yo
 way CLF.POSS-1sg.PERT 3sg like LIG-DEM.INT
 '**My** way is like that [as opposed to his].' (field notes 07/05/2011)

3.2.2.4.2 Possessor as Patient or Stimulus

Some direct constructions have a Possessor functioning as a Patient or Stimulus of a transitive event. Verb nominalisations are of this type and typically occur with a third singular Possessor, but some are found in the corpus with a first or second person singular suffix. Those that are refer to attributes of humans, or personified characters in a story. Some examples of third person nominalisations are given in Section 3.3.2.2.3.2; examples of nouns occurring with first or second person are *malala-* 'clearness', *ning(i)a-* 'look, appearance', derived from the verb *ning*, and *sipia-* 'scatteredness', from the verb *(sip)sipi*.

3.2.2.4.3 Attributes of non-human entities

Nouns describing attributes of non-human entities fall into two categories. Nouns in the first category are always suffixed, but only with a third person *-n* suffix, never with a first or second person suffix. These are therefore similar to the plant and object parts that were discussed in Section 3.2.2.2; examples are typically nouns describing physical properties of objects, such as *molea-* 'colour', *kuli-* 'depth' and *namne-* 'taste'.

Then there is a category of attributes that consists of nouns that essentially appear in their unpossessed forms, or that are indirectly possessed. They are much less commonly used in a direct possession construction. Examples include *wumwa-/wum* 'house' and *sale-/sal* 'road'. These items have a long and a short form, and there is a large group of them (see Table 15, which is far from

Table 15: A selection of nouns that can be both directly and indirectly possessed.

Form	Meaning
kamou / kamou-	words, speech
kapu / kapua-	gift of food or money at bride price
kulut / kulusu-	rubbish
lapan / lapana-	chief
mosap / mosape-	bride price ceremony
pêng / pêngi-	day
pwapwa / pwapwae-	story
sal / sale-	way, "path" of e.g. a story (how and by whom it is owned)
wum / wumwa-	house
yamat / yamase-	people, inhabitants

exhaustive). When they do appear in a direct possession construction, they are only attested with the default -n suffix.

The semantic relations represented by many of these constructions are maybe better characterised as "associative" rather than "possessive" relations. However, since grammatical possession already encompasses a much broader set of relations than just possession in the strict sense, any line drawn between possessive and associative constructions is rather arbitrary. Lichtenberk (2006: 24) argues for an analysis that distinguishes between possessive and associative relations, in which the latter are non-specific and non-referential, in contrast to the former: "Associative constructions are not relational in the way possessive constructions are". The modifying noun represents a type rather than an entity, as in e.g. *kulusu-n kei* 'sawdust' (lit. 'rubbish-3sg.PERT tree'). For Proto-Oceanic, particles **qi* and **ni* have been reconstructed which indicated this associative relationship (cf. Lichtenberk 2006). In many modern Oceanic languages, the associative construction is still formally different from possessive constructions.

In Paluai, however, there is only one formal construction indicating both possessive and associative relations: the -n suffix. It is therefore likely that the reflexes of POc **ni* and **-ñ* '3sg possessive' have merged at some stage. It could be argued that if a form is only attested with the third person -n suffix and not with a first or second person suffix, it can only appear in an associative construction and cannot be directly possessed. But the absence of other forms could also just be due to a gap in the data. Therefore, the difference between possessive and associative constructions can only be established on semantic or referential grounds.

3.2.2.5 Fluidity between possessive constructions

The phenomenon of fluidity in possessive systems was already mentioned in Section 3.2.2.4.1 above, and an example was given involving the noun *mamarou* 'way'. The same fluidity occurs in other places as well. Example (26) includes the noun *kamou* 'words, speech', which falls into the category of nouns discussed directly above.

(26) *ngapwa kopul kamoun kun*
 nga=pwa ko-pul kamou-n kun
 1sg=want.to IRR.1sg-tell speech-PERT small.basket
 'I am going to talk about baskets [i.e. tell "the story of the basket" in a generic sense].' (MK050311_0001)

(27) *tapkano yong kamou ran lapan tararap*
 tap=ka-no yong kamou ta-n lapana
 1pl.INCL=IRR.NS-IPFV hear speech CLF.POSS-PERT chief
 ta-tap
 CLF.POSS-1pl.INCL
 'We will obey the words of our chief [i.e. specific words that he just said].' (LL300511_1_0044)

Examples (26) and (27) typify the use of many items similar to *kamou*, like the ones mentioned in Table 15. Thus, even though there may not be a possessive relation in these cases, the semantic distinction between alienable and inalienable still seems somehow valid. The inalienable relation is represented by direct suffixation, and stands for a relation that is upheld across several specific occasions, a generic relationship. The alienable relation is represented by suffixation of a possessive classifier, and refers to a specific occasion, a more volatile relationship.

Fluidity is also observed for one kinship term, and specificity seems once again to play a role, but a slightly different one. The directly possessed term *naru-* 'son, child' has an indirectly possessed counterpart *not*, probably originating from the same source. The difference between the use of the two terms is exemplified below.

(28) *wong narum*
 wong naru-m
 1sg.FREE son-2sg.PERT
 'I am your son.' [one specific individual]

(29) *som not tao*
       som   not   ta-o
       one.ANIM   son   CLF.POSS-2sg.PERT
       'one of your children/sons' [any one, non-specific, non-individuated]

In these cases, use of the indirect construction is associated with situations where the reference to the possessee is generic and not highly individuated. The indirect construction is also typically used to refer to children of both genders, whereas the directly possessed form is used (almost) exclusively for male children, at least in the contemporary language.[26]

3.2.3 Noun subclasses, type III: numeral and possessive classifiers

3.2.3.1 Numeral classifiers

A third and final way in which nouns can be divided is based on which class of numerals they take. Many languages in the region have an extensive system of numeral classifiers: among them are Loniu (Hamel 1994), Mussau (Brownie and Brownie 2007), Kele (Lynch et al. 2002) and Seimat (Wozna and Wilson 2005). The same kind of elaborate system may have been present in Paluai but, if it was, it is all but obsolete in the present-day language. The data that were gathered on this subject are mainly elicited and often inconsistent, since there is disagreement among native speakers about which classifier a given noun would take.

Numeral classifiers take the form of a suffix on a cardinal numeral modifying the noun; which suffix is added depends on the semantics of the modified noun. In addition, there are a number of nouns referring to fractions (meanings such as 'part of', 'half of', 'a piece of') or collections, which are strictly speaking not numerals, but whose distribution is dependent on the same system of semantic classification.

The only distinction that still holds firm ground in the present-day language is between animate and inanimate referents. Numerals referring to animate entities take the suffix *-mou*, those referring to inanimate entities take the suffix *-êp* (some numerals have suppletive forms – see Section 4.5 for more details). These forms are consistently used, also in the higher numbers. Otto (personal communication)

---

[26] There used to be a counterpart to refer to daughters: *naru- pein* (lit. 'female child'). However, this term took on an additional meaning, referring to a group of people in customary exchange ceremonies. Perhaps partly for that reason, the indirectly possessed term *pên* 'daughter' is now predominantly used.

mentions that during his fieldwork in the 1980s the -*mou* suffix was used exclusively for humans and pigs, but this has definitely changed, since the corpus on which this work is based contains many examples of numerals suffixed with -*mou* used for other kinds of animals, such as fish.

One other classifier that is still commonly used in addition to the animate one specifies 'elongated objects' such as trees, canoes and roads, where the numeral takes the suffix -*ei*. However, the -*ei* suffix is not attested with numbers higher than four. All other semantic distinctions are attested only for the numbers one and (sometimes) two, and native speaker intuitions as to their exact scope differ. Some of the forms have cognates in Loniu, which, at the time fieldwork was done by Patricia Hamel for her 1994 grammar, still had many noun classes that were formally distinguished by means of classifiers suffixed to numerals up to ten. This indicates that the Paluai classifier system may have been more elaborate in the past. For number 'one', the forms encountered are given in Table 16.

Table 16: Semantic distinctions involving the numeral 'one'.

Form	Can combine with a noun referring to
saya	place, area of land, village
sakam	leaf
sei	elongated object
seleuk	part of elongated object
sikil	leaf, wing, betel pepper fruit
sip	inanimate object (residue or default category)
som	animate being
sum(ut)	flat object, grass skirt
supu	heap, bundle, group of (e.g. fruit or people)
sui	heap of fruits
sêk	half of elongated object
sopol	half of round object

Some examples of the use of numeral classifiers in spontaneous speech are given below.

(30) *pureu supu reo ila ro ai*
 pureu supu te-yo i=la to a=i
 grass one.heap LIG-DEM.INT 3sg=go.to be OBL=3sg
 'There is one bunch of grass on it.' (Game1_280812_0086)

(31) *kolêp pit la sakam sakam*
  ko-lêp   pit   la   sakam   sakam
  IRR.1sg-take be.close go.to one.leaf one.leaf
  'I will collect [the coconut leaves] one by one.' (AK160411_2_0004)

These semantic distinctions made for inanimates with regard to the numeral they take (and possibly many more) have all but disappeared and are now grouped together, as reflected by the use of a single default marker *-êp*. In the present-day language, the numerals for animates are also used for inanimate objects that are somehow perceived important. This may be an innovation due to the loss of the classifier system, or it may be that nouns referring to important and abstract concepts are classed with animates. In the corpus, numerals for animates have been encountered in combination with the following nouns: *pule-n* 'thing', *nel* 'rope', *ngai* 'name', *yêkyêk* 'feeling', *pangpangai* 'thought', *puron* 'thing' and *kamou* 'speech'. *nel* used to refer to a string of dog's teeth and thus to money, which is probably the reason why it is perceived as important.

In addition to classifiers suffixed to the numerals, there are separate morphemes that often accompany a numeral and a noun and that give more information about the thing counted. They seem to have developed from generic nouns, and most of the ones encountered refer to fractions and collections. When used, they precede the noun they modify. Table 17 shows the forms encountered in the corpus. Interestingly, they all end in *-n*, and so may contain

Table 17: "Generic" nouns used for fractions and collections.

Form	Literal meaning	Used when counting
patan	stick, stem, trunk	trees, tree trunks (timber), poles/masts
kerin	year	"hands" or rows on a bunch of bananas
kuen	[unknown]	bunches of round fruits
manun	[unknown]	pieces of clay pot
matan	eye	heaps of stones, or yams; especially to be distributed at traditional ceremonies
napukun	[unknown]	heaps of stones or yams
nipen	[unknown]	parts of s.t. round, esp. soft things; pieces of s.t. sharp; areas of land
paripen	[unknown]	coins; single dog's teeth; maybe paper money
puan	fruit	heaps, bundles (e.g. of firewood), groups of people
sipen	[unknown]	parts of s.t. hard (stone, metal); areas of land
yôpôn	backbone; knuckle	parts of s.t. elongated

the third person pertensive suffix. However, due to inconsistent and scattered data, not much can be concluded about them with some degree of certainty.

Some examples of these forms in spontaneous language use are given below.

(32) *suei kila napukun sip, mwayen kila napukun sip*
   suei   ki-la   napukun sip   mwayen
   short.yam   IRR.3sg-go.to   heap   one.INANIM   yam
   ki-la   napukun sip
   IRR.3sg-go.to   heap   one.INANIM
   'The short yam will go onto one heap, the long yam will go onto one heap.' (NK290311_1_0003)

(33) *iro sap sipen sum*
   i=to   sap   sipen   sum
   3sg=be   some   part.of   one.flat
   'He was at some little part of land...' (MS250311_0045)

It appears that there are a number of important semantic distinctions on which the classifier system was (and partially still is) based. As mentioned above, the strongest distinction is animate-inanimate. Other important distinctions are round as opposed to elongated in form, and whether a noun refers to a partition, a single whole, or a collective. This is in line with cross-linguistic findings: the most common parameters with regard to classifiers are animacy and dimensionality (Aikhenvald 2003).

From a cultural point of view, it makes sense that collectives and partitions are distinguished. Highly complex gift exchanges are the core of traditional ceremonies on Baluan. There are many different ceremonies, but each of them involves a stage in which gifts (of money, cloth, pigs and garden food – depending on the kind of ceremony) from various groups aligned to the person that the ceremony is held for are gathered and put together, followed by a stage in which the collected gifts are redistributed among other stakeholders in the ceremony (see e.g. Otto [1991, 1992, 2002] and Otto and Pedersen [2005]). Thus, collecting, counting, division and (re)distribution are highly significant activities in Paluai culture, and this is reflected, among other things, by a large number of different verbs referring to these activities, and a strong focus in the classifier system.

3.2.3.2 Possessive classifier *ka-*

Oceanic languages tend to have at least one possessive classifier referring to things intended to be eaten (Lynch et al. 2002: 41). Paluai meets this expectation:

for food items possessed and intended to be eaten (or already eaten) the classifier *ka-* is used. *ka-* is a bound root suffixed with a form from the paradigm used for direct possession (see Section 4.2.3.2). This possessive classifier has been reconstructed for Proto-Oceanic (Lynch et al. 2002: 77) in exactly the same form **ka*, probably derived from the verb 'eat'. Interestingly, this form differs from most other noun modifiers because it precedes the noun instead of following it. It thus differs from *ta-*, the classifier for general alienable possession, which always follows the noun (*ka-* and *ta-* never occur together). *ka-* is also used for betel nut, cigarettes and chewing gum, but not for things intended for drinking. The following sentences provide examples.

(34) *iplêp kaip nik*
 ip$_A$=lêp [ka-ip nik]$_O$
 3pl=take CLF.food-3pl fish
 'They took their fish [for them to eat].' (LK250111_0040)

(35) *kong naluai a kong pou*
 ka-ng naluai a ka-ng pou
 CLF.food-1sg.PERT garden.food and CLF.food-1sg.PERT pig
 'My garden food and pork [for me to eat].' (LK100411_0104)

(36) *kom puan tiok*
 ka-m puan tiok
 CLF.food-2sg.PERT fruit piper.betle
 'Your *tiok* fruit [for you to chew with betel nut].' (LL300511_2_0015)

3.2.4 Noun morphology

3.2.4.1 Compounding

The most common morphological process associated with nouns is compounding. Two types can be distinguished: associative compounds, formally identical to possessive constructions, and compounds in which two nouns are juxtaposed. As mentioned in Section 3.2.2.4.3 above, when two nouns are in an associative relation the first one bears a *-n* pertensive suffix. A number of examples are given below.

(37) *naru-n* 'son-PERT' + *pou* 'pig' → *narun pou* 'piglet'
    lipa-n 'teeth-pert' + *mui* 'dog' → *lipan mui* 'dog's teeth'
    paru-n 'head-pert' + *punat* 'sky' → *parun punat* 'cloud'
    pata-n 'stem-pert' + *kei* 'tree' → *paran kei* 'tree trunk'

Compounds thus formed are left-headed; e.g. *lipan mui* is a kind of tooth, not a kind of dog. The left noun thus determines the meaning of the entire structure. Although some of them are connected to part-whole or kin relations, the compounds mentioned in (37) are probably better analysed as associative rather than possessive constructions, since their reference is generic. Thus, *lipan mui* usually refers to dog's teeth in general, not to the tooth of a specific dog (although of course, in certain contexts, it can refer to a single tooth of a single dog). As before, it can be seen that, due to their formal similarity, the distinction between associative and possessive constructions is unclear.

In the second type of compounds, two nouns are juxtaposed without an *-n* suffix. Examples are given in (38).

(38) *pei* 'stingray' + *manuai* 'eagle' → *pei manuai* 'eagle ray sp.'
    kun 'basket' + *mwat* 'bandicoot' → *kun mwat*   'basket type'
    pein 'woman' + *pwalei* 'spirit' → *pein pwalei* 'spirit woman'

This type of compound is again left-headed. The difference with the first type is that there is no relation between the two elements that could have its origin in a possessive relation. The second noun simply describes an attribute of the first one.

There is a fairly large number of compounds (many of them animal and/or plant names) involving two elements one of which is not encountered by itself in the present-day language. A few examples are given below. Moreover, even if both elements are attested, the meaning of the compound is not always compositionally derivable from the meaning of the two parts, as was also shown by e.g. *parun punat* and *kun mwat* above.

(39) *kanan* (unknown) + *pulêng* 'dawn'[27] → *kanan puleng* 'tomorrow'
    nia-n 'stomach' + *laleu* (unknown) → *nian laleu*   'fish species'

3.2.4.2 Reduplication

Noun reduplication is not a productive process in the present-day language, but it may have been in the past. If a base noun is monosyllabic, the entire

---

[27] *pulêng* has a slightly irregular vowel; cf. e.g. *kanan puleng* and *lenglengin* 'shiny, bright'.

syllable is reduplicated; if it is multisyllabic, only the first CVC-, CV- or VC- sequence is prefixed (see also Section 2.3.1.5.2). There are many inherently reduplicated nouns that are either fully or partially reduplicated, but that do not have a non-reduplicated counterpart.[28] Many of the attested forms are insect, plant and bird names. There are a few exceptions that have a non-reduplicated and a reduplicated form in the present-day language:

(40) *akon* 'root'   → ***a**kakon* 'many roots'
      pule-n 'sort of, kind of' → ***pul**pule-n* '(all) sorts of'
      puron 'thing'   → ***put**puron* 'things'
      pwapwem 'morning'   → ***pwa**pwapwem* 'early morning'
      poyep 'afternoon'   → ***po**poyep* 'late afternoon'
      kasun 'cononut cream'   → *kasu**sun*' 'thick coconut cream'[29]

The above examples, plus the fact that many inherently reduplicated nouns refer to things that usually come in large numbers (such as weeds, small insects or fish), invite the conclusion that noun reduplication is a strategy to indicate plurality. At the same time, it seems that a reduplicated noun often refers to many *small* things. It is a well-established fact that reduplication can function both as a marker of plurality and as a diminutive device in the language family that Paluai belongs to (Kiyomi 1995). Exceptions are the abovementioned *pwapwem*, *poyep* and the irregular *kasun*: reduplication here functions more like an intensifier. A number of inherently reduplicated nouns with plural or mass meaning are shown in (41).

(41) *langlang* fly, flies
      loyloy   ant, ants
      pweipwei caterpillar species
      latlat   gravel (made of crushed coral)
      kotkot   smoke
      lumlum   moss
      somsom   cloud, clouds; mist
      yeuyeu   star, stars

---

[28] In addition, there are many nouns which are derived from verbs by reduplication. For more on this, see Section 3.3.2.2.1.
[29] This is the only case encountered of a reduplicated and a non-reduplicated form in which there is no prefixing reduplication.

3.2.4.3 Suffixation

The major form of suffixation associated with nouns is the marking of pertensive in direct possessive and associative constructions. For more details, see Sections 3.2.2 and 5.5. In addition, there is a small number of nouns that seem derived, i.e. they bear a *-n* pertensive suffix, and are suffixed with *-an*, which normally derives nouns from verbs or adjectives (see Section 3.3.2.2.3.2 below). The forms encountered are:

(42) *koropu-n* 'base-PERT' → *koropu-n-an* 'reason, origin'
      napu-n   'taboo-PERT' → *napu-n-an*   NEG.IMP

Because of the limited number of instances and the idiosyncratic meaning difference between the suffixed and the non-suffixed forms, *-an* is not analysed as a productive suffix on nouns.

3.2.4.4 Derivation of other parts of speech from nouns

There are no productive derivational processes used to derive other parts of speech from nouns. There may be a minor process to derive (ad)verbs from nouns, but this seems not to be productive. The *-ek* suffix, which has a productive variant that forms applicatives on verbs (see Section 8.3.2), is also found as a frozen form on a number of verbal and adverbial forms (for the criteria to distinguish adverbs from verbs, see Section 3.6). Interestingly, some of these forms include a root shared with a nominal.

(43) *kasu-*   'bad behaviour'   *kasiek* 'not well' (Adv)[30]
      arona- 'way, character'   *aronek* 'accordingly' (Adv)
      molea- 'colour'   *moleyek* 'decorate' (V)
      sopea- 'wish'   *sopeyek* 'welcome' (V)
      sonea- 'near (in time)'   *soneyek* 'draw near (in time)' (V)

Of course, the central question is whether these forms are (ad)verbs derived from nouns or vice versa, or whether perhaps they share transcategorical roots (see Section 3.5). Since both V-N derivation and applicative derivation are very productive processes, this question is not easy to answer.

---

[30] In the case of *kasu-/kasiek*, there is an alternation involving a back and a front high vowel. Nevertheless, the two forms are considered to share the same root. The same /i/ ~ /u/ alternation shows up in other parts of the lexicon as well, for instance with *panirasip* ~ *panurasip* 'first'. It does not seem to be conditioned and is not productive.

3.3 The verb

A few hundred items make up the verb class, an open class that is relatively large, albeit smaller than the noun class discussed in Section 3.2. Verbs are typically the head of a predicate; they are the only word class allowed to take bound subject pronouns and TAM particles.[31] Some verbs also appear, without any overt morphological marking, as noun modifiers or as heads of NPs. There is no category of tense, but there are many particles (often grammaticalised from verbs) expressing aspectual and modal meanings. These will be discussed in detail in Chapter 6.

The subject, regardless of whether and how it is expressed elsewhere in the clause, is cross-referenced on the verb by a bound pronoun (see Section 4.2 for forms). Direct objects are cross-referenced on the verb only when animate and not co-expressed by a full NP. Oblique constituents are not cross-referenced at all. There are very few ditransitive verbs; the majority of verbs are either intransitive (with only a subject argument S), or transitive (with a transitive subject argument A and a transitive object argument O). Alignment is nominative-accusative. Oblique arguments, i.e. arguments which are not S, A or O, are flagged by special marking and often a specific type of Serial Verb Construction (SVC). Sections 8.1 and 9.1 discuss this. Since direct objects are generally omitted in narratives, especially when they are established discourse referents, it is hard to determine whether there are any truly ambitransitive verbs, i.e. verbs that can be either transitive or intransitive without the need for a morphological derivation.

Verbs can be subcategorised into lexical transitive verbs, lexical intransitive verbs, aspectual verbs (always intransitive) and directionals (also always intransitive). A large subclass of lexical intransitives is made up of stative verbs. Aspectual and directional verbs developed from full lexical verbs and may be on their way to losing verb status and becoming adverbials, particles or preposition-like items. However, they almost all have (often formally identical) full-verb counterparts, which can function by themselves as predicate heads.

Although property-concept predicates (Stassen 1997) generally show a zero copula, in certain cases the verbs *to* 'be, stay, exist' and *la* 'go' appear to function as copulas (see Section 7.7). SVCs, both the symmetrical and the asymmetrical type (Aikhenvald 2006), are very common. Verb serialisation is extremely productive, and both lexicalisation of verb combinations and grammaticalisation of lexical verbs into pre- and postverbal aspectual and directional particles are found in Paluai. Chapter 9 discusses SVCs in detail.

---

31 The pronominal system is discussed in Section 4.2. There is a formal difference between free and bound pronouns only for the 1sg subject form.

There are only a few valency-changing operations available. Many transitive verbs can be reduplicated in order to intransitivise them, but this operation can also derive a nominalisation. Intransitive verbs can be reduplicated as well, with reduplication generally indicating continuous or iterative aspect (or, again, a nominalisation). There is no passive. In addition to a periphrastic causative, there is a morphological causative formed by the prefix *pe-*, which only applies to stative verbs. In addition, an applicative suffix *-ek* is found. See Chapter 8 for further discussion.

Verb to noun derivation is very common (in contrast to noun to verb derivation) and is done by reduplication, suffixation or a combination of both. An adjectival form can be derived from many stative verbs, by means of suffixation. Verbs can be modified by a number of forms that are analysed as adverbs (see Section 3.6), although many "adverbial" meanings are expressed by other verbs in a SVC with the main verb. The line between SVC and verb-adverb sequence is often hard to draw.

3.3.1 Subclassification of verbs

3.3.1.1 Transitive and active intransitive verbs

Lexical verbs are of two valencies: transitive or intransitive. The great majority of simplex transitive verbs is monosyllabic. They can be grouped in a number of semantic subclasses: verbs of affect (hit, break), perception and cognition (see, hear), consumption (eat, drink), et cetera. They take an A (transitive subject) and O (transitive object) argument.[32] Full discussion of grammatical functions can be found in Chapter 8.

Intransitive verbs, on the other hand, only take an S (intransitive subject) argument and can firstly be subdivided into active and stative verbs. Active intransitive verbs fall into semantic subclasses such as verbs of posture (sit, lie), motion (dance, paddle) and bodily actions (urinate, laugh). There are a small number of existential verbs as well: *tok* 'be, exist', *toktoai* 'be, stay', and the Tok Pisin loan *gat* (*lit.* 'have').

Furthermore, there is a subclass of stative intransitive verbs subcategorised for an S argument and a further argument taking marking that differs from "regular" O marking. Most of the verbs in this subclass are verbs of emotional sensation,

---

[32] The labels S, A and O (cf. Dixon 2010a: 122) will henceforth be used as convenient abbreviations for referring to the core arguments of verbal predicates (subject of intransitive and transitive verb, and object of transitive verb, respectively), not to semantic-syntactic relations. More discussion of grammatical relations can be found in Chapter 8.

which take an optionally expressed Stimulus argument; examples are *kaêrêt* 'be afraid (of)' and *mwamwasêk* 'be ashamed (of)'. When the argument is expressed, it takes the same marking as the third argument of a ditransitive clause: the preposition *a=*, or (in case of an animate referent) the possessive/locative *ta-*.

There are very few simplex ditransitive verbs in Paluai; in most cases, three-participant events are expressed by means of SVCs. There is no difference in marking between S and A (the language has nominative-accusative alignment), but the third argument of a ditransitive clause is marked differently from an O argument. The latter is cross-referenced by an enclitic on the verb complex in case it is animate and receives no further marking, whereas the former is never cross-referenced, but marked by the preposition *a=*, or possessive/locative *ta-* in case of an animate referent. Again, a more elaborate treatment can be found in Chapter 8.

3.3.1.2 Stative verbs

An important semantic subclass of intransitive lexical verbs is formed by stative verbs expressing certain property concepts, which I call "transient states". Paluai is a split-adjective language in the sense defined by Stassen (1997: 165), as it encodes some property concepts by means of verbal predication and others by means of non-verbal predication.[33] A "transient state" can be defined as a state that is not inherent to an object or a person but is likely to vary over time. Examples are for instance 'be ripe', 'be cold', 'be afraid', etc. Stative verbs are often ambiguous between a "static state" and a "dynamic state" reading; they can either mean 'be X/have property X' or 'become X/increasingly have property X'. Stative verbs function just like all other verbs: they can take TAM particles and bound subject pronouns and typically head a predicate. They differ from active verbs in a number of significant ways, however: 1) they can undergo a morphological causative operation (see Section 8.3.1), 2) they can be modified by the adverb of degree *paran* 'very', just like adjectives, and 3) many of them allow suffixation with *-n* to form an adjective (see Section 3.4.1 below). These three operations are not allowed in the case of active verbs. Typically, stative verbs also occur as second verbs in certain types of SVCs (see Section 9.1.1), but this is not a sufficient or necessary criterion.

There is a difference in interpretation and usage between a stative verb and its derived adjective. A good example of this is the verb *yamyam* 'redden; ripen'. Sentences (44) and (45) denote the process of ripening of a particular taun fruit

---

[33] For the concepts expressed by those stative verbs for which a derived adjective has been attested, both strategies are obviously available.

(*Pometia pinnata*), pointing out the stage it is at. Sentence (46) only refers to a characteristic of the fruit, namely, that it is red-coloured; no reference is made to the process of ripening and the time span that is involved. This sentence would for instance be used to contrast this piece of fruit with another, yellow- or green-coloured taun fruit.

(44) *nau i(ro) yamyam*
 nau   i=(to)   yamyama
 taun.fruit 3sg=(CONT)   redden
 'The taun fruit is reddening [i.e. getting ripe].'

(45) *nau in yamyam*
 nau   i=an   yamyama
 taun.fruit 3sg=PRF   redden
 'The taun fruit has reddened [ripened].'

(46) *nau yamyaman*
 nau   yamyama-n
 taun.fruit redden-ADJ
 'A red-coloured taun fruit.'

Sometimes, the same form is attested both as a stative and an active verb. This is the case for e.g. *tet*, which can mean both 'ebb (of tide)' and 'move (a short distance)'. Thus, sentence (47) is grammatical whereas (48) is not.

(47) *met in paran tet*
 met   i=an   paran tet
 tide 3sg=PRF   very   ebb
 'The tide has become really low.'

(48) **Sion in paran tet*
 *Sion i=an   paran tet
 John   3sg=PRF   very   move
 Intended: 'John has really left.' (compare *Sion i=an tet* 'John has left')

All stative verbs that have no attested derived adjective fall into the semantic domains of Human Propensity (e.g. 'jealous' or 'tired'), Physical Property (e.g. 'cracked' or 'loose'), or Induced Motion/Position (e.g. 'hang' or 'sink'). The Physical Property domain in particular contains many stative verbs with derived adjectives, with meanings such as 'wet', 'dry', 'rotten', etc, but also a large number of adjectives for which no verbal source has been found. This is therefore the domain that shows the greatest variation in formal expression. The domains of Dimension, Colour and Form are split between stative verbs with derived adjectives and adjectives with no attested verbal source. Terms referring to Value (with the exception of *palak* 'bad'), Age, Material and Gender are exclusively adjectives without attested verbal counterparts, or nouns. This seems to confirm the Adjective Hierarchy as proposed by Stassen (1997: 169).

3.3.1.3 Directionals

Directionals form a closed subclass of active intransitive verbs, which have become grammaticalised to function as direction markers for or related to the action expressed by the main verb. However, they also act as the head of a predicate. A feature that distinguishes them from regular active intransitive verbs is that they cannot be nominalised. Table 18 shows the directional paradigm.

Table 18: Directional paradigm.

Form	POc root(s)	Paraphrase
la, lak	*lako 'go, thither'	go to, motion away from DC, thither; not specified for absolute FoR
me	*mai 'come, hither'	come, motion towards DC; not specified for absolute FoR
si	*sipo 'go down'	come seawards (down), towards DC
sa, sak	*sake 'go up'	come inland (up), towards DC
wot	*ua[tu] 'go to addressee'	go parallel to the shore (horizontally), away from DC
suwot	*sipo + *ua[tu]	go seawards (down), away from DC
sot	*sake + *ua[tu]	go inland (up), away from DC
wen	*pano 'go, walk (away)'	move parallel to the shore (horizontally), not deictically anchored
suwen	*sipo + *pano	move seawards (down), not deictically anchored
sen	*sake + *pano	move inland (up), not deictically anchored

Each motion event includes "a Figure in Motion along a Path oriented with respect to one or more Grounds" (Wilkins and Hill 1995: 217). Sometimes, but not always, the source (the location from where the motion starts) and the goal (where the motion terminates) are explicitly stated. When referring to motion in Paluai, two dimensions are important: whether or not the motion is oriented or grounded with regard to a deictic centre, and whether or not the motion is oriented or grounded with regard to an absolute Frame of Reference (FoR) based on fixed bearings (Levinson 2003).

Paluai has eight directionals that are specified with regard to an absolute FoR; a three-way distinction is made between 1) movement seawards, 2) movement inland, and 3) movement parallel to the shore. As in the case of other Austronesian languages, the land-sea axis is an important concept within spatial reference. Baluan is a cone-shaped volcanic island; its highest point is the crater of the now dormant volcano located more or less in the middle. Therefore, going inland always means going up, and going towards the shore always means going down. In addition, since motion parallel to the shore (i.e. intersecting the land-sea axis) usually means moving on more or less the same level, this has obtained the secondary meaning of 'moving on a horizontal level'. At sea, the same system is used by extrapolation: thus, for moving towards the shore the same directionals are used as for moving inland, and for moving out to sea the same directionals are used as for moving towards the shore when on land.

Of the ten directionals, seven are specified for deixis, indicating motion either towards or away from the deictic centre; the other three are specified with regard to the absolute FoR, but are not deictically anchored. With the deictically anchored verbs, the speaker can most often be understood as the DC, but this can be varied for pragmatic reasons. Five of the directionals specified for deixis are additionally specified for the absolute FoR. In the case of the three verbs that are not deictically anchored, both the source and the (intended) goal of the motion are located at points removed from the DC, and thus the motion is directed neither towards nor away from the DC. These terms are often used for motion transverse to the DC. For instance, when speaker and addressee, from their DC slightly uphill, are commenting on a third person walking along the shore, the form *wen* could be used. No distinction is made between motion from left to right or the reverse. Two directionals, *la* and *me*, are deictically anchored, but not specified for the absolute FoR.

The interrelation described above is illustrated in Table 19. The first and the third row show forms all ending in -*ot* and -*en*, respectively, and it is possible that the forms in the first column start with the same formative *si*- ~ *su*-. It is likely that some of the directionals were in fact morphologically complex forms, and the presumable POc roots are indicated for each of them in Table 18. For two of the directionals, there are two different forms attested: *la/lak* and *sa/sak*. The long form is used when the directional is used by itself (either as the main verb or in a SVC), whereas the short form is used when the directional introduces a constituent and is thus followed by an adverbial or NP. Note that there are two omissions in the paradigm. There is no antonym of *wot*, a directional indicating motion towards the DC, parallel to the shore. This slot is filled by *me*, which is

Table 19: Organisation of the directional paradigm along dimensions of absolute FoR and deixis.

Absolute FoR Deixis	down, seawards (on land); out to sea (on water)	up, inland (on land); towards the shore (on water)	parallel to the shore, level, horizontally	not specified
away from deictic centre	suwot	sot	wot	la, lak
towards deictic centre	si	sa, sak	-	me
not deictically anchored	suwen	sen	wen	-

discussed below. In addition, there is no term that is neither specified with regard to the absolute FoR nor deictically anchored. This is no surprise, as such a term would add no meaningful information to a lexical motion verb.

The directional paradigm provides a precise reference structure with ample use in discourse. For virtually all actions that in some sense involve motion (including perception-based actions such as seeing/looking, speaking or listening), the direction of the action has to be specified with a directional, a phenomenon that is very common in Oceanic languages. In Paluai, this is done either by a directional used as a preverbal particle, or by a SVC in which the directional follows the main verb. See Sections 6.3 and 9.1.5 for more details; also cf. Levinson and Wilkins (2006); Schokkin (2013, forthcoming).

Some semantic differences between the directionals are not specified above. For instance, *la* is telic, with an endpoint to the motion inherent in its meaning; it is therefore also systematically translated with 'go to'. *wot*, on the other hand, seems to lack an inherent endpoint. Forms indicating motion towards the DC are slightly different again: these motion events are telic in nature, since it is implied that the object's arrival at the DC will terminate the motion. This may be part of the reason why, in many instances, *me* is used as an antonym to *wot*. Firstly, the semantics of *me* does not clash with this reading, and secondly, with 'come' forms, the arrival at the DC may be the meaning facet that speakers focus on most, rather than the exact direction the object came from.

As in other Oceanic languages, the deictic centre does not necessarily have to be the current position of the speaker, but it can be varied in order to put a certain constituent or discourse participant in focus (cf. Hamel 1994 for Loniu). In addition, directionals are used to keep track of who did what to whom, because they specify the locations of speech act participants relative to each other.

3.3.1.4 TAM verbs or particles

TAM verbs or particles give information about the grammatical categories of Tense, Aspect and Modality (TAM) of a main verb. Most of them probably grammaticalised from full verbs in SVCs, but they are at different stages of grammaticalisation; some still carry a large amount of lexical meaning whereas others are almost completely devoid of it. Therefore, although they may form a closed class, it is not a class that is homogeneous (and some forms can be said to be "more" part of it than others). Most of the forms appear in preverbal position, but some follow the main verb. Others can both precede and follow the main verb. The forms are shown in Table 20 below.

Table 20: TAM verbs and particles.

Form	Aspectual meaning	Potential lexical source / related form
pwa	Desiderative; Inchoative	pwa 'say, think'
no	Imperfective	no 'only, just'
pe	Perfective	pe 'do, make'
an	Perfect	unknown
to	Continuative; Habitual	to 'be, stay, remain'
yen	Progressive	yen 'lie'
tu	Stative continuative	tu 'stay, remain'
sa	Ability; Apprehension	tai 'look at'
nêm	Completive	nêm 'be finished'
wot	Durative	wot 'move across'

It looks as if there have been successive "waves" of grammaticalisation of verbs into TAM particles. *yen* and *wot*, for instance, occur just as often as lexical verbs as they do as modifiers to other verbs, whereas *an* has completely lost the connection with its lexical source. The former development therefore seems to have taken place much more recently than the latter. Chapter 6 discusses these and other elements of the verb complex (VC) in detail.

3.3.1.5 Existential verbs

Truly existential predicates seem to be quite uncommon, since a predicate headed by *tok* 'be' or *toktoai* (lit. 'stay-be-at') could often have an alternative locative interpretation. Two examples of *tok* with a clear existential interpretation are given below.

(49) *urêrok tepwo*
 wurê=tok te-pwo
 1pc.EXCL=be LIG-DEM.INT
 'We remain/exist today.' (OL201210_0099)

(50) *pian te irok o, inêm*
 pian te i=tok yo i=nêm
 good REL 3sg=be DEM.INT 3sg=be.finished
 'The goodness that existed is finished.' (PK290411_3_0046)

For humans and abstract concepts, existential predicates such as the above are still used. With regard to inanimate, concrete objects, however, an existential meaning is more likely to be expressed by the Tok Pisin verb *gat*:

(51) *tepwo pun igat yuêp kain mun te ipto pwa panuatu*
 te-pwo pun i=gat yuêp kain mun te
 LIG-DEM.PROX INTF 3sg=have two.INANIM kind banana REL
 ip=to pwa panuatu
 3pl=HAB say Vanuatu
 'Nowadays, there are two kinds of banana that are called "Vanuatu".' (KS030611_1_0044)

3.3.1.6 Light verb constructions

Light verbs are verbs with very little semantic content that join forces with an additional item, usually a noun, to form a predicate. In Paluai this type of construction is only found with the verb *pe* 'make, do'. Table 21 gives an overview of some frequent light verb constructions.

Table 21: Examples of light verb constructions with *pe*.

Form	Meaning
pe net / pe kou	fish (*lit*. 'make ocean' / 'make hook')
pe kui	cook (*lit*. 'make pot')
pe kanum	work in the garden (*lit*. 'make garden')
pe nin	fight (*lit*. 'make fight')
pe yep	make fire, light a fire
pe lik / pe kun	make a basket
pe mangat	do work
pe nayei	lie (*lit*. 'make/do lie')

These constructions could be cases of noun incorporation (Mithun 1984), a process commonly encountered in the Oceanic language family. However, based on grammatical and semantic criteria it is not very likely that this is indeed the case. Most importantly, another element such as an adverb can intervene between the two components, just as in a regular verb-object sequence. This is shown in the examples below.

(52) *kipe pe ansê kui*
 ki-pe pe an-sê kui
 IRR.3sg-PFV make PART-DIM pot
 'He will do a bit of cooking.' (SP190311_0046)

(53) *urêno ro pe liliu kou wot*
 wurê=no ro pe liliu kou wot
 1pc.EXCL=IPFV CONT make again fishing go.level
 'We were fishing again.' (MK060211_0017)

Secondly, these constructions are not nominalised as complete entities. Instead, only the verb is nominalised, resulting for instance in a form such as *pe-i-nan kui* (make-3sg=NOM pot 'cooking'). Finally, the verb *pe* is encountered with a plethora of different nouns and also with nominalisations; Table 21 provides no more than a very small selection. It thus seems more likely that *pe* is a light verb with very general semantics that can be used as a transitive verb on many occasions, instead of another more specific verb. The abovementioned combinations are therefore regular VO sequences.

3.3.2 Verb morphology

Several morphological operations on verbs can be distinguished: compounding, reduplication and affixation. Prefixation (with a prefix that expresses irrealis) is considered an inflectional operation and will not be discussed here but in Chapter 6. Reduplication can have different functions, one of which is nominalisation. Suffixation always leads to a change of word class: either nominalisation or, for stative verbs, adjective derivation.

3.3.2.1 Deriving another verb

3.3.2.1.1 "Compounding"

Whether compounding is a relevant process for verbs is much harder to establish than for nouns, since serialisation is extremely productive. Most likely, there is a continuum ranging from fully lexicalised to fully productive verb sequences.

At the most lexicalised end of the continuum, we find lexical verbs that consist of a sequence the meaning of which cannot be compositionally derived from the two parts; they can be said to be fully idiomatic. Probably, these are cases of lexicalisation of a SVC; see Chapter 9. In many of these cases, one of the parts is not attested by itself. Some examples are given in Table 22.

Table 22: Some lexicalised verb sequences.

Sequence	Consists of	Meaning
antek yut	antek 'put away' + yut 'split'	collect branches after burning a garden
lup san	lup (unknown) + san 'cut'	compare, judge
tou put	tou 'give' + put (unknown)	regret
wui antek	wui 'grow; surround' + antek 'put away'	push out by force using a lever; forget about
soksok yit	sok(sok) 'hit' + yit 'chip off'	hatch (of egg)

However, there are also many "semi-productive" verbal collocations where, either for the first or for the second verb slot, a choice can be made from within a small range of verbs that are semantically related to each other. These collocations are not fully idiomatic, since their meaning is still to some extent reconstructible from the meanings of the parts; in addition, they are productive, at least to some extent. These and other aspects of verb serialisation are discussed in Chapter 9.

3.3.2.1.2 Reduplication

There are two types of verbal reduplication: one derives a noun from a verb (making it a word-class-changing derivation; see below for discussion), the other derives another verb. The latter form of reduplication has a different effect for transitive and intransitive verbs: with transitives, it is a valency-reducing strategy, whereas with intransitives it adds aspectual meanings.

Nominalising reduplication can be distinguished from other forms of reduplication based on distributional differences: whether the resulting form appears in a nominal or a verbal slot. In addition, disyllabic reduplicated forms that are verbal in nature tend to carry stress on the final syllable, whereas nominal forms tend to be stressed on the penultimate syllable. However, these are tendencies at best; they cannot be solely relied upon to determine word class of a particular form.

3.3.2.1.2.1 Transitive verbs

Reduplication is often mentioned as an intransitivising device in Oceanic languages (see Evans 2003 and references therein) and it is also encountered as such in Paluai. The former transitive subject A will be the S of the intransitive construction. Both full and partial reduplication are attested. If a form is monosyllabic, the entire structure will be reduplicated; if it is multisyllabic, only the first CVC-, CV- or VC- sequence will be prefixed (see also Section 2.3.1.5.2). Examples are given below for the transitive verbs *ngan* 'eat' and *lomêek* 'plant (yams)'.

(54) *epnganngan nêm*
 ep=ngan.ngan nêm
 1pl.EXCL=REDUP.eat be.finished
 'We finished eating.' (KM060111_0075)

(55) *ipkalomêek suei le mwayen, kalolomêek nêm...*
 ip=ka-lomêek suei le mwayen ka-lo.lomêek nêm
 3pl=IRR.NS-plant short.yam or yam IRR.NS-REDUP.plant be.finished
 'They will plant yams. (When) they will finish planting... (KM190211_0035)

3.3.2.1.2.2 Intransitive verbs

Reduplication of intransitive verbs is not as common as reduplication of transitive verbs. With intransitive verbs, reduplication does not change the valency of the verb. For both active and stative intransitive verbs, reduplication indicates a durative state or iterative action, which has (already) been going on for a stretch of time. The reduplicated form is often preceded by the continuative particle *to*. Some examples are given below, for the stative verb *porok* 'be painful' and the active verb *aluk* 'paddle'.

(56) *niong iro (paran) potporok*
 nia-ng i=to (paran) pot.porok
 stomach-1sg.PERT 3sg=CONT (very) REDUP.be.painful
 'My stomach is being (very) painful [and has been for a while].'
 (field notes 15/09/2012)

(57) *taman no ro alaluk*
       tama-n   no   to   al.aluk
       father-PERT   IPFV   CONT   REDUP.paddle
       'His father was paddling.' (NP210511_1_0025)

3.3.2.2 Word class-changing derivations

3.3.2.2.1 Reduplication

There are a number of frequently used nouns that only have a reduplicated form. Most of them can be traced back to a corresponding verb. Both transitive and intransitive verbs can be reduplicated to form nominalisations.

Some examples are given in Table 23. They can be regarded as Agentive nominalisations in the case of intransitive verbs. For transitive verbs, they can be analysed as Result or Instrument nominalisations, where the reduplicated form refers to an object or abstract concept that is the result of the action described by the verb, or an instrument used while carrying out the action. Reduplication, however, can also yield Action nominalisations. Their nominal status is evidenced for example by the fact that they can be followed by a possessive classifier, as in (58), or follow a preposition, as in (59). Examples are shown below for the transitive verb *san* 'cut' and the intransitive verb *wau* 'move'. In particular cases, reduplication may have negative overtones (do s.t. for too long, overdo it), as shown in example (60).

Table 23: Some nominalising reduplications.

Base form	Meaning	Reduplicated form	Meaning
angou	arrive (intr)	*angangou*	stranger
wop	fly (intr)	*wopop*	flying fish
ngan	eat (tr)	*nganngan*	food
ning	see (tr)	*ningning*	view, opinion
ngayup	spit (intr)	*ngangayup*	spittle
pangai	think of/about (tr)	*pangpangai*	thoughts, mindset
soyek	push, move (tr)	*soysoyek*	change(s)
tuk	beat (tr)	*tuktuk*	beat (N); masher
yuêt	ask (tr)	*yuyuêt*	question

(58) *tareo nêmnêmti nêm, i reo sansan tan ngoyai*
 ta=te-yo nêmnêmti nêm
 SPEC.COLL=LIG-DEM.INT all be.finished
 i te-yo san.san ta-n ngoyai
 3sg LIG-DEM.INT REDUP.cut CLF.POSS-PERT cuscus
 'All of that, it was [the result of] the cutting of/by the cuscus.'
 (LL010711_0083)

(59) *...lalon um pari ai wauwau pit tararap*
 lalo-n wumwa pari a=i wau.wau pit
 inside-PERT house belonging.to OBL=3sg REDUP.move be.close
 ta-tap
 CLF.POSS-1pl.INCL
 '...inside our meeting house.' [*lit.* 'inside our house for coming together']
 (LL300511_1_0043)

(60) *wono ro pe no roktok si*
 wo=no to pe no tok.tok si
 2sg=IPFV CONT do only REDUP.sit come.down
 'You are just sitting around.' [*lit.* 'you are just doing sitting']
 (field notes 02/10/2012 – said as a reprimand to a child)

3.3.2.2.2 A note on inherently reduplicated verbs

There is a fairly elaborate list of verbs that seem to be inherently (partly) reduplicated: no non-reduplicated counterpart was attested. Interestingly, most of these verbs are intransitive (and among them there are a large number of stative verbs), many describing bodily functions. Most of these denote actions where no or little volition on the part of the S argument is involved; typical examples are e.g. yawning, sneezing, dreaming etc. The verb forms encountered are listed in Table 24.

It is possible that these forms developed diachronically from forms derived by valency-reducing reduplication as discussed above. Semantically, however, for many verbs this does not seem to make sense. For most of the meanings expressed by these verbs, the intransitive is the semantically unmarked form. If more than one constituent were expressed, it would rather be an optional locative or instrumental constituent (e.g. 'vomit *on*', 'drift *towards*', 'be ashamed *about*') expressed with a SVC or prepositional phrase, instead of a direct object.

Table 24: Inherently reduplicated verbs.

Verb	Valency	Meaning
alalmau	Act intr	yawn
amamsi	Act intr	sneeze
ilili	Tr	stand up (on), lead (a clan)
kulpopot	Act intr	whistle
kupkup	Stat intr	turn grey (of hair)
loklok	Stat intr	hang
lôlôt	Stat intr	cool down
lulum	Tr	take off s.t. that is hanging (laundry, fruit)
lulut	Tr	undress
mamat	Stat intr	be awake
mayai	Stat intr	be quick
memeyek	Tr	spoil, ruin
mêpmêp	Act intr	dream
mimi	Act intr	urinate
mumut	Act intr	vomit
mwamwanget	Stat intr	be lazy, tired
mwamwasêk	Stat intr	be ashamed
ngolngol	Act intr	grunt (of pig)
nopnop	Stat intr	be jealous
nunuau	Stat intr	be keen, energetic
pelpelek	Stat intr	shiver, tremble
pepek	Act intr	defecate
pêtpêt	Act intr	cough
pipilek	Tr	insult, abuse s.t. or s.b.
pitpit	Act intr	weep, sob
pweipwei	Stat intr	settle down, be quiet
sisi	Tr	sweep
sosol	Stat intr	be sorry
sosou	Stat intr	be cold, feel cold
tenten	Act intr	speak to an ancestor, pray
têktêk	Act intr	hop
wokok	Act intr	float, drift
yamyam	Stat intr	redden
yangyang	Act intr	love
yatyat	Act intr	jump up and down and wave the arms [by a toddler]
yauyau	Stat intr	stoop, bend down
yaya	Act intr	swim
yepyep	Stat intr	be itchy
yuyut	Act intr	fall off

3.3.2.2.3 Suffixation

3.3.2.2.3.1 The -a and -o suffix
Formally, three suffixes deriving nouns from verbs can be distinguished: *-(n)an* (see below), *-a* and *-o*. *-a* is encountered in forms that can be described broadly as Result nominalisations, with a few exceptions. This suffix seems to be used with intransitive verbs more than with transitive verbs, but firm conclusions are hard to draw because of the limited number of forms and the fact that the suffix seems to be no longer productive. The same is true for the *-o* suffix. Some examples are given in Table 25 below. Suffixation also derives adjectives from stative verbs; this process will be discussed in Section 3.4 on adjectives.

Table 25: Nominalisations formed by adding an *-a* or *-o* suffix.

V	Meaning	N	Meaning	Type
mumut	vomit	*mumur-a*	vomit	result
sisi	sweep	*sisi-a*	broom	instrument
mimi	urinate	*mimi-a*	urine	result
ngui	snore	*kongui-a*	snoring (the sound)	result
sosol	feel sad, sorry	*sosol-a*	mourning, sorrow	result/action
pepek	defecate	*pepek-a*	defecation	action
mamat	be awake	*mamar-a*	wake, being awake	result/action
kol	wait for	*kolkol-o*	place to sit and wait	instrument

3.3.2.2.3.2 The -(n)an suffix
In contrast to *-a* and *-o*, the suffix *-(n)an* is very productive. *-(n)an* can only be applied to transitive verbs and derives an Action nominalisation, with the former direct object of the verb modifying the derived verbal noun. This suffix, as its form already indicates, is again linked to ideas of association and/or possession. The resulting noun X heads an NP with the former direct object Y in an associative or possessive construction, which globally means 'the X-ing of Y'. Some examples are given below. In (62), there are two nominalisations: one involves the verb-noun combination *pe kui* 'cook' (*lit.* 'make pot'), the other one the verb *sui* 'fry'.[34]

---

[34] Light verb constructions with *pe* 'do, make' are very common and always refer to so-called "name-worthy activities". The construction is not analysed as a form of noun incorporation. One reason is that it does not form one phonological word, as evidenced by the fact that the combination is not nominalised as a whole: only the verb is, as shown in example (62).

(61) *worou porok pari ai awuian paralan*
    wo=tou   porok   pari   a=i   awui-an   paralan
    2sg=give strength belonging.to OBL=3sg weave-NOM   relatives
    'You give [him] strength for keeping together the clan.'
    [*lit* 'weaving-of the clan'] (YK290411_2_0026)

(62) *a peinan kui reo i suian yapi*
    a   pe-i-nan   kui te-yo   i   sui-an   yapi
    and   make-i-NOM   pot LIG-DEM.INT 3sg   fry-NOM   sago
    '...and that [way of] cooking is frying sago.' [*lit*. 'frying-of sago']
    (CA120211_1_0014)

Its use with derived forms and with loans from Tok Pisin demonstrates that *-an* suffixation is very productive. For instance, the transitive verb *pemat* 'kill', a causative construction derived from the stative verb *mat* 'die', can be nominalised to form *pemat-an yamat* 'murder' (*lit*. 'make die-of person'). Another example is the somewhat curious compound *toktoai* 'dwell, exist at', which can be glossed as 'remain-be-OBL=3sg' and consists of two verbs plus a preposition. This compound verb can be nominalised, forming *toktoai-an* 'life at/of'. An example involving a loan word is *soim-an* 'showing', formed with the Tok Pisin loan *soim* 'show'.

Sometimes the suffix has the form *-nan* instead of *-an*: both *tou-nan* and *tou-an* 'giving' and *pei-nan* and *pe-an* 'doing, making' can be found. There does not seem to be a meaning difference between the two forms. In addition, sometimes an element *-i* is found in a nominalisation, as was already shown for *peinan* in (62), and this element is never seen on the base verb. Other examples encountered are *ngan-i-an* 'eating' (in addition to *ngan-an*) and *ning-i-an* 'looks, appearance' (in addition to *ning-an*). This is possibly a remnant of a once-productive transitivising suffix retained in one variant of a frequent nominalisation, but no firm conclusions can be drawn in the matter.

3.3.2.2.3.3 The -n suffix
An adjective can be derived from most stative verbs by adding a *-n* suffix. The final vowel of the verb root will surface in the process, and thus the final vowel of the derived adjective will be unpredictable. Adjective derivation is not productive; it is discussed in more detail in Section 3.4.1.

3.3.2.2.4 Zero derivation
Verbs are also encountered in nominal slots without any overt morphological operations, a phenomenon that can be treated as zero derivation. Stative intransitive

verbs in particular appear in nominal slots with zero derivation. These forms often modify nouns referring to human beings and are used to describe a personal characteristic. A few examples are given below, involving the forms *mwamwanget* 'be lazy', *nopnop* 'be jealous' and *palak* 'be bad'. *palak* is slightly irregular because it is apparently unable to take a bound subject pronoun: **nga=palak* (1sg=be.bad) is ungrammatical.[35] It is often encountered in the adverbial construction *la palak* 'badly' (see Section 3.6.2.2). The verbal status of *palak* is therefore somewhat questionable. However, because it is allowed to take TAM particles it is primarily analysed as a verb, with a zero-derived nominal and adjectival variant (from which a phrasal adverb can be derived).

(63) *i reo kinan mwamwanget tang*
i   te-yo   kina-n   mwamwanget   ta-ng
3sg   LIG-DEM.INT mark-PERT   be.lazy   CLF.POSS-1sg.PERT
'It is a sign of my laziness.' (WL020611_0048)

(64) *yamtan nopnop*
yamta-n   nopnop
owner-PERT be.jealous
'A jealous person.' (field notes 30/04/2011)

(65) *umaro pe palak pwên*
u=ma=to   pe   palak pwên
3du=NEG₁=HAB do   evil   NEG₂
'They did not do bad things.' (LL010711_0010)

3.3.3 Functions of nominalisations

Nominalisations function like nouns: they can be an argument to a verb or else follow a preposition. Grammatically they have more in common with nouns than they do with verbs: there is for instance no TAM morphology found on nominalisations, but they do exhibit nominal categories such as possessive and

---

35 Possibly, this is due to the fact that the stative verb *palak*, when used in reference to humans, can only mean 'angry', and in spontaneous speech only occurs in the construction *nia-TAM palak* 'stomach TAM bad', with the pertensive suffix on *nia-* referring to the person experiencing the emotion. Direct attachment of a 1sg subject bound pronoun to the stative verb (under elicitation conditions) may be rejected by native speakers for this reason. The adjective *palak* can be used with human reference, but only in combination with a free pronoun, and with different meanings: *wong palak* 'I am unwell' or 'I am evil'.

number. Nominalisations show no differences in syntactic behaviour compared to other (abstract) nouns.

3.4 Adjectives: overview and semantic classes

There is a separate class of adjectives, although many members of this class are derived from stative verbs. The forms for which no stative verb counterpart is attested seem to encompass in particular Dixon's Set A: Dimension, Age, Value and Colour, but there are also many terms referring to Physical Property from Set B (Dixon 2010b: 73–74). Adjectives derived from stative verbs mostly refer to Physical Property, with a few Dimension and Colour terms and one Speed term. Human Propensity and Speed (from set B) are almost exclusively expressed by stative verbs that have no derived adjective. Terms from Set C are generally not expressed by means of adjectives (or stative verbs) but in other ways. In Table 26 below, some examples are given of adjectives and other forms in each semantic class. For non-adjectival forms, word class is indicated.

Table 26: Semantic classes of adjectives, with some examples.

Set	Semantic type	Examples
A	Dimension	*menengan* 'big'; *aloen* 'long, tall' (from V *alau* 'grow')
	Age	*salai* 'old (of people)'
	Value	*pian* 'good; goodness'
	Colour	*pilipil* 'yellow'; *yamyaman* 'red' (from V *yamyam* 'be red')
B	Physical Property	*konun* 'heavy'; *kôkin* 'hot' (from V *kôk* 'be hot')
	Human Propensity	*peruruek* 'ignorant'; *nopnop* 'be jealous' (V)
	Speed	*mayayen* 'quick' (from V *mayai* 'be quick')
C	Difficulty	metaphorical extension of Physical Property terms such as 'hot' or 'light'
	Similarity	*ngonomek* 'be corresponding' (V); adverbs
	Qualification	*nuanan* 'true, thruth'; adverbs
	Quantification	Quantifiers (see Section 4.6)
	Position	Adverbs (see Section 3.6); Spatial nouns (see Section 3.2.2.3)

Cross-linguistically speaking, the two grammatical functions of adjectives are 1) modifier to a noun within the NP (attributive function), and 2) head of a verbless predicate, or copula complement (predicative function). Paluai adjectives fulfil both these functions. Adjectives can be distinguished from verbs because they are not allowed to take TAM particles and bound pronouns. They can be distinguished from nouns because 1) they can be modified by the adverb of degree *paran* 'very', 2) reduplication of adjectives functions as a mitigating rather than an augmentative device, and 3) adjectives are the only word class that can function as Parameter in comparative constructions (see Section 7.8). In addition, the nominalising suffix *-an* can also be used on adjectives, as it can on verbs.

3.4.1 Formal characteristics of adjectives

Derived adjectives always end in *-Vn*; the same is true for a large number of adjectives with no attested verbal counterpart. The vowel in derived adjectives is unpredictable and is probably part of the verb root. Thus, as is the case with nouns, many verbs seem to have underlying long forms, whose final vowel only surfaces in an adjectival, suffixed, form. The vowel can be /a/, /ɛ/, /u/, /i/ or /ɔ/. /o/ and /e/ are not attested. Some examples of derived adjectives are given in (66).

(66) *kai* 'darken'   → *kaya-n* 'dark, black'
   kôk 'be hot'   → *kôki-n* 'hot'
   mat 'die'   → *mare-n* 'dead'
   mut 'break'   → *muri-n* 'broken'
   sakôi 'dry (up)' → *sakôyu-n* 'dry, dried up'

Examples of adjectives ending in *-Vn* that have no attested verbal counterpart include *mwalolon* 'fragrant', *kawin* 'crooked' and *menengan* 'big'. It is possible that these adjectival forms are diachronically derived from verbs that have been lost.

3.4.2 Adjectives within the NP

Within the NP, an adjectival modifier normally follows the noun. There are four exceptions: *kurun* 'small', *pwakpwak* 'fat, big', *tinawayen* 'huge; very many' and *lapan* 'excellent'. These can all be considered adjectives of dimension, except for *lapan*, which may be used ahead of common nouns to distinguish it from the homophonous noun meaning 'chief'. Four examples of attributive adjectives within the NP (the first two postposed, the following preposed) are given below.

(67) panu   menengan
   land   big
   'Manus mainland' [*lit.* 'the big place']

(68) pêng pian
    night good
    'good night'

(69) kurun not
    small child
    'a small child'

(70) tinawayen yoy
    huge   stone
    'a huge stone'

3.4.3 Predicative adjectives

Like nouns, adjectives can be heads of a predicate, but they cannot take bound pronouns, indicating their non-verbal status (there is a formal difference between free and bound pronouns only for first person singular; see Section 4.2). Compare (71) with (72) and (73) with (74). For more on non-verbal predicates, see Chapter 7.

(71) wong   pian
    1sg.FREE good
    'I am alright.'

(72) *nga=pian
    1sg=good

(73) wong   aloen
    1sg.FREE long
    'I'm tall.'

(74) *nga=aloen
    1sg=long

3.4.4 Adjectival morphology

3.4.4.1 Reduplication
Semantically, reduplication of an adjective reduces its denotational strength; this is one way in which adjectives can be distinguished from nouns. Examples

are *menmenengan* 'quite big' and *pipian* 'quite good'. The process is not very frequent. Periphrastic constructions with the adverb of degree *ansê* 'a bit' (derived from a quantifier *sê*) are a more common strategy to express reduced denotational strength; see Section 3.6.1.2 for further details.

3.4.4.2 Suffixation

The suffix *-an*, which functions as a deverbal nominalisation suffix (see Section 3.3.2.2.3.2 above), is also occasionally used to derive a noun from an adjective. Examples are given in (75) and (76).

(75) *lou repwo, ma konunan sa?*
lou te-pwo ma konun-an sa
living.person LIG-DEM.PROX EMP heavy-NOM what
'[My goodness,] what is the weight of this person!'
(LM190611_0023)

(76) *ino ro ai aloenan maranip*
i=no to a=i aloen-an mata-n ip
3sg=IPFV be OBL=3sg long-NOM eye-PERT 3pl
'It was [dependent] on how far they could see [*lit.* 'the length of their eyesight'].' (LK100411_0127)

3.5 Forms that appear in more than one word class

Several forms can occur as both nouns and verbs, or as both nouns and adjectives, based on distributional criteria. In addition, there are forms whose status as either verb or adverb is problematic; they will be discussed below, in Section 3.6 on adverbs. Most forms for which word class is hard to determine refer to abstract concepts.

3.5.1 Forms that appear as nouns and verbs

A number of forms appear both as nouns and as verbs. Zero derivation of intransitive verbs to nouns was already discussed in Section 3.3.2.2.4 above, as were other roots that appear both as nouns and verbs but with different suffixes, and for which it is also difficult to determine which form is primary and which one is derived.

There is also a transitive verb that occurs as a noun with zero derivation: *puk* 'open; reason, cause'. Examples of its nominal and verbal use are given below; it is noteworthy, though, that there is an unpredictable semantic shift between the two uses. In (77), *puk* is modified by a noun that is itself a zero derivation from a verb.

(77) *eppuk koukou*
ep=puk   koukou
1pl.EXCL=open fence
'We opened the fence.' (KM060111_0067)

(78) *puk tan kaêrêt*
puk   ta-n   kaêrêt
cause CLF.POSS-PERT be.afraid
'[This snake is] a reason for fear.' (Game1_021012_0533)

3.5.2 Forms that appear as nouns and adjectives

The following forms are encountered both as nouns and adjectives: *nuanan* 'true; truth', *parayoyop* 'power; strong' and *pian* 'good; goodness'. Because nouns and adjectives are distributionally and formally very similar, the only way to distinguish between them is whether or not they can be modified by an adverbial of degree such as *paran*. Examples involving *pian* are given below. *pian* can function as a core argument and be modified by a relative clause, as in (79), where it therefore has the hallmarks of a typical noun; it can also occur modified by *paran*, as in (80), which turns it into a typical adjective. It may be safely concluded that *pian* has therefore lexical entries in both word classes.

(79) *pian te irok o inêm*
pian te i=tok   yo   i=nêm
good REL 3sg=stay DEM.INT 3sg=be.finished
'The goodness that existed, it is finished.' (PK290411_3_0046)

(80) *panu reo i paran pian*
panu te-yo   i   paran pian
place LIG-DEM.INT 3sg very good
'That place is very good.' (MS250311_0076)

3.6 Adverbs

Adverbials modify word classes other than nouns, or else modify clauses. In Paluai, there is a marked difference between adverbs that modify verbs and adverbs that modify other word classes and clauses. Adverbs that modify verbs are themselves often verbal in nature and origin, and sometimes it is hard to determine whether a form is fully verbal or adverbial. Because of the fluid boundaries between verbs and verb-modifying adverbs, this subclass of adverbs can be regarded as an open class.[36] Adverbs that modify word classes other than verbs, or that modify clauses, are either nominal in origin or their origin is unclear. They form closed subclasses within the adverbial word class.

There are no clear formal criteria for adverbs. Distributionally, (verb-modifying) adverbs can be distinguished because they are the only lexical word class other than verbs occurring inside the verb complex, occupying a slot between the verb and the object enclitic. In contrast to verbs, on the other hand, they can never head a predicate. Many of the verb-modifying adverbs end in *-ek* and are thus formally similar to derived applicatives. However, since no underived counterparts are attested synchronically, the *-ek* in these verb-modifying adverbs is a frozen suffix.

3.6.1 Types of adverbs

Semantically and distributionally, several types of adverbs can be distinguished:
- Manner adverbs, which modify verbs
- Adverbs of degree, which modify stative verbs, adjectives and possibly other adverbs
- Temporal adverbs, which modify clauses
- Spatial adverbs, which modify verbs
- Modal/epistemic adverbs, which modify clauses

3.6.1.1 Manner adverbs
Manner adverbs modify a verb within the verb complex, giving information about the manner in which the action described by the verb is carried out. Table 27 shows a number of the forms encountered. Since the manner of an action can also be expressed with a SVC (see Chapter 9), strict criteria have been

---

[36] Alternatively, this subclass of adverbs could instead be regarded as a subclass of verbs that have undergone grammaticalisation and incomplete reanalysis.

Table 27: Some manner adverbs.

Form	Meaning
apurek	quickly
kasiek	not well
keleyek	turning
lilisek	forgetting about
liliu	again
nayek	around, here and there
neyek	following
nonot	understanding; clearly (used with verbs of perception and cognition)
nonosi	exactly
pelpelek	shivering
pulêek	too, as well, also
pwotpwot	exactly
sapeyek	not in a straight line
sapuluek	around, in a circle
seniek	frequently
sou	leaning
tiyek	completely; nicely
tuliek	confused
tut	completely
yokosek	mixed
wuliek	slowly, tardily

employed to distinguish manner adverbs from verbs. The most important criterion is that the form is never encountered as a full verb, i.e. as predicate head. One exception is made here. One form, *liliu*, occurs both as a full verb meaning 'return' and as an adverb meaning 'again'. However, the two roles can be distinguished based on distributional grounds; this is discussed in Section 9.2.1.

Further criteria related to semantics and distribution state that the prospective adverbial form must refer to the manner of the entire action (and not, e.g., to the manner of a kind of "result" of the action) and that it is allowed to intervene between a transitive verb and its direct object.

Some example sentences with manner adverbs are given in (81) to (83) below. All main verbs are transitive and shown in boldface; the manner adverb always directly follows the verb and thus appears between the verb and its direct object.

(81) *wong, ngasi rou pulek Korup*
      wong   nga=si   **tou**   pulêek   Korup
      1sg.FREE 1sg=come.down   bring   too   K.
      'Me, I came too, in order to bring Korup.' (ANK020995_0015)

(82) *irêno rêk sapuluek kôlôlôi menengan*
      irê=no   **têk**   sapuluek   kôlôlôi   menengan
      3pc=IPFV build   going.round gathering.place big
      'They built [their houses] in a circle around a big gathering place.'
      (OL200111_0094)

(83) *kapwa epkape kasiek napun ep teo...*
      kapwa   ep=ka-**pe**   kasiek   napu-n   ep
      if   1pl.EXCL=IRR.NS-make   not.well   food.taboo-PERT   1pl.EXCL
      te-yo
      LIG-DEM.INT
      'If we do not follow our food taboos properly...' (NP260511_0003)

Sentence (82) shows that *-ek* adverbs could still be somewhat related to applicatives. It could indeed be interpreted as 'they built-on_{APPL} [the gathering place]_O in a circular manner'. However, *kôlôlôi* is a local noun, and thus does not need to be preceded by a preposition. It is therefore hard to establish whether this is a "proper" applicative construction (for more on these, see Chapter 8). The difference between *sapuluek* and *nayek* becomes clear when sentence (82) above is compared to sentence (84) below.

(84) *ngapa ro ningning nayek*
      nga=pa   to   ning.ning   nayek
      1sg=yet CONT   REDUP.see   around
      'I am still looking around.' (Game1_021012_0096)

3.6.1.2 Adverbs of degree

Adverbs of degree are hard to delimit in almost any language. In Paluai, forms that modify stative verbs or adjectives, giving information about the extent of the property or state described by them, are considered adverbs of degree. Table 28 gives an overview.

Adverbs of degree can modify other word classes beside stative verbs and adjectives, such as quantifiers and other adverbs. *Pun* is a general intensifying or augmentative particle; it is used, among others, to form comparative and

Table 28: Adverbs of degree.

Form	Meaning
paran	very, really
pun	extremely (intensifier)
ansê	somewhat, quite

superlative constructions. These will be discussed in more detail in Section 7.8. *Ansê* is derived from the quantifier *sê*, and is also itself used as a quantifier. Quantifiers are discussed in Section 4.6. Just as interesting are the historical sources of *paran* and *pun*. They seem to be nominal in origin and are related to the terms for 'stem' and 'base' of a tree, respectively. The latter were metaphorically extended to mean 'most important part', 'core', and probably over time grammaticalised into adverbial modifiers.

(85) *nirasip pun kala alilêt*
 panurasip pun ka-la alilêt
 first INTF IRR.NS-go.to forest
 'First of all, one has to go to the bush.' (CA120211_2_0004)

(86) *paran menot teo irok*
 paran menot te-yo i=tok
 very many LIG-DEM.INT 3sg=stay
 'Very many are there.' (LL300511_2_0004)

Adverbs of degree can also be used with loan words:

(87) *mangat teo i paran expensive pun*
 mangat te-yo i paran expensive pun
 work LIG-DEM.INT 3sg very INTF
 'That ceremony is very expensive/the most expensive.'
 (PK290411_3_0041)

In a small number of cases, *paran* modifies a noun. Interestingly, all these cases show a directly possessed noun, mostly *kane-n* 'its meat', heading a non-verbal predicate. It seems that *paran kanen* is a set expression, meaning something like 'the real thing'. This is a further indication of the nominal origin of *paran*. Since this is probably a less grammaticalised sense of *paran* with a limited distribution, it does not change the analysis of *paran* as an adverbial modifier. An example of its use is given below.

(88) *i reo i paran kanen yamat pun*
    i   te-yo   i   paran kane-n   yamat   pun
    3sg   LIG-DEM.INT   3sg very   meat-PERT   person INTF
    'This here, it is really a person / a real person [i.e. not a spirit].'
    (OL200111_0039)

3.6.1.3 Temporal adverbs
Temporal adverbs modify clauses or verb complexes; an overview is given in Table 29.

Table 29: Temporal adverbs.

Form	Meaning/paraphrase
kanan puleng	tomorrow, the next day
mapêng	three days from now
minak	today, the present day
mino	yesterday, the previous day
monokin	afterwards, after X (behind-PERT)
nipêng	before, the other day
pa	yet, still
palosi	before, in the past
papwên	not yet (yet-NEG)
panuan	before X (front-PERT)
panurasip	(at an) earlier (moment in time)
pempom	last night, the previous night
perelian panu	(at) noon, (at) midday
pêng	(at) night; day of 24 hours; occasion
pêngino	the day before yesterday, two days previously
poyep	(in the) afternoon
pururuan pêng	(at) midnight
pwa in	first (before something else); for the time being
pwapwem	(in the) morning
pwotnan	at that time
teloan	now
tepwo-om	right now
wulian	finally
yupêng	the day after tomorrow, two days from now

A number of observations can be made with regard to temporal adverbs. First of all, they either modify a clause or a verb complex. They usually appear as the first constituent in the clause they modify. An exception is *pa* 'yet', which usually appears as the first element in a verb complex, just after the bound pronoun. Secondly, temporal adverbs either have nominal origin or their origin is unknown. Many, in particular the terms for parts of the day and 'yesterday', 'tomorrow' etc., are also attested as nouns.

In addition, Paluai seems to have had a "numeral classifier-like" counting system for days. Lele, a language spoken on Manus mainland, seems to have retained many more of the terms for days, even counting up to 'ten days from now' (Juliane Boettger, personal communication). Moreover, temporal operators such as *pempom* 'last night, previous night' are not deictic shifters as, e.g., in English. When English temporal adverbs are used in an indirect quotation, for instance, they have to be adjusted based on the moment the sentence is uttered. It is impossible to say, for instance: *?Last Tuesday she told me she went to the movies last night*, where *last night* refers to Monday night. Instead, one has to say *Last Tuesday she told me she went to the movies the night before*. In contrast, Paluai *pempom* could felicitously be used in this instance.

A few examples of the use of temporal adverbs are given below.

(89) *taim te wong pa namwi a tamong i mat tu niong*
 taim te wong pa namwi ya tama-ng i mat
 time SUB 1sg.free yet small then father-1sg.PERT 3sg die
 tu ni=ong
 stay away=1sg
 'When I was still little, my father died and left me.' (SY100411_0005)

(90) *minak wuirok tepwo*
 minak wui=tok te-pwo
 present.day 1du.EXCL=stay LIG-DEM.PROX
 'Nowadays, the two of us live here.' (KM060111_0009)

3.6.1.4 Spatial adverbs

An overview of spatial and directional adverbs is given in Table 30. The first three instantiations are complex forms consisting of the preposition *a=*, the ligature *te-*, and a member of the demonstrative paradigm; they form spatial deictics (for a discussion of the demonstrative system, see Section 4.3). Adverbials are only directional when combined with a directional verb in a SVC (see Section 3.3.1.3 and Chapter 9).

Table 30: Spatial adverbs.

Form	Meaning/paraphrase
atepwo	(right) here (spatial deictic)
ateyo	there (spatial deictic)
atelo	over there (far) (spatial deictic)
masayen	outside
maso	away; apart, separate
ni	away
nok	far away
paye	down, low; beneath
wat	up, high; above

A number of examples of the use of spatial adverbs are given below.

(91) *i rapari monok telo iret wen nok nansê*
 i ta=pari monok te-lo
 3sg SPEC.COLL=belonging.to behind LIG-DEM.DIST
 i=tet wen nok nan=sê
 3sg=move move.level far.away PART=DIM
 'The one at the back is leaning away a little.' (Game3_280812_0176)

(92) *ipnu ru wau naynayek masayen*
 ip=no tu wau nay.nayek masayen
 3pl=IPFV STAT.CONT move REDUP.around outside
 'They were walking around outside.' (LL300511_1_0038)

(93) *wosa yen arepwo pwên*
 wo=sa yen a=te-pwo pwên
 2sg=MOD lie OBL=LIG-DEM.PROX NEG2
 'You cannot lie here.' (NP210511_2_0058)

3.6.1.5 Modal/epistemic adverbs

Modal or epistemic adverbs modify a clause or a sentence. They give information about speaker attitude towards the information represented, for instance about the certainty of the speaker's knowledge about or the (non-)desirability of the state of affairs represented in the clause or sentence. Some members of this category of adverbs have an affinity with subordinating conjunctions that reflect, for instance, a concessive or causal relationship between two clauses.

Because the terms mentioned here cannot function by themselves as subordinators, but must be accompanied by the general subordinator *te*, they are considered adverbs. Forms are shown in Table 31 below.

Table 31: Modal/epistemic adverbs.

Form	Meaning
aronan	consequently, therefore
kanopa	like, for example
konan	like, such as
longoan	consequently, therefore
naman	perhaps, maybe
punan	luckily
ta po	indeed; actually
tangoan	consequently, therefore
tole	for sure
(la) temenin	like (that); as follows

Most modal and epistemic adverbs form the first constituent in the clause. An exception is *tole*, which usually appears clause-finally. *kanopa* and *konan* occur frequently, but in contexts where they seem to function more as interjections; they will therefore be discussed in more detail in Section 4.9. A few examples of the use of modal or epistemic adverbs are given in (94) to (96).

(94) *i reo pwên tararap tole*
    i   te-yo   pwên   ta-tap   tole
    3sg   LIG-DEM.INT   NEG   CLF.POSS-1pl.EXCL   for.sure
    'This [thing] is not ours [i.e. is not native to Baluan Island] for sure.' (Game1_021012_0090)

(95) *pein teo, rabo ila ran narun Kireng*
    pein   te-yo   tapo   i=la   ta-n   naru-n   Kireng
    woman LIG-DEM.INT   indeed   3sg=go.to   CLF.POSS-PERT   son-PERT   K.
    'This woman, indeed she belongs to Kireng's son.' (KW290311_0016)

(96) *naman nian ip palak ai*
    naman   nia-n   ip   palak   a=i
    perhaps   stomach-PERT   3pl   be.bad   OBL=3sg
    'Perhaps they got angry about it.' (NP190511_2_0023)

naman is used in a variety of contexts where a speaker does not want to vouch for the truth value of a proposition. The proposition can refer to something that could have happened in the past, or something that could happen in the future. Modality as part of the verbal TAM system will be discussed in more detail in Section 6.5.

3.6.2 Derivational processes related to adverbs

3.6.2.1 Reduplication
As with verbs, adverbs can be reduplicated to indicate prolonged and/or iterative action. Example (92), repeated below as (97), showed a reduplicated form of *nayek* 'around'. As with verbs, reduplication of adverbs can have slightly negative overtones ('do s.t. for too long, overdo it').

(97) *ipnu ru wau naynayek masayen*
  ip=no   tu   wau   nay.nayek   masayen
  3pl=IPFV STAT.CONT move REDUP.around outside
  'They were walking around outside.' (LL300511_1_0038)

3.6.2.2 Periphrastic derivation of adverbials with *la*
Adverbial phrases can be formed with the directional *la* 'go to' followed by an adjective. In these cases, the motion and direction semantics of *la* has usually fully disappeared. It is probably the copula variant of *la*, meaning 'become', that has grammaticalised in this way; *la* as a copula is discussed in more detail in Chapter 7. A few examples are given below.

(98) *tapan pemat ip parian tap la mangsilan*
  tap=an   pe-mat   ip paria-n   tap   la   mangsilan
  1pl.INCL=PRF CAUS-die 3pl wife-PERT 1pl.INCL go.to meaningless
  'We have killed our wives for no reason.' (LK100411_0088)

(99) *ngamasan la pian pwên*
  nga=ma=san   la   pian pwên
  1sg=NEG$_1$=cut go.to good NEG$_2$
  'I haven't cut [it] well.' (LL010711_0031)

3.7 Overview

Tables 32 and 33 summarise the distinguishing criteria for the word classes discussed in this chapter. Table 34 shows derivational morphology.

Table 32: Distinguishing criteria for word classes N, V, A and Adv.

I. Criteria to distinguish nouns
- Cannot take TAM particles and prefix
- Cannot take bound subject pronouns
- Cannot take a modifier of degree
- Can be modified by demonstrative or possessive classifier
- Can follow the preposition *pari*
- Reduplication indicates plurality
- Most typical function: head of a NP that is an argument to a verb

IIA. Criteria to distinguish verbs
- Can take TAM particles and prefix
- Can take bound subject pronouns
- Most typical function: head of a predicate

IIB. Criteria to distinguish stative verbs (as for verbs, but additionally:)
- Are always intransitive (some can take an E argument)
- Can take causative prefix *pe-*
- Can take a modifier of degree
- A noun-modifying form (A) can be derived from some, but not all

III. Criteria to distinguish adjectives
- Cannot take TAM particles and prefix
- Cannot take bound subject pronouns
- Can take a modifier of degree
- Reduplication indicates mitigation
- Can be Parameter of comparison in comparative constructions
- Most typical function: modifier of a N within the NP

IV. Criteria to distinguish (manner) adverbs
- Cannot head a predicate
- Are the only elements that can appear between a transitive V and its O argument, and between a stative V and its TAM elements
- Most typical function: modifier of a V

Table 33: Comparison of word classes with respect to distinguishing criteria.

Function	N	V	Vstat	A	Adv
head predicate	x	x	x	x	
TAM		x	x		
bound subject pronoun		x	x		
causative			x		
transitive		x			
mod. of degree			x	x	
parameter of comparison				x	
modifies N	x			x	
modifies V			(x)		x
modifies A					x
modifies Adv					x
appears in verb complex		x	x		x

Table 34: Word class-changing derivational morphology.

to \ from	N	V	Vstat	A	Adv
N		redupl. -(n)an -a / -o	–	–an	–
V	–		–	–	–
Vstat	–	–		–	–
A	–	–	–n		–
Adv	-ek	–	–	la A	

4 Word classes II: closed classes

4.1 Preliminaries

This chapter discusses the closed word classes found in Paluai, those that have a limited number of members, usually restricted to functional or grammatical morphemes. In what follows, the closed classes of pronouns, demonstratives, adpositions, numerals, quantifiers, interrogative words, negation and mood markers, conjunctions and other clause connectors, and interjections will be discussed. Some members of closed classes, such as TAM particles, were already discussed in the previous chapter together with the word class they function as modifiers for; this information will not be repeated here.

For some of the classes discussed in this chapter, such as conjunctions and interjections, it is unclear whether they are really closed as such. There are, for instance, a number of loans found in these classes. In any case, the distinction between open and closed classes is probably better characterised as gradient rather than absolute. In addition, functional categories sometimes have affinities with forms in other word classes. This is especially true for items whose form is dependent on the semantics of the noun that they modify, such as numerals, quantifiers or classifiers. This issue will be discussed where relevant. In the final section, a clitic formative *ta=*, which is encountered with forms from a range of open and closed word classes, will be discussed in more detail.

4.2 Pronouns

Pronouns form a closed class of grammatical items. Paluai distinguishes between free and bound personal pronouns, and two sets of suffixes that are attached to the possessive forms *ta-* and *ka-*; the latter also attach to directly possessed nouns. Free pronouns are free morphemes that can be used independently, as head of a NP, and can be modified by deictic demonstratives. They function in possessive constructions with non-singular Possessors to refer to person/number of the Possessor (see Section 5.5). Bound pronouns are clitics that attach to the first element (in case of the subject clitic) or the last element (in case of the object clitic) in the verb complex, in order to index the S/A and sometimes the O argument of the verb. Verbal predicates are discussed in detail in Chapter 6. The third person non-singular free pronouns are also used to indicate number on nouns; see Section 5.2 for more on this. Free pronouns can carry sentence accent, whereas bound pronouns are never stressed. Both free and bound pronouns can be used for anaphoric reference; see Chapter 12.

Among personal pronouns, a distinction is made between first person (referring to the speaker), second person (referring to the addressee) and third person (referring to neither speaker nor addressee but to an additional speech act participant or some person or entity not present at the speech act). Four numbers are distinguished: singular (referring to one entity), dual (two entities), paucal (several entities) and plural (many entities). Paucal number refers to any quantity greater than two and smaller than about ten, but it has no strict upper limit. This is in line with observations made for the paucal in other languages (Dixon 2010b: 191–192): it basically refers to a relatively small number compared to a larger number, rather than to any specific number. In one text, for example, the population of one village is contrasted with the population of Baluan Island as a whole. The village population (probably numbering several hundred people) is referred to by the paucal, whereas the entire population of Baluan (well over a thousand individuals) is referred to by the plural.

For the non-singular first persons (i.e. dual, paucal and plural), a distinction is made between inclusive (referring to two or more people including the addressee) and exclusive (referring to two or more people excluding the addressee). There is no gender or animacy distinction in the pronominal system. Naturally, first and second person pronouns generally refer to human beings or personified animals or objects, but a third person subject clitic can refer to either a human, an animal or an item, without a change in form.

Free pronouns can replace nouns and function as core arguments of the verb (i.e. as transitive and intransitive subject, and transitive object), or as subject constituent in a verbless clause. Free-standing pronouns are often followed by demonstrative forms, which function as definiteness and topic markers (see Section 4.3 below and Chapter 12). However, unlike a noun, they cannot be followed by adjectives or numerals. Non-core arguments are never expressed by a free pronoun used in isolation, but will be introduced by either the preposition *a=*, the possessive/locative *ta-* or the possessive classifier *ka-*; they are often preceded by a directional (see Chapter 8). Subject bound pronoun clitics can only attach to verbs and preverbal particles, not to other word classes.[37] This bound pronoun will often be a pronoun "copy" of a pronominal or full NP subject that precedes it in the same clause. See Chapters 6 and 8 for more on verbal categories, grammatical relations and the order of elements in the clause.

---

[37] Since adverbs can appear between the verb and its object, the object bound pronoun enclitic can attach to an adverb.

Pronouns also have multiple functions within the noun phrase, such as indicating number on nouns, and marking the comitative, e.g. *taman u tinan* 'father and mother' (*lit.* 'father 3du mother'). These functions are discussed in Chapter 5.

4.2.1 The free pronoun paradigm

The free pronoun paradigm is given in Table 35.

Table 35: The free pronoun paradigm.

		1st person	2nd person	3rd person
Singular		wong	wo	i
Dual	inclusive	tau	au	u
	exclusive	(w)ui		
Paucal	inclusive	tarê	arê	irê
	exclusive	(w)urê		
Plural	inclusive	tap	ap	ip
	exclusive	ep		

The pronouns are complex forms. Clearly, the dual pronouns share a formative [u], which may go back to Proto-Oceanic *rua 'two'. The paucal forms share a formative [rê] ~ /tê/, which may go back to POc *tolu 'three' (Lynch et al. 2002: 72). Thus, the paucal may originally have been a trial. The plural forms share a formative [p], probably from POc *pat 'four', and are thus likely to have developed out of a quadral. The inclusive paradigm starts with a formative *ta-*, which is most likely a reflex of POc *kita 'we (incl.)'. Thus, the inclusive paradigm reflects an older state of affairs whereas the exclusive paradigm may be an innovation.

In everyday conversation, inclusive pronouns are used as hortative markers without any additional sentence material:

(1) *kay, tarê!*
    kay   tarê
    okay   1pc.INCL
    'Okay, let's [go ahead and do what we want to do].'

4.2.2 The bound pronoun paradigm

4.2.2.1 Subject forms

The forms of the subject bound pronouns, preceding the verb complex, are given in Table 36. The table shows that the forms are identical to the free pronouns, except for the first person singular; in addition, word-initial [w] and [j] tend to be dropped more frequently for the bound pronouns. They are analysed as clitics that attach to the first element in the verb complex.

Table 36: The bound pronoun paradigm (subject forms).

		1st person	2nd person	3rd person
Singular		nga=	(w)o=	i=
Dual	inclusive	tau=	au=	u=
	exclusive	(w)ui=		
Paucal	inclusive	tare=	arê=	irê=
	exclusive	(w)urê=		
Plural	inclusive	tap=	ap=	ip=
	exclusive	ep=		

4.2.2.2 Object forms

The forms of the object bound pronouns, following the verb complex, are given in Table 37. Again, forms are nearly identical to the free pronouns. In contrast to bound subject forms, there is no optional [w] that is dropped in the majority

Table 37: The bound pronoun paradigm (object forms).

		1st person	2nd person	3rd person
Singular		=ong	=o	=i (animate) =Ø (inanimate)
Dual	inclusive	=tau	=au	=u (animate) =Ø (inanimate)
	exclusive	=ui		
Paucal	inclusive	=tarê	=arê	=irê (animate) =Ø (inanimate)
	exclusive	=urê		
Plural	inclusive	=tap	=ap	=ip (animate) =Ø (inanimate)
	exclusive	=ep		

of cases. The forms are analysed as enclitics that attach to the last element in the verb complex. For the third person, there is an animate/inanimate distinction: animate O arguments are formally overt, whereas inanimate O arguments have a zero form. No such distinction exists for the first and second person, which (normally) refer to animate beings.

4.2.3 Possessive forms

Pronouns also appear as suffixes on the classifier for general alienable possession *ta-* and on the alimentary classifier *ka-* (and on directly possessed nouns). Because the resulting forms are slightly irregular in some cases, the entire paradigms are given here, with the surface realisation forms placed between square brackets.

4.2.3.1 The possessive/locative form *ta-*

The bound root *ta-* is suffixed with a pronominal form as shown in Table 38. It is used to mark the Possessor of an indirect possession construction (see Section 5.5) or as a marker of an Oblique argument with animate reference (see Sections 4.4.3 and 8.1). When it occurs as a postposed element to a noun, it forms one phonological word with it. This is evident from a phonological process of assimilation of the /t/ when the noun it modifies ends in /l/ (see Section 2.2.2.1.2), for instance in /samɛl=ta-ŋ/ > [samɛl laŋ] 'my knife'. This type of assimilation only occurs within a phonological word. The possessive classifier *ta-* is therefore best analysed as a clitic.

Table 38: The possessive paradigm (with *ta-*).

		1st person	2nd person	3rd person
Singular		*ta-ng* [taŋ]	*ta-o* [taɔ] ~ [tɔ]	*ta-i* [tai] *ta-n* [tan]
Dual	inclusive	*ta-tau* [tarau]	*ta-au* [tau]	*ta-u* [tau]
	exclusive	*ta-ui* [tui]		
Paucal	inclusive	*ta-tarê* [tararẹ]	*ta-arê* [tarẹ]	*ta-irê* [tairẹ]
	exclusive	*ta-urê* [turẹ]		
Plural	inclusive	*ta-tap* [tararap] (expected: [tarap])	*ta-ap* [tap]	*ta-ip* [taip]
	exclusive	*ta-ep* [tɛp]		

For the third person singular, there are two possibilities: *ta-* is suffixed by *-i* when it is not followed by a NP constituent, and by *-n* when it is. The suffixes that *ta-* takes are predominantly the same ones that are taken by directly possessed nouns, except for the second and third person singular: the *-o* and *-i* suffixes are formally identical to the second and third person object bound pronouns, respectively. Because it takes a mixture of nominal and verbal dependent morphology, *ta-* seems to be straddling the boundary between the nominal and verbal realms.

In addition, there is some degree of syncretism in surface realisations within the possessive paradigm, and also between the personal and the possessive paradigms. The surface form *tarê*, for example, can either refer to first person paucal inclusive, or to second person paucal possessive.

4.2.3.2 The possessive classifier *ka-*

Table 39 shows the forms that occur with the possessive classifier *ka-*.

Table 39: The possessive paradigm (with *ka-*).

		1st person	2nd person	3rd person
Singular		*ka-ng* [kɔŋ]	*ka-m* [kɔm]	*ka-n* [kan]
Dual	inclusive	*ka-tau* [karau]	*ka-au* [kau]	*ka-u* [kau]
	exclusive	*ka-ui* [kui]		
Paucal	inclusive	*ka-tarê* [kararẹ]	*ka-arê* [karẹ]	*ka-irê* [kairẹ]
	exclusive	*ka-urê* [kurẹ]		
Plural	inclusive	*ka-tap* [kararap]	*ka-ap* [kap]	*ka-ip* [kaip]
	exclusive	*ka-ep* [kɛp]		

In contrast to *ta-* discussed above, the classifier *ka-* only takes suffixes from the nominal paradigm. An interesting formal difference between *ta-* and *ka-* is the different vowel in the first and second person singular forms. The change from /a/ in the *ka-* form (*kong* vs. *tang*) is probably due to a historical version of the pertensive suffix, which contained a vowel that caused alternations to the stem vowels. The same process is at work in the suffixation of directly possessed nouns; for more on this, see Section 2.2.7.2.

4.3 Demonstratives

Demonstratives are deictic expressions; in situational deixis, they "indicate the location of referents along certain dimensions, using the speaker (and time and place of speaking) as a reference point or 'deictic centre'" (Cruse 2006: 44). With discourse deixis, demonstratives function as anaphors to refer to previously mentioned discourse participants or stretches of discourse, or they act as cataphors and refer forward. In many languages, demonstrative forms do double duty as situational and discourse deictic elements. Paluai has an elaborate system for the expression of discourse deixis, definiteness, topic and focus, discussed in more detail in Chapter 12.

There is a three-way demonstrative system, which distinguishes 1) a position at or very near the deictic centre, 2) an intermediate position, somewhat removed, and 3) a position at significant distance from the deictic centre. The deictic centre is usually the speaker, and it appears to be very small: the proximate demonstrative seems almost exclusively used for objects held in one's hands or on one's body. The terms are used relative to each other: when talking about two objects at different distances, the intermediate and distal demonstratives will be used to establish a contrast between them. Relative distance from the addressee is not a factor in the demonstrative system. Visibility may be a factor, as the distal demonstrative tends to be used when something is handed to a person when this person doesn't see the item. Thus, the distal demonstrative may more generally be used for objects that are not visible, but it is also used for objects that are still visible, but further away than something that was referred to with the intermediate form.

There is no gender or animacy distinction in the demonstrative paradigm. The paradigm is quite large, with at least twelve basic and complex demonstrative forms. They have different functions, but all have one of the three basic units as their main formative. Demonstratives always follow the element they modify and usually appear last in the NP (see Section 5.1 for more on the order of elements within the NP).

4.3.1 Basic demonstrative forms

The three basic forms on which demonstratives are built are the following:

(2) *pwo* proximate demonstrative, 'this'
    yo intermediate demonstrative, 'that'
    lo distal demonstrative, 'that (far)'

These basic demonstrative forms can modify a noun or pronoun, but they are most often encountered with the 3sg pronoun. As a situational deictic device, *i pwo* is often used when pointing at an object or offering somebody an object: 'here it is'. *i yo* often occurs as a discourse marker at the end of a narrative ('okay', 'that's it'). When modifying a noun, the basic form probably functions as a marker of definiteness; this is similar to its function when combined with the ligature *te-* (see below). *lo* is not encountered as often as the other two, but is used to point at a distant object. The examples below show some uses of the basic demonstrative forms. In (3), a situational deictic use of *pwo* is shown, and in example (4) a discourse deictic use of *yo*.

(3) *i pwo, moni reo pepa sangal*
 i pwo moni te-yo pepa sangal
 3sg DEM.PROX money LIG-DEM.INT ten.kina ten
 'Here. The money is one hundred kina.' [said when the money is handed over] (PK290411_3_0036)

(4) *i o. naman kamou rang teo inêm*
 i yo naman kamou ta-ng
 3sg DEM.INT perhaps speech CLF.POSS-1sg.PERT
 te-yo i=nêm
 LIG-DEM.INT 3sg=be.finished
 'That's it. Perhaps my talk is finished.' (OL201210_0195)

4.3.2 Demonstratives with the formative *te-*

Demonstratives do not occur all that often in their basic form. In most cases, the forms are used with a formative *te-* preceding them. The exact meaning of *te-* is not entirely clear. It may have an emphatic function, or it may predominantly be used as a ligature between the (pro)noun and basic demonstrative form. There is also a general dependent clause marker *te* (see Chapter 11), which may be related to the formative. The demonstrative with *te-* forms one phonological word with the element that precedes it. This is evident from a phonological process of assimilation of /t/ to a preceding /l/ (see Section 2.2.2.1.2), for instance in /silal=tɛ-jɔ/ > [silal lɛʲɔ] 'the spirit'. This type of assimilation only occurs within a phonological word. Demonstratives, both with and without the formative *te-*, are therefore best analysed as clitics.

The complex demonstrative forms *tepwo*, *teyo* and *telo* can modify both nouns and pronouns, and again can be used either for situational or discourse deixis. *i teyo* is used in a way very similar to *i yo*, as a discourse marker at the end of a story. *tepwo* and *telo* function as anaphoric and cataphoric devices.

A number of important observations should be made with regard to the *te*-demonstratives. First of all, the proximate form *tepwo* in particular does not only refer to place, but also to time. It can thus mean 'this, here', but also 'now'. Secondly, these demonstratives can modify proper nouns, place names and first and second person pronouns as well as common nouns. For instance, when beginning a story, speakers usually introduce themselves with something similar to the following phrase, which is quite difficult to translate into proper-sounding English:

(5) *wong tepwo, ngayong Lorat*
 wong te-pwo ngaya-ng Lorat
 1sg.FREE LIG-DEM.PROX name-1sg.PERT Lorat
 'I here, my name is Lorat.' [Tok Pisin: *Mi nau, nem bilong mi Lorat.*]
 (OL201210_0014)

Thirdly, the intermediate demonstrative *teyo* in particular is used to indicate definiteness. Newly introduced elements, which are indefinite (regarded as unidentifiable for the hearer), lack a demonstrative modifier. For more on definiteness and information structure, see Chapter 12. Some examples of *te*-demonstratives are given below. All sentences come from recordings of the *Man and Tree* game (Levinson et al. 1992) in which pictures are described, so they predominantly have situational deictic functions. At the same time, however, they also seem to mark definiteness of discourse entities that have been mentioned before.

Basic demonstratives and the demonstrative forms with *te-* do not occur by themselves, but always modify another element (a noun or pronoun). Grammatically, they can best be analysed as clitics: they cannot occur independently, and can attach to more than one word class. Phonologically, however, they show behaviour that is atypical of clitics, since they can be stressed. These demonstratives are different from the ones starting with *a=* and *ta=* discussed below, which can occur as independent constituents.

(6) *on pwa mun tepwo iro Paluai?*
 wo=an pwa mun te-pwo i=to Paluai
 2sg=PRF think banana LIG-DEM.PROX 3sg=be P.
 'Do you think this kind of banana grows on Baluan?'
 (Game3_280812_0230)

(7) *mun teo i makerin sip pwên*
 mun te-yo i ma=kerin sip pwên
 banana LIG-DEM.INT 3sg NEG₁=bunch one.INANIM NEG₂
 'Those bananas, they are not in a bunch.' (Game3_280812_0229)

(8) *kei raywei relo ila ro monokin irê*
 kei ta=yuei te-lo i=la to monoki-n irê
 tree SPEC.COLL=two.long LIG-DEM.DIST 3sg=go.to be behind-PERT 3pc
 'Those two trees are behind them.' (Game4_280812_0178)

4.3.3 Spatial demonstratives with *a=*

A further set of demonstratives is created by adding the preposition *a=* (see Section 4.4.1) as a proclitic to the forms with *te-*. The resulting form deictically refers to the location where the action described by the verb takes place, and thus acts as an adverbial modifier to a verb. In the case of *ateyo*, this is usually a location mentioned previously in the discourse. In sentence (10) it is a garden located high on the mountain, mentioned just before. The proximate and distal spatial demonstratives usually deictically refer to a location, either the location at which the speech act takes place, as in (9), or a location removed from it, as in (11).

(9) *wosa yen arepwo pwên*
 wo=sa yen a=te-pwo pwên
 2sg=MOD lie OBL=LIG-DEM.PROX NEG
 'You cannot lie here.' (NP210511_2_0058)

(10) *wuisot kunawayut areo*
 wui=sot kunawayut a=te-yo
 1du.EXCL=go.up take.rest OBL=LIG-DEM.INT
 'We went up to take a rest there.' (KM050995_0017)

(11) *ola lêp kong payanpôl sip te ila ro arelo me*
 wo=la lêp ka-ng payan.pôl sip
 2sg=go.to take CLF.food-1sg.PERT dry.coconut one.INANIM
 te i=la to a=te-lo me
 REL 3sg=go.to be OBL=LIG-DEM.DIST come
 'You go and take my coconut [for me to eat] that is over there, and bring it here.' (LK100411_0063)

4.3.4 Free demonstrative forms formed with *ta=*

Lastly, there is a complex demonstrative made up of a formative *ta=* (discussed in more detail in Section 4.11), the aforementioned ligature *te-* and a basic demonstrative form. This demonstrative can be used as an independent form that can substitute for a noun and be head of a NP, for instance as an S/A or O argument to a verb. These forms can be compared to Indo-European demonstrative pronouns, such as English *this* in 'This is rubbish' or 'I don't like this'. Interestingly, they are independent forms all parts of which can be analysed as clitics.

Again, the proximate and distal forms are usually encountered in situational deictic uses, whereas the intermediate form is often used as a discourse deictic. Their usage is exemplified below. In (12) and (14), the proximate and the distal demonstrative, respectively, function as subjects in a verbless clause. In (13), the intermediate demonstrative functions as O argument.

(12) *Nulik, tarepwo ran sê?*
 Nulik ta=te-pwo ta-n sê
 Nulik SPEC.COLL=LIG-DEM.PROX CLF.POSS-PERT who
 'Nulik, whose is all this?' (WendyLawan020611_0044)

(13) *irouek nêm tareo la ran pein teo*
 i=touek nêm ta=te-yo la
 3sg=show be.finished SPEC.COLL=LIG-DEM.INT go.to
 ta-n pein te-yo
 CLF.POSS-PERT woman LIG-DEM.INT
 'She showed all that [what had been talked about just before] to the woman.' (KS030611_1_0017)

(14) *tarelo yeuyeu*
 ta=te-lo yeuyeu
 SPEC.COLL=LIG-DEM.DIST star
 'Those are stars.' (052b_0169)

4.3.5 Overview of demonstrative forms

The complex demonstrative forms do not differ from the basic forms in meaning, but they clearly have different functions. Table 40 gives an overview. Across all functions, the intermediate demonstrative has a definiteness marking and discourse deictic function more often than the other two.

Table 40: The demonstrative paradigm.

Form	Modifies noun	Modifies pronoun	Modifies verb	Independent form
pwo 'this'	sometimes	yes	no	no
yo 'that'	sometimes	yes	no	no
lo 'that (far)'	sometimes	yes	no	no
tepwo 'this'	yes	yes	no	no
teyo 'that'	yes	yes	no	no
telo 'that (far)'	yes	yes	no	no
arepwo 'here'	no	no	yes	no
areyo 'there'	no	no	yes	no
arelo 'over there'	no	no	yes	no
tarepwo 'this'	no	no	no	yes
tareyo 'that'	no	no	no	yes
tarelo 'that (far)'	no	no	no	yes

4.4 Adpositions

Spatial relations are often expressed by spatial nouns (see Section 3.2.2.3) or directional verbs (Section 3.3.1.3). A small number of forms, often with generic semantics, could be analysed as prepositions; Table 41 gives an overview. There are no postpositions.

Table 41: Prepositions and related forms in Paluai.

Form	Meaning/paraphrase
a=	inanimate Oblique argument marker
pari	from, belonging to
ta-	possessive, locative; animate Oblique argument marker

4.4.1 The preposition *a=*

When used to introduce a prepositional phrase, *a=* is always cliticised to the third person singular pronoun *i*. *a=* is considered a clitic because it attaches to other forms as well, without the *i*, for example to the demonstrative forms with *te-* (as discussed in Section 4.3.3 above), the locative interrogative word *pa* 'where' (see Section 4.7) and the dependent clause marker *te* (see Section 4.9.2).

a= has a very general semantics, comparable to the Tok Pisin form *long*. It functions as an obligatory Oblique argument marker with common nouns, and

also serves to introduce the Standard of comparison in comparative constructions (see Section 7.8). When local nouns and directly possessed nouns (often spatial nouns) form a locative constituent, *a=* is not required. When expressing movement, *a=* is preceded by a directional verb indicating the direction of motion. At least the following uses of *a=* can be distinguished:
- Location;
- Motion towards a goal or from a source;
- Moment in time / period of time;
- Instrument;
- Stimulus;
- "Theme" (often followed by nominalised verb).

An example of each is given below.

(15) *ipat mun to ai kanum teo*
    i$_A$=pat   [mun]$_O$ to   a=i   [kanum te-yo]$_{LOC}$
    3sg=plant banana be OBL=3sg garden LIG-DEM.INT
    'He planted bananas in the garden.' (WL020611_0029)

(16) *wope yen piy me ai yapi*
    wo$_A$=pe   yen   piy=Ø$_O$ me   a=i   [yapi]$_{LG}$
    2SG=PFV CONT squeeze=3sg.ZERO come OBL=3sg sago
    'You will continue squeezing [it] into the sago.' (CA120211_1_0036)

(17) *wuirok ai kerin sip*
    wui$_S$=tok   a=i   [kerin sip]$_{TIME}$
    1du.EXCL=stay OBL=3sg year   one.INANIM
    'The two of us stayed there for one year.' (KM060111_0039)

(18) *ngalêp pein, nga lêpi ai kokon*
    nga$_A$=lêp [pein]$_O$ nga$_A$=lêp=i$_O$ a=i[kokoni]$_{INSTR}$
    1sg=take woman 1sg=take=3sg OBL=3sg money
    'I took a woman, I took her with money.' [i.e. I paid my bride price]
    (KM060111_0100)

(19) *maloan no wop ai kamou rai*
    [maloa-n]$_S$ no   wop   a=i   [kamou ta-i]$_{STIMULUS}$
    spirit-PERT IPFV fly OBL=3sg speech CLF.POSS-3sg
    'She was surprised [*lit.* 'her spirit flew'] by his words.' (WL020611_0038)

(20) *nganêm ai nganngan*
 nga_S=nêm a=i [ngan.ngan]_{THEME}
 1sg=be.finished OBL=3sg REDUP.eat
 'I am done (with) eating.' (Game1_021012_0476)

4.4.2 The preposition *pari*

The form *pari* can be used by itself, heading a predicate, or as a modifier within the NP. When followed by a common noun, it needs to be accompanied by *a=* (see above) cliticised to *i*. It then usually introduces a constituent with a purposive meaning, or meaning 'about'. When followed by a local noun, it is not accompanied by *a=*, and it can be translated in English as 'from'. However, it can only be used to refer to a person's or an object's origin and it does not have a directional or ablative use. *Pari* has an affinity with possessive constructions and is therefore glossed as 'belonging to'. One important difference between *pari* constructions and possessive constructions "proper" is that the former are not used to refer to human beings as possessors. Examples of the various uses of *pari* are given below. *pari* is also part of the complex interrogative word *pari ai sa* 'why' (see Section 4.7) and the complex subordinate clause marker *pari ai te* 'because' (see Chapter 11).

Sometimes, *pari* heads the predicate of a verbless clause. When it is the head of a predicate with a 3sg subject, it is usually preceded by the 3sg subject pronoun *i*. In these cases, *pari* could be analysed as a copula (see Section 7.5). It is not a full verb, since it cannot take bound pronouns.

In (21), an example is given of *pari* indicating a person's origin. In this example, it heads a verbless predicate. (22) shows *pari* as a NP modifier with a purposive meaning. *lalon* is a spatial noun, so *a=* is not needed.

(21) *taman i pari Ulput*
 tama-n i pari Ulput
 father-PERT 3sg belonging.to Ulput
 'His father was from Ulput.' (OL201210_0019)

(22) *kei nangin pari lalon kanum*
 kei nangin pari lalo-n kanum
 herbs scent belonging.to inside-PERT garden
 'Nice-smelling herbs [meant] for inside the garden.' (NK290311_1_0013)

Sentences (23) and (24) show examples of possessive or purposive uses of *pari* as a noun modifier. In (23), it refers back to the common noun *matmat*, the name of a particular kind of ceremony.

(23) *ikipe si pe mangat pari ai*
 i=ki-pe si pe mangat pari a=i
 3sg=IRR.3sg-PFV come.down do work belonging.to OBL=3sg
 'He will come and do the work for it [a ceremony].'
 (SY100411_0025)

(24) *...pari ai wauwau pit taip*
 pari a=i wau.wau pit ta-ip
 belonging.to OBL=3sg REDUP.move be.close CLF.POSS-3pl
 '[a house] for (the purpose of) their meetings.' (LL300511_1_0024)

pari as a noun modifier is also encountered describing a typical activity or proclivity of a certain person or group of people. In these cases, *X pari ai* could be replaced by *yamta-n*, which literally means 'owner'.

(25) *ip yamat pari ai peinan nin*
 ip yamat pari a=i peinan nine
 3pl person belonging.to OBL=3sg making.of fight
 'Troublemakers, people inclined to fighting.' [TP: *man bilong pait*]
 (NP220611_2_0009)

4.4.3 The "preposition" *ta-*

Ta- functions as a possessive classifier in indirect possessive constructions (see Section 5.5), but is also used, often in combination with a directional, to mark Oblique arguments that have animate reference; it always bears a suffix cross-referencing person and number of this argument. In this function, *ta-* is in complementary distribution with *a=*. *ta-* can mark Oblique arguments with a variety of semantic roles, but since the argument refers to an animate being, it will often be a Goal or Recipient (see Chapter 8 for more on grammatical relations).

Paluai *ta-* is probably a reflex of the Proto-Oceanic root **ta*, analysed as a form with both possessive and locative/benefactive functions (Ross 1988: 103).

As in its possessive use, *ta-* receives a pronominal suffix, as indicated in Table 38 above, when it is an argument marker. An example of *ta-* marking a Stimulus argument to the Experiencer subject verb *kaêrêt* 'be afraid (of)' is given below in (26). (27) shows the same verb with an inanimate stimulus introduced by *a=*. Many examples of the use of *ta-* in ditransitive constructions are given in Chapters 8 and 9.

(26) *ngaru kaêrêt tan muyou*
 nga$_S$=tu   kaêrêt   ta-n   [muyou]$_{STIMULUS}$
 1sg=STAT.CONT be.afraid CLF.POSS-PERT snake
 'I was afraid of the snake.' (Game1_021012_0562)

(27) *ipto kaêrêt ai aronan kauwat taip*
 ip$_S$=to   kaêrêt   a=i   [arona-n   kauwat   ta-ip]$_{STIMULUS}$
 3pl=HAB be.afraid OBL=3sg way-PERT tradepartner CLF.POSS-3pl
 'They used to be afraid of the customs of their tradepartners.' (MS250311_0046)

4.5 Numerals

Cardinal numerals follow a decimal system, with numbers seven to nine expressed by subtractive numerals. This is a commonly encountered feature in Eastern Admiralties languages (Ross 1988: 342). Numerals bear suffixes depending on the noun class they are modifying (see Section 3.2.3). The numbers one, four and five, and their multiplications by a factor 10 or 100, have suppletive forms.

Numerals modify nouns and can be the head of a non-verbal predicate. They are morphologically complex forms. What follows will discuss their analysis in as much detail as possible. There are no ordinal numbers.

4.5.1 The numerals one to ten

In Tables 42 and 43 below, two paradigms containing the numerals one to ten are given. The first table shows the numerals suffixed with *-mou*, used for animates, and the second one those suffixed with *-êp*, used as a "residue" category for inanimates. A tentative morphological structure is given in the second column of

Table 42: Numerals one to ten for animates.

Numeral	Morphological structure	Translation
som	sV-m[ou]	one
yumou	yu-mou	two
tulumou	tulu-mou	three
pamou	pa[t]-mou	four
ngunan	ngV-nan	five
ngonomou	ngV-onom-mou	six
nganorulumou	nga-no-tulu-mou	seven (ten minus three)
nganoyumou	nga-no-yu-mou	eight (ten minus two)
nganosom	nga-no-som	nine (ten minus one)
sangal	sV-nga-l	ten

Table 43: Numerals one to ten for inanimates.

Numeral	Morphological structure	Translation
sip	sV-êp	one
yuêp	yu-êp	two
tulêp	tulu-êp	three
talot	unknown	four
ngunan	ngV-nan	five
ngunêp	ngV-onom-êp*	six
nganorulêp	nga-no-tul-êp	seven (ten minus three)
nganoyuêp	nga-no-yu-êp	eight (ten minus two)
nganosip	nga-no-sip	nine (ten minus one)
sangal	sV-nga-l	ten

*Tentative analysis: the POc root *onom* is not reflected by the current form.

each table where possible, but most of the numeral formatives are not attested independently in the present-day language.

The tables show that the morphological makeup of numerals is the same for both paradigms, except for the final suffix. A number of other remarks can be made. Firstly, it is clear that some of the numerals reflect Proto-Oceanic forms. *ta ~ *sa 'one', *rua 'two', *tolu 'three', *pat 'four' and *sa[ŋa]-puluq 'ten' can be recognised in the Paluai forms (Lynch et al. 2002: 72). It is possible that *onom 'six' is part of the paradigm for animates, but this is less clear for inanimates. Another Proto-Oceanic root whose reflexes are widespread in Oceanic languages, *lima (meaning 'five' and also 'hand') is only reflected in the form for 'fifty', shown below. The inanimate form for 'four' is also irregular. The formative -nga-, which is probably derived from the form for 'ten' (cf. POc *sa[-ŋa]-puluq 'ten'), is the base for the subtractive numerals seven to nine. Ross (1988: 344) has reconstructed the subtraction morpheme as the PEAd form *(a)nto-. Clearly, the formative -no- in the forms for seven to nine is a reflex of this.[38]

4.5.2 Higher numerals

The forms for multiples of ten and hundred are given in Tables 44 and 45 below. These forms do not show the numeral classifier suffix, but they will mostly be followed by a numeral one to nine, which bears the suffix. Rounded tens and hundreds for some reason do not bear the suffix. The system on which the tens and hundreds are formed is practically the same as that for the numbers one to ten, but with a formative -nga- and -ngo- respectively.

Table 44: Multiplications of ten.

Numeral	Morphological structure	Translation
yungal	yu-nga-l	twenty
tulungal	tulu-nga-l	thirty
pawoy	pa[t]-woy	forty
limlim	lim-lim	fifty
wolongal	wolo-nga-l	sixty
nganorulungal	nga-no-tulu-nga-l	seventy
nganoyungal	nga-no-yu-nga-l	eighty
nganosangal	nga-no-sV-nga-l	ninety
songot	sV-ngo-t	one hundred

Subtractive numerals that are multiplications of ten have the same formative nga- as a base, as observed for the units above. This may be an old form: for PMP, *-ŋa is attested as a linker for multiples of ten (Blust and Trussel 2017). There are irregular forms for fifty and five hundred, and numbers 600 to 900 are made up of the numbers for 60 to 90 plus a formative kasip, which is not encountered anywhere else. Its exact meaning is not clear. Complex numbers are built by simple combination. A few examples:

---

38 The -no- formative may also be related to the adverb no 'only, just'.

Table 45: Multiplications of hundred.

Numeral	Morphological structure	Translation
yungot	yu-ngo-t	two hundred
tulungot	tulu-ngo-t	three hundred
pangot	pa[t]-ngo-t	four hundred
misimin	unknown	five hundred
wolongal kasip	numeral 'sixty' plus unknown formative	six hundred
nganorulungal kasip	numeral 'seventy' plus unknown formative	seven hundred
nganoyungal kasip	numeral 'eighty' plus unknown formative	eight hundred
nganosangal kasip	numeral 'ninety' plus unknown formative	nine hundred
mwason	unknown	one thousand

(28) *sangal a sip*   eleven (*lit.* 'ten and one')
    tulungal a (kulan) nganosip thirty-nine (*lit.* 'thirty and (plus) nine')
    mwason yuêp   two thousand

Numbers exceeding a couple of thousands are not attested, but they could theoretically be formed combining the numerals given above. In the past, the higher numerals would be used when performing traditional exchange ceremonies, but nowadays they have been replaced by English and/or Tok Pisin numerals. Because of this, many speakers, especially from younger generations, are losing or have already lost proficiency in Paluai numerals. One could therefore say that the Paluai numeral system is more critically endangered than the language as a whole (cf. Comrie 2005).

4.5.3 Morphology and syntax of numerals

4.5.3.1 Numerals as noun modifiers

Like most modifiers, numerals usually follow the noun. However, the forms for 'one' sometimes precede the noun. This may be a strategy to express non-specific, categorial reference, which will be discussed in more detail in Section 5.2; see sentence (29) for an example. Similarity is also expressed by the numeral for one: *no sip* (*lit.* 'only one') can, depending on the context, be translated with 'the same'; see sentence (30).

(29) *ope rou lai sip kui menengan*
 wo=pe tou la a=i sip kui menengan
 2sg=PFV put go.to OBL=3sg one.INANIM pot big
 'You will put [it] into a big pot.' (CA120211_2_0020)

(30) *woning parun teo? mano sip pwên*
 wo=ning patu-n te-yo ma=no sip pwên
 2sg=see head-PERT LIG-DEM.INT NEG$_1$=only one.INANIM NEG$_2$
 'You see their heads? They're not the same.' (Game3_280812_0310)

The formative *ni-* 'other' can precede the numerals for one, the resulting form meaning 'another'. The form of the numeral will depend on the semantics of the head noun: *nisaya* refers to another place, *nisopol* to another side, and *nisip* and *nisom* to another inanimate and animate entity, respectively. Usually, forms with *ni-* refer to an indefinite entity. They usually follow the noun, but can also precede it, or be used by themselves, as in (32).

(31) *pou nisom ila ro ilili la ro naêmwan*
 pou ni-som i=la to ilili la to naêmwa-n
 pig other-one.ANIM 3sg=go.to CONT stand.up go.to be backside-PERT
 'Another pig is standing behind him.' (Game2_021012_0246)

(32) *pian, ngan akêp nisip*
 pian nga=an akêp ni-sip
 good 1sg=PRF pick.up other-one.INANIM
 'Alright, I picked up another one.' (Game1_021012_0094)

Lastly, the sequences *(a=)sip (a=)sip* or *(a=)som (a=)som* are used to express 'one by one, one after the other'; see sentence (33).

(33) *kipe la ro yil asip asip*
 ki-pe la to yil a=sip a=sip
 IRR.3sg-PFV go.to CONT dig OBL=one.INANIM OBL=one.INANIM
 'She will be digging [them] up one by one.' (KM190211_0047)

Numerals are also often encountered as noun modifiers preceded by the formative *ta=*, which was already briefly mentioned in the discussion of complex demonstratives in Section 4.3.4 above. It will be discussed in more detail in Section 4.11.

4.5.3.2 Numerals heading a predicate

A numeral can be the head of a non-verbal predicate. It is possible that only numerals referring to animates have this property. Examples are given in sentences (34) and (35).

(34) *urê ramwen urê ngonomou, a irê rabein, irê ngunan*
     wurê   ta=mwen   wurê   ngonomou a   irê
     1pc.EXCL SPEC.COLL=man 1pc.EXCL six.ANIM and 3pc
     ta=pein   irê   ngunan
     SPEC.COLL=woman 3pc five
     'We men, we are six, and them women, they are five.' (OL201210_0042)

(35) *pou re kope yiuek kup tan tamong nganoyumou*
     pou   te   ko-pe   yiu-ek=Ø   kup
     pig   REL IRR.1sg-PFV pull.in-APPL=3sg.ZERO pigs.lined.up
     ta-n   tama-ng   nganoyumou
     CLF.POSS-PERT father-1sg.PERT eight.ANIM
     'The pigs with which I will pull the rope for my father [as part of a traditional ceremony] are [i.e. number] eight.' (YK290411_2_0049)

4.5.4 Counting money

There is an interesting way of counting money. The Tok Pisin loan *pepa* (lit. 'paper') refers to a ten kina note. When money is counted, in particular at traditional ceremonies, this is done in units of ten kina referred to by *pepa*, and modified by the numerals for animates, as in example (36):

(36) *moni pepa yungal a kulan som*
     moni   pepa   yungal a   kulan som
     money ten.kina twenty and plus   one.ANIM
     'The money is two hundred and ten kina.' (PK290411_1_0010)

It is possible that this convention is a relic from the time when customary payments were made with dog teeth. These were generally counted in units of ten or one hundred, strung on a line. This may also be the reason why *nel* 'rope' is modified by a numeral for animates. *nel som* conventionally referred to one string of dog teeth and was as such counted with the numeral for animates.

4.6 Quantifiers

Quantifiers usually modify nouns and indicate quantity or scope. In Paluai, they form a small closed class, members of which are shown in Tables 46 and 47. Paluai quantifiers show a fair bit of variation with regard to their position in the NP or clause (see also Section 5.1). Some quantifiers can be used by themselves, in which case they function as a core argument of a verb. Some are also used in an adverbial phrase introduced by the directional verb *la*, meaning 'become much; do many times' (see Section 3.6.2.2).

Table 46: Quantifiers referring to large quantities.

Form	Meaning	Position in NP/clause
menot	many; much	usually follows N (as a modifier within a NP); also used adverbially with *la* and as a predicate head
menton[39]	many; much	unknown (possibly only appears in adverbial phrases with *la*)
mwason	many (*lit.* thousand)	precedes N (as a modifier within a NP)
naringiai	many	mostly follows N (as a modifier within a NP), also used adverbially with *la*
nêmnêmti	all	follows N or a pronoun (as a modifier within a NP); sometimes used by itself
tasom / tasip	whole, (al)together	follows N (as a modifier within a NP); also used by itself
wut	every, all (land)	precedes N (as a modifier within a NP)

Table 47: Quantifiers referring to small quantities.

Form	Meaning	Position in NP/clause in base form
no	only, just	precedes N
sôkôm	some (indefinite, countable)	follows N or a pronoun (as a modifier within a NP)
sê	diminutive	follows N (as a modifier within a NP); occasionally follows V (as an adverbial modifier)
sut	some, a small amount (land, paper money)	follows N (as a modifier within a NP)

---

[39] *menot* and *menton* are formally very similar, and may be variants of the same root.

Quantifiers referring to small quantities in particular have affinities with numerals and numeral classifiers. There are at least three members of this class whose use seems to depend on the semantics of the noun they modify. *sut* (and a probably related form *wut*) is only used for areas of land and for money. *sê*, in its base form, is a general diminutive device, while *sôkôm*, the indefinite quantifier, seems to be used for people and countable objects. The forms *tasip* and *tasom* are based on forms for the numeral 'one' and will be discussed in Section 4.11.

nan= and *an-* can be attached to quantifiers referring to small quantities, in order to yield partitive meanings. *nan=* can also be attached to numerals. The resulting complex forms can modify verbs, adjectives or nouns; they can either precede or follow the noun they modify. The forms with *nan=* can also be used by themselves; this seems not to be the case for forms with *an-*. In what follows, quantifiers indicating large quantities ('a lot', 'many') are discussed first, followed by quantifiers indicating small quantities ('some, a few') and the complex forms that can be built from them.

4.6.1 Quantifiers referring to large quantities

The use of quantifiers referring to large quantities is usually quite straightforward. *naringiai* seems to be used predominantly for countable objects and in particular animate beings, while *menot* and *mwason* have a more general use. Sentences (37) to (39) show examples. *mwason* is also the numeral for 'thousand'. Its use as a quantifier can usually be readily recognised because of its position ahead of the noun.

(37) *epan lêp naringiai ngoyai*
 ep=an lêp naringiai ngoyai
 1pl.EXCL=PRF take many cuscus
 'We had caught many cuscus.' (NP210511_2_0013)

(38) *ipting antek puron menot*
 ip=ting antek puron menot
 3pl=check put.away activity many
 'They abolished many ceremonies.' (SP190311_0068)

(39) *iplêp mwason nik*
 ip=lêp mwason nik
 3pl=take thousand fish
 'They caught many fish.' (PN100411_0018)

menot can also head a non-verbal predicate:

(40) *numun parun mamenot pwên*
    numun   patu-n   ma=menot   pwên
    hair   head-PERT NEG$_1$=many NEG$_2$
    'He doesn't have much hair.' [*lit.* 'his hair not much'] (Game3_280812_0368)

wut seems only to be used in combination with *panu*; meaning 'everywhere', it is the universal quantifier referring to place:

(41) *kinan irok wut panu*
    kina-n   i=tok   wut   panu
    mark-PERT 3sg=stay every.land place
    'Its mark remains everywhere.' (LM260511_1_0066)

For other nouns, including those referring to time, the universal quantifier *nêmnêmti* 'all, every' is used. When following a pronoun, it usually refers to people and means 'everybody'. Two examples of *nêmnêmti* are given below.

(42) *koayit nêmnêmti*
    ko-ayit   nêmnêmti
    IRR.1sg-separate all
    'I will separate [them] all.' [i.e. separate the inner layer of the bark of tree branches, for basket weaving].' (MK050311_0013)

(43) *ip nêmnêmti iptet sak*
    ip   nêmnêmti   ip=tet   sak
    3pl   all   3pl=move come.up
    'All of them came up.' (LL300511_1_0039)

4.6.2 Quantifiers referring to small or indefinite quantities

Quantifiers referring to small or indefinite quantities are slightly more complex. The particle *no* means 'only, just'; it has a homophone marking imperfective aspect (see Chapter 6). A phrase combining *no* with *tasip* or *tasom* means 'only one (of)'; a phrase combining *no* with the bare numeral *som* or *sip* means 'the same'. Some examples of the use of *no* are given below.

(44) *woro ning no kukuminan a parei*
 wo=to ning no kuku.mina-n a parei
 2sg=CONT see only wrist-PERT and pole
 'You see only his wrist and the pole.' (Game4_280812_0458)

(45) *maran no rasip teo iro pei*
 mata-n no ta=sip te-yo i=to pei
 eye-PERT only SPEC.COLL=one.INANIM LIG-DEM.INT 3sg=CONT appear
 'Only one of his eyes is showing/visible.' (Game3_280812_0365)

(46) *menenganan taywêp nêm no sip*
 menengan-an ta=yuêp nêm no sip
 big-NOM SPEC.COLL=two.INAMIM be.finished only one.INANIM
 'They are of the same size [*lit.* 'the bigness of the two is only one'].'
 (Game2_280812_0201)

sôkôm refers to a few, usually non-specific, members of a collection and can thus be regarded as a (categorial) indefinite quantifier. It is therefore typically used with countables, and very often refers to animate beings, in particular when following a pronoun. It is attested only once with the form *nan=* attached to it.

(47) *ip numun sôkôm teo kosan antek*
 ip numun sôkôm te-yo ko-san antek
 3pl fibre some LIG-DEM.INT IRR.1sg-cut put.away
 'Some of the fibres I will cut off.' (MK050311_0024)

(48) *ip sôkôm pe, ip sôkôm mape pwên*
 ip sôkôm pe ip sôkôm ma=pe pwên
 3pl some do 3pl some NEG$_1$=do NEG$_2$
 'Some [people] did, some didn't.' (LM260511_1_0042)

4.6.2.1 The forms *an-* and *nan=* with quantifiers

Of all quantifiers, *sê* and *sut*, and the forms *an-* and *nan=* that often accompany them, are the most complex in their distribution and analysis. *nan=* is also attested with numerals. In their base forms, *sê* and *sut* are used as diminutive particles to nouns, always following the noun they modify; they can't occur independently.[40]

---

[40] Interestingly, the numerals for 'one' and these quantifiers all begin with *s-*. They may share a formative.

In a small number of cases, *sê* is used as an adverbial modifier, meaning 'do X a little'. A complex form is encountered more often in this function. The following sentences give some examples of noun modification.

(49) *i mapari sap nipen sut pwên*
     i   ma=pari   sap nipen   sut pwên
     3sg NEG_1=belonging.to any part.of.round DIM NEG_2
     'He is not from just any old small piece of land.'
     (PK290411_3_0088)[41]

(50) *monmon sê, suisuinot, ipe yong suisuinot*
     monmon sê   suisuinot   i=pe   yong suisuinot
     bird   DIM sunbird.sp 3sg=PFV hear sunbird.sp
     'A small bird, the *suisuinot*, she heard a *suisuinot*.' (LK100411_0094)

The forms *an-* and *nan=* can be attached to *sê* and *sut* to express partitive meanings. *an-* is attested in the following functions (Table 48):

Table 48: Functions and distribution of *an-*.

Function	Position	Meaning/paraphrase
1. Modifier to a mass noun	Preceding N	a bit of N
2. Modifier to an adjective	Following A	somewhat, quite A
3. Modifier to a verb	Following V	do V a bit, a while (affirmative clause) not do V at all (negative clause)

nan= is attested in the functions shown in Table 49. Examples of each use of *an-* and *nan=* are given below. The quantifier and the element it modifies are placed between square brackets. Sentences (51) to (53) show examples of *an-* forms modifying a noun (51), an adjective (52) and a verb (53), respectively.

(51) *worou ansê pau namwi lai*
     wo=tou [[an-sê   pau]   namwi] la   a=i
     2sg=put PART-DIM coconut.oil small   go.to OBL=3sg
     'You put a small bit of coconut oil into it.' (NK290311_2_0020)

---

[41] *sap* is an interrogative word that can also function as an indefinite pronoun meaning 'any, which(ever)'. It will be discussed in Section 4.7.1 below.

Table 49: Functions and distribution of *nan=*.

Function	Position	Meaning/paraphrase
1. Independent form	Argument to a predicate	a bit, a small piece
2. Modifier to a mass noun	Preceding N	a bit of N
3. Independent form (attached to a numeral) or modifier to a noun	Following N	[unclear]
4. Modifier to a verb	Following V	do V a bit, a while (affirmative clause) not do V at all (negative clause)
5. Temporal adverbial phrase (preceded by *no*)	Clause-initial	in a while, almost

(52) *iro aloen ansê ai Paluai*
   i=to   [aloen an-sê]   a=i   Paluai
   3sg=be long   PART-DIM   OBL=3sg P.
   'It is quite far from Baluan.' (LL030611_0005)

(53) *ngamapwa koning ansêo pwên*
   nga$_A$=ma=pwa   ko$_A$-[ning an-sê]=o   pwên
   1sg=NEG$_1$=want.to   IRR.1sg-see   PART-DIM=2sg   NEG$_2$
   'I don't want to see you ever again.' (WL020711_0082)

Sentences (54) to (58) show examples of forms that have *nan=* attached to them. The forms are used autonomously (54) or as a mass noun modifier (55), a verbal modifier to a stative verb ((56) and (57)) or a temporal adverbial modifier (58).

(54) *worou palsi nansê sot*
   wo=tou   palosi   [nan=sê]   sot
   2sg=put   first   PART=DIM   go.up
   'You put a little bit [of the sago] first [into the frying dish].' (CA120211_1_0026)

(55) *wope lêp nansê yon*
   wo=pe   lêp   [nan=sê   yanu]
   2sg=PFV take   PART=DIM   water
   'You will take a bit of water.' (CA120211_1_0032)

(56) *ino ret nali nansê wot*
    i=no   tet   [nali   nan=sê]   wot
    3sg=IPFV move be.lost PART=DIM go.level
    'It is going away a little bit to the side.' (Game4_280812_0201)

(57) *kino ru, ma kilôlôt nansê...*
    ki-no   tu   ma   ki$_S$-[lôlôt   nan=sê]
    IRR.3sg-IPFV   stay and IRR.3sg-be.cool PART=DIM
    'It should remain [inside the pot] and when it will cool down a bit...'
    (CA120211_2_0042)

(58) *no nansê, kola ai sou reo*
    [no   nan=sê]   ko-la   a=i   sou   te-yo
    only   PART=DIM IRR.1sg-go.to   OBL=3sg   reef LIG-DEM.INT
    'In a little while, I will go to that reef.' (LK100411_0131)

4.6.2.2 The form *nan=* with numerals

When occurring with a numeral, *nan=* is either attached to the bare numeral or to the form with *ta=* (see Section 4.11). It is not clear why numerals modifying nouns sometimes occur with *nan=*, and what the semantic difference is (if any). When both *nan=* and *ta=* are attached to the numeral 'one' for animates, i.e. *som*, this yields the autoreflexive meaning 'on his own'. Some examples of numerals with *nan=* are given below.

(59) *maran nansip menengan a nansip namwi*
    mata-n   nan=sip   menengan a   nan=sip
    eye-PERT   PART=one.INANIM   big   and PART=one.INANIM
    namwi
    small
    'One of his eyes is big and one is small.' (Game3_280812_0296)

(60) *i ranisip telo, u not nantaymou liliu*
    i   ta=ni-sip   te-lo   u   not
    3sg   SPEC.COLL=other-INANIM   LIG-DEM.DIST 3du child
    nan=ta=yumou   liliu
    PART=SPEC.COLL=two.ANIM   again
    'The next one, (of) the two boys again.' (Game3_280812_0346)

(61) *Ngat i no nantasom o*
    Ngat   i   no   nan=ta=som   yo
    N.   3sg   only PART=SPEC.COLL=one.ANIM   DEM.INT
    'Ngat is just on his own [an only child].' (LM240611_0045)

Thus, although the functions of *an-* and *nan=* seem to partially overlap, they serve in different domains. *an-* forms never occur independently, but always as modifiers to other forms. *an-* does not attach to numerals either. Definiteness does not seem to play a role in the selection of these forms, since both *an-* and *nan=* are encountered in forms modifying elements whose referents have not been introduced into the discourse before. It is possible that specificity or individuatedness plays a role, which could be a partial explanation as to why *nan=* occurs with numerals, whereas *an-* does not. Finally, there may be interspeaker differences in the use of these forms, but this cannot be confirmed or otherwise on the basis of the data in the corpus.

4.7 Interrogative words

Table 50 shows an overview of the Paluai words marking content questions. In what follows, interrogative words will be discussed in turn, starting with words questioning identity, and followed by forms questioning time, relations, purpose and manner.

Table 50: Interrogative forms.

Form	Translation	Morphological makeup	Relates to	Affiliated word class
pa	where; wherever	–	Location	N, Dem, Adv
kapi	when	unknown	Point in time	Adv
la sa	how	*la* 'go (to) + *sa* 'what'	Manner; Quality	Adv
pari ai sa	why; for what	*pari* 'belonging to' *a=i* OBL=3sg, *sa* 'what'	Purpose; Reason	–
sa	what	–	General; Identity of object	N
samai-	what (relationship) of	unknown; contains *sa* 'what'	Relationship	N

Table 50 (continued)

Form	Translation	Morphological makeup	Relates to	Affiliated word class
samnon	how many	unknown; contains *sa* 'what'	Quantity	Numeral
sap	which(ever); something	unknown, contains *sa* 'what'	Identity of object	N
sê	who(ever); someone	unknown	Identity of person	N
tenepa	how	unknown, maybe contains *pa* 'where'	Manner	A, Adv

4.7.1 Questioning identity: *sa*, *sap* and *sê*

The syntax of interrogative clauses is discussed in Chapter 9. The interrogative form remains in situ, occupying the same grammatical slot as the element it questions. *sa* is the most general interrogative; as can be seen above, some of the complex forms contain *sa* as well. *sap* questions entities ('which'); it usually modifies a noun such as *pule-n* 'thing' or *kanai* 'kind' and then questions the identity of a certain object. *sap* is also used in declarative clauses, where it refers to a non-specified entity: 'something' (or, in the complex form *sesap*, 'anything', for a negated clause). The same is true for *sê*, questioning persons. Examples of *sa*, *sap*, and *sê* are given below; (64) and (65) exemplify the indefinite use of *sap* and *sê* in a declarative clause. Note that the use of *sa* for questioning a person's name is unusual in Melanesia; a more commonly encountered strategy is to use the interrogative for 'who'.

(62) *ngayom sa?*
 ngaya-m   sa
 name-2sg.PERT what
 'What's your name?'

(63) *i reo i maloan sap pulen, kolpanu le?*
 i   te-yo   i   maloa-n   sap   pule-n   kolpanu   le
 3sg LIG-DEM.INT 3sg photo-PERT which thing-PERT green.tree.snake or
 'This is a picture of which thing, a green tree snake or?'
 (Game1_021012_0016)

(64) *naman sap palawêk in pei la ro panu*
naman   sap   palawêk i=an   pei   la   to   panu
perhaps something badness 3sg=PRF happen go.to be   home
'Perhaps something bad has happened at home.' (WL020711_0105)

(65) *ma kapwa wo sê re wopwa wolak...*
ma   kapwa wo   sê   te   wo=pwa   wo=lak
but   if   2sg who   REL   2sg=want.to   2sg=go
'But if you are someone who wants to go...' (LL030611_0109)

4.7.2 Questioning time and place

kapi 'when' and *pa* 'where' are the interrogative words used to ask questions relating to time and place, respectively. When used in a sentence, the preposition *a=* has to be added to the form *pa*, as in (66). However, in daily conversation *pa* is also often used elliptically, as in (67).

(66) *Jema i ro apa?*
Jema   i=to   a=pa
J.   3sg=be   OBL=where
'Where is Jema?'

(67) *Jema pa?*
Jema pa
Jema where
'Where is Jema?'

4.7.3 Questioning relations with *samai-*

Perhaps not surprisingly for a language in which direct possession and association play such an important role (see Sections 3.2.3 and 5.5), Paluai has a separate interrogative form for direct possessive relations. This form, which is itself directly possessed, can question both kinship and part-whole relations; see examples below. In (69), the speaker is referring to a body part of the pig in the picture, but is unsure what body part it is exactly and therefore uses the interrogative form. However, the clause in its entirety is declarative.

(68) *Pokut i samaim?*
 Pokut i samai-m
 P. 3sg what.of-2sg.PERT
 'How is Pokut related to you?' (052b_0015)

(69) *puan ngusun le samain to ai nipêng teo*
 puan.ngusu-n le samai-n to a=i nipêng te-yo
 upper.lip-PERT or what.of-PERT be OBL=3sg other.time LIG-DEM.INT
 'His upper lip or what of him was (visible) on it just before.'
 (Game4_280812_0292)

4.7.4 Questioning purpose or reason

The form *pari ai sa* can be translated as 'why', but is more accurately described as an interrogative form referring to a purpose relation. The expression *kipe la sa*, which can be characterised as a rhetorical question ('now what?'), could be used in a situation when someone's actions or motives are questioned. An example of *kipe la sa* is given below, followed by an example of *pari ai sa*.

(70) *ikipe la sa? imaarei ngonomek i re ngaakêp pwên*
 i=ki-pe la sa i=ma=arei ngonomek i te
 3sg=IRR.3sg-PFV go.to what 3sg=NEG₁=say corresponding 3sg REL
 nga=akêp pwên
 1sg=pick.up NEG₂
 'Now what? He didn't correctly describe the one that I picked up.'
 (Game1_021012_0023)

(71) *woro pe yep pari ai sa?*
 wo=to pe yep pari ai sa
 2sg=CONT make fire what.for
 'Why are you making fire?' / 'What are you making fire for?'
 (field notes 12/10/2012)

4.7.5 Questioning manner

la sa and *(la) tenepa* both act as manner interrogatives, but in different ways. *la sa* is only attested as a modifier to a main verb; it questions the manner of the action described by this main verb, e.g. *ngan* 'eat' in (72) and *pe* 'do' in (73).

tenepa is attested as a modifier to a main verb in an adverbial phrase with *la*, as in (74), but can also head a predicate, as it does in (75) and (76). It questions more generally the manner of a situation, procedure or appearance.

(72) *kapwa kongan, kope nganan la sa?*
kapwa ko-ngan ko-pe ngan-an [la sa]
if IRR.1sg-eat IRR.1sg-do eat-NOM go.to what
'If I would eat [this], how would I do the eating?' (Game1_021012_0302)

(73) *kanopa i siai, le ipe la sa?*
kanopa i=siai le i=pe [la sa]
like 3sg=arms.bent or 3sg=do go.to what
'Like he is standing with his arms bent, or how?' (Game2_280812_0086)

(74) *a ngaakêpi la renepa?*
a nga=akêp=i [la tenepa]
and 1sg=pick.up=3sg go.to how
'And how do I pick it up?' [i.e. what is the right procedure]
(Game1_021012_0020)

(75) *tare kanopa kaywun le i renepa?*
ta=te-yo kanopa kaywun le i tenepa
SPEC.COLL=LIG-DEM.INT like white or 3sg how
'This thing, it is like white or how [does it look]?' (Game2_280812_0142)

(76) *kay, wope pul ma wo renepa?*
kay wo=pe pul ma wo tenepa
okay 2sg=PFV tell EMP 2sg how
'Okay, tell me. What's your story?' [TP: *yu olsem wanem?*] (LK100411_0054)

4.8 Negation and mood markers

TAM markers were already listed in Table 20 in the previous chapter. Reality status is indicated by zero for realis, and by a *kV-* prefix for irrealis (see Chapter 6). In addition, there are markers giving information about the mood and polarity of a clause. Declarative mood is unmarked. Interrogative mood is marked by a distinct intonation pattern and, in the case of content questions, by one of the interrogative forms listed in Section 4.7 above. Imperative mood is generally only marked by the absence of a bound subject pronoun or by the irrealis; the imperative meaning can be deduced from the context. Negative imperatives, however, are marked by *napunan* (*lit.* 'it is forbidden').

Negative polarity of both verbal and non-verbal predicates is marked by double negation: the marker *ma=* is inserted before, and the marker *pwên* after the material over which the negation has scope. *pwên* can be used by itself, e.g. as a negative answer to a question. Mood and polarity are discussed in more detail in Chapter 10.

4.9 Conjunctions and clause connectors

Clauses can be subdivided into two types: main and dependent. Main clauses can stand on their own, whereas dependent clauses usually cannot; an exception to this are complement clauses of the verb *pwa* 'say' reflecting a direct quotation. Main and dependent clauses are not distinguished by constituent order.

Paluai has a dependent clause marker *te* that introduces relative clauses (which modify nouns) and complement clauses (which are arguments to a verb). They are discussed in Sections 11.1.1 and 11.1.2, respectively. In addition, the marker *te* can follow a form in order to identify an adverbial subordinate clause, specifying e.g. time, manner or reason. The forms that can be followed by *te* will be discussed below. Generally speaking, *te* indicates that the element it follows is modified by a clausal instead of a phrasal constituent.

4.9.1 Coordination

The markers linking main clauses are shown in Table 51. Two of the coordinating clause connectors, *a* and *le*, can also connect two NPs as a conjunctive and a disjunctive marker, respectively. The markers *ma* and *onga* are only attested as clause connectors. *ma* can mean 'and' but can usually be translated as 'but',

Table 51: Markers for the coordination of two main clauses.

Form	Meaning/function
a	and (conjunctive)
ma	but, and
le	or (disjunctive)
onga	(and) then, (and) so

in particular when it occurs in the form *ma i te*, which presumably contains the subordinate clause marker *te*. *ma* is also attested as an emphatic marker, and in the complex form *ai sa? ma*, which can be translated as 'because'.

ma is used not only for relations of contrast (in which case it translates as 'but'), but also to mark additivity. Most instances of *ma*, however, have a clear contrastive meaning. Example (77) below shows a relation of contrast with *ma*, which is absent from (78). A contrastive reading of (78) would be strange, because the storyteller is talking about returning home from the garden, which people usually do in the afternoon. For a more detailed discussion, see Section 11.2.

(77) *ngamaakêp nganngan pwên, ma ngano akêp muyou*[42]
 nga=ma=akêp nganngan pwên ma nga=no akêp muyou
 1sg=NEG₁=pick.up food   NEG₂ but 1sg=IPFV pick.up snake
 'I didn't pick up the food, but I picked up the snake.' (Game1_021012_0510)

(78) *ila poyep, ma uliliu si*
 i=la   poyep   ma u=liliu   si
 3sg=go.to afternoon and 3du=return come.down
 'It had become afternoon and they returned home.' (LM190611_0006)

The phrase *ai sa? ma* may be a calque from Tok Pisin *(bi)long wanem* 'because'. *ai sa* literally means 'for what?' and is used as a rhetorical question, the answer to which is given in a subsequent clause introduced by *ma*. In example (79), question and answer are translated into English as a single clause introduced by *because*. It is unlikely that *ma* means 'and' in this context. It may be the case that *ma* is used here as an emphatic marker; for more information, see Chapter 12.

(79) *ai sa? ma nik in pwak ai lau*
 a=i sa   ma nik i=an   pwak   a=i   lau
 OBL=3sg what EMP fish 3sg=PRF be.stuck OBL=3sg fishing.net
 '[The net went under] because fish had filled up the net.' (NP210511_1_0016)

The disjunctive coordinator *le* connects main clauses and can be translated as 'or'. It indicates a choice between two options. Often, *le* is repeated after the second of the two clauses it connects.

---

[42] *ma* is formally identical to the first element of the clausal negation *ma=* (which occurs in combination with *pwên*). They may be diachronically related. Negation is discussed in Section 10.2.

(80) *i reo, iro patan nan le iro patan yoy, le?*
 i te-yo i=to pata-n nan le i=to
 3sg LIG-DEM.INT 3sg=be on.top-PERT ground or 3sg=be
 pata-n yoy le
 on.top-PERT stone or
 'Now what? He didn't correctly describe the one that I picked up.'
 (Game1_021012_0023)

Both *le* 'or' and *a* 'and' can be used when connecting two NPs. In (81), two different words for 'spirit' are used, connected by *le*.

(81) *silal le pwalei, ipe pung nangin sasawan nik*
 [silal] le [pwalei] i=pe pung nangin sasawan nik
 spirit or spirit 3sg=PFV smell scent strong.smell fish
 'A spirit, he smelled the strong smell of the fish.' (NP210511_1_0028)

(82) *kokonin teo, moni a tiap*
 kokoni-n te-yo [moni] a [tiap]
 money-PERT LIG-DEM.INT money and cloth
 'His money [to be distributed at the *pukan kokon* ceremony] consists of paper money and cloth.' (SY100411_0045)

The marker *(te) onga* is somewhat enigmatic. It is used so frequently by some speakers, in particular in narratives, that it can almost be regarded as a filler. It basically means 'and so, and then' and thus may indicate a closer connection between two clauses than the much shorter form *a*. However, due to its frequent use it seems almost devoid of any meaning in present-day Paluai. An example of *onga* is given below; for more detailed discussion, see Section 11.2.1.1.

(83) *epsi ret onga epsi panu, epsi panu a sin ilol...*
 ep=si tet onga ep=si
 1pl.EXCL=come.down move and.so 1pl.EXCL=come.down
 panu ep=si panu a sin i=lol
 home 1pl.EXCL=come.down home and sun 3sg=be.dark
 'We went and so we came home, we came home and the sun had set...' (LL030611_0078)

4.9.2 Subordination

Main and dependent clauses are not differentiated syntactically (see Chapter 11 on clause relations for more details). There are, however, a number of conjunctions in the corpus that mark a subordinate relation between one clause and another. Most of them contain a dependent clause marker *te*, which marks an adverbial subordinate clause. Temporal subordinate clause markers are discussed first, followed by other types of subordinate clause markers.

4.9.2.1 Temporal subordinate clause markers

Temporal subordinate clause markers are shown in Table 52 below. All subordinate clause markers indicating temporal relations contain the marker *te*. Many of them have a nominal origin. Most indicate simultaneity of two events on the time axis: 'at the time X happened, Y happened (also)'. Others indicate a sequence of events ('before/after X happened, Y happened') or a sequence of time stretches ('X happened until Y happened').

Table 52: Temporal subordinating conjunctions.

Form	Translation	Origin
no te	(just) when	possibily *no* 'only'
panua-n te	before	*panu* 'front'
monoki-n te	after	*monok* 'back'
pwotna-n te	when	possibly *pwotwpot* 'exactly'
pêng te	when	*pêng* 'day, occasion'
taim te	when	TP loan *taim* 'time'
inap te	until	TP loan *inap* 'until'

Most of these markers are used not only for past events, but also for future or hypothetical ones. When a temporal subordinator is used, the main clause will usually be marked with the particle *ya*. This particle plays an important role in discourse and information structure, and is also attested with the emphatic marker *te-* (discussed in Section 4.3 on demonstratives) attached to it. This is the case e.g. in example (84) below. A more elaborate discussion of temporal subordinate clauses can be found in Section 11.1.3.1. Example (84) shows the use of *taim* for a hypothetical event.

(84) *taim te pên tao ila nurupui rea, wope lêpi...*
    taim te pên ta-o i=la nurupui te-ya
    time SUB daughter CLF.POSS-2sg.PERT 3sg=go.to mature LIG-then
    wo=pe lêp=i
    2sg=PFV take=3sg
    'When your daughter has her first period, then you will take her...'
    (NK290311_2_0005)

4.9.2.2 Other types of subordinate clauses

Other subordinate clause markers are shown in Table 53. All forms show the marker *te*, except the conditional clause markers and *kanpwên*.

Table 53: Other subordinating conjunctions.

Form	Translation	Type of relation
arona-n te	consequently	Consequence
a=i sa? ma	because	Reason
(a=)te	for, because	Reason
kanopwên	if	Counterfactual conditional
kanpwên	as if	Manner
kapwa	if, when	Conditional
longoa=n te	consequently	Consequence
ma i te	but, well	Concessive
pari ai te	because	Reason
tangoa=n te	consequently	Consequence
(te) onga	so that	Result
te ... sa	lest, for fear of	Possible consequence

The markers *(a=)te* and *pari ai te* appear to be very similar in meaning. Sometimes, *(a=)te* is used in an elliptical sentence. Example (85) below was used as a warning to a child. What is clearly meant is 'cover your head, lest the sun burn you!'. However, the subordinate clause (with apprehensive meaning) is almost entirely elided. All types of subordinate clauses mentioned in this section are discussed in detail in Chapter 11.

(85) *wopolpol, te sin!*
    wo=polpol te sin
    2sg=cover.head SUB sun
    'Cover your head, because of the sun!' (field notes 25/09/2012)

4.10 Interjections and formulaic words and phrases

A residual word class is formed by interjections, i.e. words or phrases that usually give information about the attitude of the speaker towards what is said, but that fall outside of the syntax of a sentence. Ameka (1992) distinguishes expressive interjections, "the vocal gestures which are symptoms of the speaker's mental state", conative interjections, "those expressions which are directed at an auditor", and phatic interjections, "used in the establishment and maintenance of communicative contact" (Ameka 1992: 113–114). Formulaic words (and phrases) such as *sorry* and *thank you* are slightly problematic: they are "intentional and (socially) expected reactions to situations" (Ameka 1992: 109), whereas interjections are spontaneous. However, since formulaic words often cannot be readily assigned to another word class, they are put together with interjections.

An overview of frequent interjections and formulaic words and phrases is given in Tables 54, 55 and 56. Although interjections and formulas are usually regarded as elements outside of grammar in a narrow sense, there are clearly several types that can be distinguished. Interjections and formulas also have various functional slots in grammar and discourse. On the one hand, there are

Table 54: Formulaic words and phrases.

Form	Paraphrase	Usage
nok mwe- (directly possessed)	sorry	sympathy, concern (not apology)
wuro	thank you	appreciation
pwapwem (pian)	(good) morning	greeting
poyep (pian)	(good) afternoon	greeting
pêng (pian)	(good) night	greeting
wo / au / arê / ap teyo (2nd person pronoun plus DEM)	hello	greeting
wo=tu tet au= / arê= / ap=ka-tu tet	goodbye (*lit.* 'you will be going')	greeting with leave-taking (party who stays)
wo=no tok au= / arê= / ap=ka-no tok	goodbye (*lit.* 'you will be staying')	greeting with leave-taking (party who goes)
arê no / arê rebo-ong	my goodness	surprise, indignation, dismay
yi tinang	mother!	alarm, dismay
konan	never mind	dismissal
ma in	don't know	uncertainty

Table 55: Expressive/conative interjections.

Form	Usage
eh	general expression of dismay, disgust, disagreement
oh	surprise, sign of comprehension

Table 56: Phatic/conative interjections.

Form	Usage
ah	affirmation of statement, comprehension
au	when hesitating
eh	after mistake or as self-repair
hm	backchanneling device
i lou	when hesitating (filler, when speaker can't think of word)
kanopa	hedging device, filler
kay	resolution, agreement
uu	affirmation of statement, agreement; general backchanneling device

those interjections that can be used to reply to an utterance of another speaker, for instance *ah*, *kay*, *oh* and *uu*. They have a communicative function and a potential illocutionary force. Formulas always have a communicative function.

On the other hand, there are those interjections that mainly function as expressions of attitude (expressive interjections) or that function as "self-regulating" devices (a subtype of phatic/conative interjections). *eh* is used in situations of self-repair, while *au*, *ilou* and *kanopa* are used as fillers and when hesitating. *kanopa* is also used as a hedging device. The primary function of the latter type of interjections is not communicative, but they play a facilitating role, helping the speaker along in her utterances. Of course, expressive interjections do have a secondary communicative function, giving the listener information about the emotional state of the speaker. They can therefore be characterised as conative as well as expressive (Ameka 1992).

4.11 The formative *ta*=

A formative *ta*=, briefly mentioned in Section 4.5.3, is added as a proclitic to members from several word classes. This complex form then occurs either by itself or is preceded by a pronoun, most commonly the 3sg pronoun *i*. The historical origin of *ta*= may be POc **ta* 'indefinite common nonhuman article'

(Lynch et al. 2002: 71), although Paluai *ta=* is used to refer to humans. When used together with a pronoun, the constituent seems to indicate specificity, or, to be more precise, to single out a specific member or members of a (more amorphous, but identifiable) set. Budd (2014) discusses a number of instances where reflexes of **ta* or of the POc numeral 'one' are used to express partitives. Paluai *ta=* would partially fit an analysis as a partitive, in particular the "one/other" subtype described by Budd (2014: 533), which he defines as "one of a pair (or small number) of identifiable entities".

In what follows, the use of the *ta=* formative will be discussed in more detail. It derives independent nominal heads from forms that cannot otherwise be used independently (e.g. adjectives, demonstratives, prepositions, numerals and quantifiers), but it is also attested on forms that can be used as such, i.e. nouns. For each word class, the use of *ta=* without an accompanying pronoun will be discussed first, followed by its use with a pronoun. *ta=* forms one phonological word with the element preceding it. This is evident from the alternation of /t/ to [l] when *ta=* follows a form ending in /l/: /sal ta=talɔt/ > [sal laralɔt] 'four roads'. This alternation only occurs within phonological word boundaries.

4.11.1 Use of *ta=* with nouns

Ta= by itself, without a pronoun, is not often encountered with nouns other than personal nouns. With personal nouns, however, it is quite common; its main use here is to introduce the name of a person who was immediately before referred to by means of a kinship term. Since kin terms are classificatory, they can refer to an identifiable set (e.g. not only a person's biological father, but also father's brothers, etc.). The personal name could thus be used, introduced by *ta=*, to specify a member of this set.

(86) *sopol lan tupung ta Kiralue*
      sopol   ta-n   tupu-ng   ta=Kiralue
      one.half.round CLF.POSS-PERT   grandparent-1sg   SPEC.COLL=Kiralue
      'The side of my grandfather Kiralue.' (KW290611_0004)

(87) *tamom ta Yêp Ponaun*
      tama-m   ta=Yêp   Ponaun
      father-2sg.PERT SPEC.COLL=Yêp   Ponaun
      'Your father Yêp Ponaun.' (PK290411_1_0005)

Ta= in combination with *i* is often encountered with the nouns for 'left' and 'right', which refer to spatial orientation. In (88), the *ta=* phrase acts as a modifier to the head of the NP, *minan*.

(88) *kei ila ro minan i raalmaru*
 kei i=la to mina-n i ta=almaru
 tree 3sg=go.to be hand-PERT 3sg SPEC.COLL=right
 'The stick is in his right hand.' [*lit.* 'The stick is in his hand [that is] the right one.'] (Game2_021012_0017)

Secondly, *i ta=* is often used with *not* 'child' or *manak* 'elder' to refer to a younger or older individual, respectively, often a sibling.

(89) *iro yik i ranot teo*
 i=to yik i ta=not te-yo
 3sg=CONT search.for 3sg SPEC.COLL=child LIG-DEM.INT
 'He was searching for the younger one.' (WL020711_0162)

Thirdly, *i ta=* is often used with the forms *nisip* 'other', *panurasip* 'first' and *monok* 'after', in which case it refers to an instance or object that came first, came later, or will come next.[43]

(90) *ino pwa kingonomek i rapanurasip*
 i=no pwa ki-ngonomek i ta=panurasip
 3sg=IPFV want.to IRR.3sg-correspond.to 3sg SPEC.COLL=first
 'It is likely corresponding to the first one.' (Game1_021012_0384)

Occasionally, *ta=* is encountered with pronouns other than *i* to refer to a subset of a bigger set of humans, in this case a subset of siblings:

(91) *urê ramwen urê ngonomou, a irê rabein, irê ngunan*
 wurê ta=mwen wurê ngonomou a irê
 1pc.EXCL SPEC.COLL=man 1pc.EXCL six.ANIM and 3pc
 ta=pein irê ngunan
 SPEC.COLL=woman 3pc five
 'We men, we are six, and them women, they are five.' (OL201210_0042)

---

[43] Word class membership of *panurasip* is not entirely clear. It is mostly used as clausal adverbial, but probably has nominal origins; it is related to the noun *panua* 'front'.

4.11.2 Use of *ta=* with adjectives

With adjectives, *ta=* in combination with *i* derives an independent form meaning 'the one with property X'. Some examples are given below. The phrases *i takaywun*, *i tayamyaman* and *i tapilipil* each form the head of a NP that functions as an argument in the clause (in this case, the subject of a verbless clause). Again, the *ta=* forms specify an entity from a set, or one of a pair, as e.g. in (93).

(92) som i menengan, i rakaywun i menengan
      som   i   menengan i   ta=kaywun   i   menengan
      one.ANIM 3sg big   3sg SPEC.COLL=white 3sg big
      'One is big, the white one is big.' (Game1_021012_0580)

(93) i rayamyaman teo isot wat a i rapilipil leo isi paye
      i   ta=yamyaman te-yo   i=sot   wat   a   i
      3sg SPEC.COLL=red LIG-DEM.INT 3sg=go.up up.high and 3sg
      ta=pilipil   te-yo   i=si   paye
      SPEC.COLL=yellow LIG-DEM.INT 3sg=come.down down.below
      'The red one is up high and the yellow one is down below.'
      (Game2_280812_0153)

Ta= is occasionally used without *i* to derive an independent form from an adjectival base. Again, the derived form heads a NP functioning as a core argument. Sentence (94) below describes an overripe banana, with black spots all over it.

(94) takayan ino pei muyan o
      ta=kayan   i=no   pei   muya-n   yo
      SPEC.COLL=black 3sg=IPFV appear skin-PERT DEM.INT
      'Black spots are / "blackness" is appearing all over its skin.'
      (Game3_280812_0224)

4.11.3 Use of *ta=* with demonstratives

There is a complex demonstrative that is made up of *ta=*, the ligature *te-*, and a basic demonstrative form, discussed in Section 4.3.4. This demonstrative can be used as an independent form that can substitute for a noun, for instance as a subject or object argument to a verb or a subject in a non-verbal predicate. An example is given in (95).

(95) *ipwa, "Nulik, tarepwo ran sê?"*
i=pwa   Nulik   ta=te-pwo   ta-n   sê
3sg=say Nulik   SPEC.COLL=LIG-DEM.PROX CLF.POSS-PERT who
'She said, "Nulik, whose is all this?"' (WendyLawan020611_0044)

4.11.4 Use of *ta=* with prepositions

ta=, with and without *i*, is also encountered preceding the preposition *pari* 'belonging to'. The construction usually has spatial reference. The phrase starting with *ta=* modifies the element that precedes it, regardless of whether this is a full NP or the 3sg pronoun *i*.

(96) *wai rai rapari paye relo*
wai   ta-i   ta=pari   paye
loincloth CLF.POSS-3sg.PERT SPEC.COLL=belonging.to down.below
te-lo
LIG-DEM.DIST
'The lower part of his loincloth.' (Game3_280812_0110)

(97) *i rapari monok telo*
i   ta=pari   monok   te-lo
3sg SPEC.COLL=belonging.to behind   LIG-DEM.DIST
'The one at the back.' (Game3_280812_0110)

4.11.5 Use of *ta=* with numerals

Used with numerals, *ta=* is furthermore used to specify an entity or entities from a limited, identifiable set. When counting or tallying objects, the bare numerals will be used. However, when a numeral refers to a specified number of objects from a set, it is introduced by *ta=*. Example (98) was used during a game of cards, to tell someone to pick two cards (specific entities) from the stack (a limited, identifiable set). Example (99) shows the use of the numeral for two from the animates paradigm, *yumou*, both with the formative *ta=* and without it. A married couple consists by definition of two people, husband and wife, which is a limited set. In this instance, the numeral refers to the entire set. In contrast, the number of children of a married couple is not limited. Hence, the formative *ta=* is not used.

In addition, *tasip* and *tasom*, consisting of *ta=* with the forms for 'one', have the meaning '(the) whole (of)'. An example is shown in sentence (100).

(98) *lêp taywêp*
    lêp   ta=yuêp
    take   SPEC.COLL=two.INANIM
    'Take two.' [context: playing cards] (field notes 2015)

(99) *u pet parian taymou reo, ugat not yumou*
    u   pet   paria-n   ta=yumou   te-yo   u=gat
    3du   dyad   wife-PERT   SPEC.COLL=two.ANIM   LIG-DEM.INT   3du=have
    not   yumou
    child two.ANIM
    'The married couple, they had two children.' (WL020711_0004)

(100) *kanen tasom nêm kila yamyaman...*
    kane-n   ta=som   nêm   ki-la   yamyaman
    body-PERT   SPEC.COLL=one.ANIM be.finished   IRR.3sg-go.to   red
    'Its [the snake's] whole body is like red...' (Game1_021012_0532)

In addition, numerals with or without *ta=* can take the proclitic *nan=*, which also occurs on quantifiers; see Section 4.6.2.2.

4.11.6 Use of *ta=* with quantifiers

The formative *ta=* is encountered with the diminutive form sê, as in (101) below. This form follows the head noun. It is not entirely clear what exactly the semantic difference is between the bare diminutive and the variant with *ta=*. It could be that, similar to the use of *tasip* and *tasom* above, its meaning is 'whole', but with a diminutive overtone. (101) is a concluding remark at the end of a narrative, so it is possible that the speaker is referring to the entire story in this way.

(101) *pwapwa rasê reo, a inêm*
    pwapwae   ta=sê   te-yo   a   i=nêm
    story   SPEC.COLL=DIM   LIG-DEM.INT   and   3sg=be.finished
    'This is the whole (small) story, and it's finished.' (WL020711_0165)

4.11.7 Use of *ta=*: summary

Despite its wide-ranging uses, it appears that a common semantic core can be established for *ta=*. A summary is given in Table 57 below.

Table 57: Overview of uses of ta=.

Part of speech	With pronoun	Derives independent form?	Core meaning
Noun	no	no	one of a set of kin
	yes	no	one or several of a set
Adjective	no	yes	'X-ness'
	yes	yes	'the X one'
Demonstrative	no	yes	'(all) this/that'
Preposition	no	no	spatial noun modifier
	yes	yes	'the one at X'
Numeral	no	no	number X of a set
Quantifier / numeral one	no	no	'whole'

It appears that *ta=* contributes a meaning component indicating a more encompassing whole or set, whereas the 'specific partitive' meaning is contributed by the pronoun that often accompanies the forms with *ta=*. When there is no pronoun or personal name (which is also providing specific reference), the forms with *ta=* have a collective interpretation, but always in a limited and never in a universal sense. This is especially evident in its uses with adjectives, quantifiers, demonstratives and the numerals for 'one'.

5 The noun phrase

The head of a noun phrase (NP) is most often a noun, but it can also be a free pronoun or an independent demonstrative form starting with *ta=* (see Section 4.3.4). Only NPs that have a noun as their head have the full range of modifiers available to them. A NP headed by a free pronoun can only have a relative clause, emphatic *mwanenen* (*lit.* 'straight') and/or a demonstrative as a modifying element, and a NP headed by a demonstrative cannot contain any modifiers at all.

NPs typically function as core or peripheral arguments to verbal predicates (see Chapter 8). In addition, a NP can fill the predicative position in a non-verbal predicate (see Chapter 7). Categories that are associated with the noun will be discussed in this chapter, in addition to the structural properties of the NP. Nouns can be modified by pronouns, numerals, quantifiers, adjectives, other nouns, prepositional phrases, relative clauses and demonstratives. In addition, nouns can enter into a possessive construction, which is either direct or indirect.

5.1 Structural features of the NP

Word order in a maximal noun phrase (with a noun as head) is as follows:

(1) (Determiner) (Pre-head Mod)* **Noun** (Post-head Mod)† (Possessor)
 (Prepositional Phrase) (Relative Clause) (Demonstrative)
 * Pre-head Mod can be filled by either a quantifier or an adjective, but not both.
 † Post-head Mod can be filled by a quantifier, an adjective, a numeral or another noun.

Minimally, a NP consists of just a bare noun; all modifiers are optional. However, the noun is often accompanied by one or more modifiers, but never as many as the maximal noun phrase suggests. The use of multiple modifiers in a NP is extremely rare. When speakers wish to modify a noun multiple times (as in English phrases of the type *my huge new shiny house* or *two fat lazy pigs*), they will do so with a chain of predicates, or else they will use relative clauses. This practice seems to be typical for the New Guinea region (de Vries 2006); see Schokkin (2014) for a more detailed discussion of this topic pertaining to Paluai. Thus, the template given above for a "maximal" NP is somewhat misleading, since such a NP, with an adjectival modifier *and* a numeral *and* a prepositional modifier *and* a relative clause, will hardly ever be encountered. In what follows, each of the NP elements will be discussed separately.

5.2 Determiner

The first optional slot in the NP may be filled by a determiner indicating number. In addition, the determiner may say something about specificity of the NP referent (i.e., whether or not the NP referent is identifiable for the speaker). In contrast to determiners in languages like English, the Paluai prenominal determiner says nothing about the definiteness status of the NP referent (i.e., whether or not it is identifiable for the hearer). Definite reference can be added by using a demonstrative, which is discussed in Chapter 12. There are two options for the determiner:
– a third person non-singular personal pronoun, indicating non-singular number and possibly providing generic reference;
– the numeral 'one', indicating singular number and possibly providing categorial reference.

5.2.1 Personal pronoun as determiner

Third person pronouns are optionally used to indicate number on nouns. Bare nouns are neutral with regard to number, i.e. an unmodified noun can refer to either singular or non-singular. Thus, in (2), it has to be understood from the context that *yêpin* refers to many leaves, rather than one:

(2) *yep iret ai yêpin teo*
 [yep]$_S$ i=tet a=i [yêpin te-yo]$_{LOC}$
 fire 3sg=spread OBL=3sg leaf LIG-DEM.INT
 'The fire spread through the leaves.' (KM190211_0020)

To disambiguate whether a noun is referring to one or to multiple entities, speakers have a number of options. A numeral 'one' can be used to refer to a single entity (see below). Alternatively, a noun can be modified by a third person non-singular (i.e. dual, paucal or plural) pronoun, to indicate that it is referring to a non-singular referent. The following sentences give a number of examples.

(3) *Ton no lêp maloan ip kurun pusok si net*
 [Ton]$_A$ no lêp [maloa-n ip kutun pusok]$_O$ si net
 T. IPFV take picture-PERT 3pl small island come.down sea
 'Ton was taking pictures of the small islands down in the ocean.'
 (KM050995_0019)

(4) *ipe ro lêp tinawayen ip molat*
 i$_A$=pe to lêp [tinawayen ip molat]$_O$
 3sg=PFV HAB take huge 3pl dogtooth.tuna
 'He used to catch huge amounts of dogtooth tuna.'
 (KW290611_0037)

(5) *irê not no sot tet panu rang*
 [irê not]$_S$ no sot tet panua ta-ng
 3pc child IPFV go.up move front CLF.POSS-1sg.PERT
 'The boys went in front of me.' (NP210511_2_0040)

(6) *ip numun sôkôm teo kosan antek*
 [ip numun sôkôm te-yo]$_O$ ko$_A$-san antek=Ø
 3pl fiþre some LIG-DEM.INT IRR.1sg-cut put.away=3sg.ZERO
 'Some of the leftover fibres [after basket weaving] I cut away.'
 (MK050311_0024)

The 3pl pronoun *ip* is used much more often than the dual and paucal ones; a notable difference between them is that the dual and paucal are only used for human referents. As can be seen from the examples above, *ip* has no such constraints. Still, it does most often refer to human or at least animate beings, despite the significant number of its occurrences with inanimate referents. This correlates with the animacy hierarchy that is well established for the marking of number on nouns (Smith-Stark 1974). In the above examples, the referents of the NPs modified by a personal pronoun are identifiable for the speaker, except in example (4), where the referent of *ip molat* is non-specific generic: it refers to the dogtooth tuna species as a whole. This use of *ip* is common: compare for instance *ip Maput* 'the Titan people' or *ip palosi* 'the (people from the) past', which are also used with generic reference to a class.

5.2.2 Numeral 'one' as determiner

A numeral 'one' can precede the head noun in a NP to specify singular number. In addition, it indicates that the NP has non-specific categorial reference. Categorial reference indicates that the NP referent is an arbitrary member of a class, and as such a non-specific entity. Thus, class membership is the most important feature, rather than reference to an individuated entity. Some examples are given below.

(7) *ope rou lai sip kui menengan*
wo$_A$=pe tou=Ø la a=i sip kui menengan
2sg=PFV put=3sg.ZERO go.to OBL=3sg one.INANIM pot big
'You will put [it] into a big pot.' [any member of the class of cooking pots]
(CA120211_2_0020)

(8) *wola lêp som not taip*
wo$_A$=la lêp [som not ta-ip]$_O$
2sg=go.to take one.ANIM child CLF.POSS-3pl
'You go and take a child of theirs.' [any of their children]
(LM240611_0047)

(9) *aplêp som nel me*
ap$_A$=lêp [som nel]$_O$ me
2pl=take one.ANIM rope come
'You bring a (piece of) rope.' [just any odd piece of rope]
(LK100411_0077)

In example (10) below, it becomes especially clear that the numeral used in this sense has a categorial interpretation. In this example, dogs are urged by their chief to pick up a random tail. They have to make haste, since the house where the tails are stored is on fire. The consequences are dire, because the dogs end up with a tail that is not their own.

(10) *wono akêp sip youn mui*
wo$_A$=no akêp [sip you-n mui]$_O$
2sg=IPFV pick.up one.INANIM tail-PERT dog
'You will (each) pick up a dog's tail.' [any tail]
(LL300511_1_0059)

Whether the use of the numeral in this context is obligatory or optional is not entirely clear. If there is no prenominal numeral present, the referent of the NP is probably specific, i.e. identifiable for the speaker, but indefinite and introduced in the story for the first time. Nevertheless, there is reference to an individuated entity. When the numeral 'one' is included, the focus lies more on class membership of the entity than on the entity's individuatedness.

5.3 Prenominal modifier

The prenominal modifier slot can be filled with either a quantifier or an adjective. Most adjectives follow the head noun, but there are a number of exceptions to this rule, e.g. *kutun* 'small', *pwakpwak* 'big, fat' and *tinawayen* 'huge'. Most of these exceptions seem to involve adjectives of Dimension. Quantifiers either precede or follow the head noun; a more detailed discussion can be found in Section 4.6. There are no attested examples of nouns that are preceded by both a quantifier and an adjectival modifier. The quantifier for small quantities *sê* (or a slightly different form, depending on the semantics of the head noun) precedes the head noun when the formative *an-* is attached to it, but follows it when it is used by itself. Two examples of a quantifier used as a pre-head modifier are given below. In (11), *ansê* precedes the head noun *pau*; in (12), *ansut* precedes *kokon*. In (11), the entire phrase *ansê pau* is in turn modified by the adjective of dimension *namwi* 'small'.

(11) worou ansê pau namwi lai
wo$_A$=tou [an-sê   pau   namwi]$_O$ la   a=i
2sg=put PART-DIM coconut.oil small   go.to OBL=3sg
'You put a little bit of coconut oil with it.' (NK290311_2_0020)

(12) ipwa kime baim ai ansut kokon
i$_A$=pwa   ki-me   baim=Ø$_O$ a=i   an-sut   kokoni
3sg=want.to IRR.3sg-come buy=3sg.ZERO OBL=3sg PART-DIM money
'He wants to come and buy [it] with a bit of money.' (MK050311_0030)

5.4 Post-nominal modifier

Most adjectives follow the head noun, as do most quantifiers. Numerals, used within the NP, also follow the head noun. A head noun can be modified by a pre-head and a post-head modifier combined, as in (13), but seemingly not by two pre-head or two post-head modifiers.

(13) *u tinawayen yoy raywêp*
 u~S~ [tinawayen yoy ta=yuêp]~NVPRED~
 3du huge stone SPEC.COLL=two.INANIM
 'They were two huge stones.' (LK250111_0052)

5.5 Possessor

The noun referring to the Possessee (Pe) is always the head of a possessive construction, either direct or indirect. The Possessor (Po) is cross-referenced by a suffix on the noun referring to the Pe (in the case of direct possession) or on a classifier *ta-* following the NP referring to the Pe (in the case of indirect possession). In addition, the Po can be overtly expressed by a full NP, which always follows the Pe NP. This means that a possessive construction consists of two NPs, with the Pe NP acting as its head. Section 3.2.2 provides a discussion of how nouns can be classified according to the possessive construction(s) they can participate in. For alienable possession of edible objects, there is an alimentary classifier *ka-*, which precedes the head noun. This is the only case where a pronominal Po precedes the Pe. However, if the Po is also expressed by a full NP, it will still follow the Pe NP, as e.g. in the case of the set phrase *ka-n naluai pein* 'CLF.food-3sg garden.food woman', used to refer to the food given by the woman's side at a bride price ceremony. See Section 4.2.3.2 for the *ka-* paradigm.

5.5.1 Form of direct possession

Direct possession, which is usually considered to be linked to a semantic class of "inalienable possession", is expressed by a pronominal suffix added directly to the noun (hence the name). The suffix can take the following forms:

(14) first person singular   -ng
      second person singular   -m
      third person singular   -n (optionally followed by a full NP)
      all other persons and numbers -n plus free personal pronoun (see
                                     Section 4.2.1 for the entire paradigm)

An example of direct possessive constructions with the root *tama-* 'father' is given below:

(15) tamo-ng   'my father'
    tamo-m   'your (sg.) father'
    tama-n   'his father'
    tama-n Sion 'John's father'
    tama-n ip   'their (pl.) father'

The -n suffix itself is not specified for person and number; it only indicates a possessive relation. It could be argued that it is a default form that refers to third person singular, both when it is not further specified and when it is specified by a full lexical NP. For other persons and numbers, information referring to the Po is added by a personal pronoun that follows the possessive marker. This type of reference, with an invariant suffix on the Pe constituent and further specification of person/number of the Po by an independent pronoun, is labelled "construct" by various authors (cf. Lichtenberk 2009a). Many directly possessed forms show vowel alternations between first and second person forms on the one hand, and third person on the other; see Section 2.2.7.2 on this morphophonological alternation.

5.5.2 Form of indirect possession

General indirect possession is expressed by a person/number suffix added to the classifier *ta-*, which follows the head NP referring to the Pe. This form of possession is usually linked to a semantic class of "alienable possession". The paradigm of forms can be found in Section 4.2.3.1. There is a distinction in the third person singular, depending on whether or not the *ta-* form plus suffix is followed by a full NP. Different suffixes are used: *-n* if what follows is a full NP, *-i* in all other cases. The indirect possession pertensive suffixes for second and third person singular are different from the ones that attach directly to the noun, but are formally identical to the verbal object enclitics (see Section 4.2.2.2). Thus, *ta-* is a hybrid form that takes dependent forms from both the nominal and verbal paradigm. An example of indirect possessive constructions with the noun *wum* 'house' is given below:

(16) wum ta-ng   'my house'
    wum ta-o   'your (sg.) house'
    wum ta-i   'his house'
    wum ta-n Sion 'John's house'
    wum ta-ip   'their (pl.) house'

Indirect possession of edible or otherwise consumable items (but not potable items) is expressed by the alimentary classifier *ka-*, which precedes the noun. When the Po is expressed by a full NP, it still follows the Pe noun. Three examples of the use of this classifier are given below.

(17) *kong naluai a kong pou*
    ka-ng   naluai   a   ka-ng   pou
    CLF.food-1sg.PERT garden.food and CLF.food-1sg.PERT pig
    'My garden food and pork [for me to eat].' (LK100411_0104)

(18) *kom puan tiok*
    ka-m   puan tiok
    CLF.food-2sg.PERT fruit piper.betle
    'Your *tiok* fruit [to chew with betel nut].' (LL300511_2_0015)

(19) *kan naluai nei*
    ka-n   naluai   nei
    CLF.food-PERT garden.food rat
    'Rat food.' (Game1_021012_0284)

Alternatively, *ta-* and *ka-* can be heads of a non-verbal predicate, expressing predicative possession. This function is discussed in Chapter 7.

5.6 Prepositional phrase

The only prepositional phrases encountered within the NP are formed with *pari* 'from, belonging to'.[44] A prepositional phrase always follows the head noun it modifies. When *pari* is followed by a local noun, there is no additional modifier; otherwise, the preposition *a=* (cliticised to the third singular pronoun *i*) needs to be added. Examples (20) to (23) illustrate the use of *pari*, first with a local noun (20 and 21), then with a common noun (22 and 23).

(20) *ip pein pari Nauna*
    ip   pein   pari   Nauna
    3pl woman belonging.to N.
    'The women from Nauna.' (KS030611_1_0007)

---

[44] If possessive *ta-* is considered a preposition, phrases expressing the Po of an indirect possession construction also count as prepositional phrases within the NP.

(21) *epworup Lou, suk pari Lou*
 eps=worup   Lou suk   pari   Lou
 1pl.EXCL=descend Lou shore belonging.to Lou
 'We alighted on Lou, on the shore of Lou Island.' (NP210511_2_0006)

(22) *wong kope pwapwa repwo, pwapwa pari ai pang*
 wong_A   ko-pe   [pwapwae te-pwo]_O   [pwapwae [pari
 1sg.FREE IRR.1sg-make story   LIG-DEM.PROX story   belonging.to
 a=i   pang]]_{NP/O}
 OBL=3sg rain
 'I will tell this story, the story of the rain.' (LM260511_1_0004)

(23) *ip yamat pari ai peinan nin*
 ip   yamat   pari   a=i   peinan   nine
 3pl   person belonging.to OBL=3sg   making.of   fight
 'Short-tempered people, people prone to fighting.' [TP: *man bilong pait*] (NP220611_2_0009)

The prepositional phrase with *pari* follows adjectives and other post-head modifiers:

(24) *kei nangin pari lalon kanum*
 kei   nangin pari   lalo-n   kanum
 herbs   scent   belonging.to inside-PERT   garden
 'Fragrant herbs [meant] for inside the garden.' (NK290311_1_0013)

5.7 Relative clause

Nouns are frequently modified by relative clauses. There is a range of functions that the common argument (CA) can have in the dependent clause. Relative clauses are discussed in detail in Section 11.1.1.

5.8 Demonstrative

The demonstrative is always the final element in the NP, following all other modifiers. As such, it probably has a function signalling the end of the NP. The demonstrative paradigm is discussed in Section 4.3. Demonstratives with *te-* have been reanalysed as definiteness markers: they indicate that the referent of the NP they modify is identifiable to the hearer. When first introduced in discourse, a participant is usually not marked by a demonstrative, unless the speaker wants to add extra emphasis. Subsequent occurrences of the participant are likely to be marked with the intermediate demonstrative. The proximate and distal demonstrative forms are often used as discourse cataphors. See Chapter 12 for more elaborate discussion.

5.9 Coordination of NPs

As mentioned in Section 4.9, the coordinators *a* 'and' and *le* 'or' can be used to coordinate two NPs:

(25) *taman a tinan teo, ukape pul la rai la remenin telo*
 [[tama-n]$_{NP}$ a [tina-n]$_{NP}$ te-yo]$_A$ u=ka-pe pul
 father-PERT and mother-PERT LIG-DEM.INT 3du=IRR.NS-PFV tell
 la ta-i la temenin te-lo
 go.to CLF.POSS-3sg.PERT go.to like LIG-DEM.INT
 'Her father and mother, they would speak to her as follows.'
 (LK100411_0013)

In (25), the entire constituent *taman a tinan teyo* forms the topicalised A argument of the clause, ending with the intermediate demonstrative as a definiteness marker. Interestingly, two NPs can also be linked by means of a personal pronoun. In a number of cases, dual number pronouns are used in combination with one or two NPs to indicate a collective meaning 'together with, and'. These are examples of what Lichtenberk (2000) calls "inclusory pronominals", although he does not in fact discuss inclusory pronominals that are combined with two NPs. According to Lichtenberk (2000: 2), it is not the case that two NPs, or a pronoun and a NP, are coordinated, but rather that the pronominal "identifies the total set of participants" whereas the NP "identifies a subset". Adopting this analysis would mean that in Paluai, more than one subset of a total set of participants could be identified by the NPs in this construction.

(26) *iro ran taman u rinan*
 i$_S$=to ta-n [[tama-n]$_{NP}$ u [tina-n]$_{NP}$]$_{LOC}$
 3sg=be CLF.POSS-PERT father-PERT 3du mother-PERT
 'She was (staying) with her father and mother.' [*lit.* 'father them two mother'] (LK100411_0009)

(27) *wui Maiau pe yangyangek tui*
 [wui   Maiau]_A pe-yangyang-ek ta-ui
 1du.EXCL M.   RECIP-love-APPL CLF.POSS-1du.EXCL
 'Maiau and I fell in love with each other.' [*lit.* 'we two Maiau...']
 (KM060111_0009)

(28) *pêng pian, Sauka au Keket*
 pêng pian [Sauka]_NP au   [Keket]_NP
 night good S.   2du   K.
 'Good night to you two, Sauka and Keket.' [*lit.* 'Sauka you two Keket'] (OL201210_0002)

It is mostly the third person pronoun that is used in this construction; first and second person instances are quite rare. Similar usages of dual pronouns are reported for closely related languages, e.g. Loniu (Hamel 1994), Sivisa Titan (Bowern 2011) and Seimat (Wozna and Wilson 2005).

6 Predicates I: verbal predicates

This chapter discusses the various categories associated with verbal predicates and the verb complex. They include the indexing of subject and object, reality status, aspect and modality. With regard to aspect, a distinction can be made between pre- and postverbal aspectual markers. The former could be regarded as particles, whereas the latter take the form of coverbs or adverbs, or may perhaps be analysed as full verbs serialised with the main verb in a serial verb construction (SVC). The Paluai verb complex could be analysed as potentially containing a SVC, a main verb plus coverbs/adverbs, a main verb plus particles, or a combination of the above.

The term "verb complex" is commonly used in Oceanic linguistics and usually refers to a conglomerate of the main verb, its preceding TAM and/or directional particles and its postverbal elements, if present. In accordance with Pawley (2003), the Paluai verb complex (VC) is here analysed as a phonological phrase that resembles a word in certain respects (exhibiting rigid order of elements, with no pauses allowed phrase-internally), but that is at the same time unwordlike in the sense that its "peripheral functors" (Pawley 2003: 149) are usually free forms, or can at least synchronically be encountered as both free forms and more grammaticalised items. Often, the free form and grammaticalised item are formally identical, but some preverbal elements show more phonological dependency than their corresponding free forms, if applicable. This will be discussed below where relevant.

In addition to aspectual particles, many VCs contain a directional (see Section 3.3.1.3). Directionals can occur both preceding and following the main verb; in this chapter only preverbal instances are discussed (with the exception of *wot* 'move horizontally'). The reason for this is that a directional following the main verb is not analysed as a part of the VC, because it follows the object bound pronoun (if present) and introduces a Locative or Goal constituent by means of a SVC. Verb-particle or verb-verb sequences that are part of the VC, however, could potentially also be analysed as SVCs, but will be discussed separately in Chapter 9, which is dedicated specifically to SVCs.

In what follows, bound subject and object pronouns indexing core arguments S, A and O will be discussed first, followed by aspectual particles and preverbal directionals.[45] Although reality status is probably the most fundamental verbal category in Paluai, its discussion will follow that of aspectual

---

[45] As mentioned before, the labels S, A and O (cf. Dixon 2010a: 122) are used as convenient abbreviations for referring to the core arguments of verbal predicates (subject of intransitive

categories, because it interacts with the latter in various ways. Modality ties in with reality status and thus will be discussed directly following it. After discussing the various categories associated with the verb and the VC, the final part of the chapter will discuss the structural properties of the VC itself and how it can be schematically represented.

6.1 Indexing of S/A and O

In the case of declarative and interrogative clauses, the subject is normally expressed within the VC, either by a bound pronoun that forms a proclitic to the VC or by an irrealis marker containing person and number information (see Section 6.4 for discussion of the category irrealis). This means that if the irrealis marker is present, the bound pronoun is not obligatory. It is obligatory, on the other hand, if there is no irrealis marker: example (6) is ungrammatical. In addition, a free pronoun can precede the VC. Free and bound subject pronouns are formally distinct for first person singular only (see Section 4.2 for the pronominal paradigms). For other persons and numbers, it is in fact not formally evident whether the free or the bound form of the pronoun is used. In contrast, for the first person singular the following variation can be recognised:

(1) *wong ngakola um*
    wong   nga=[ko-la]$_{VC}$ wumwa
    1sg.FREE 1sg=IRR.1sg-go.to   house
    'Me, I will go home.'

(2) *ngakola um*
    nga=[ko-la]$_{VC}$ wumwa
    1sg=IRR.1sg-go.to   house
    'I will go home.'

(3) *wong kola um*
    wong   [ko-la]$_{VC}$ wumwa
    1sg.FREE IRR.1sg-go.to   house
    'I will go home.'

---

and transitive verb, and object of transitive verb, respectively), not to semantic-syntactic relations. More discussion of grammatical relations can be found in Chapter 8.

(4) *kola um*
    [ko-la]ᵥᴄ   wumwa
    IRR.1sg-go.to house
    'I will go home.'

(5) *ngala um*
    nga=[la]ᵥᴄ wumwa
    1sg=go.to house
    'I went home.'

(6) **wong la um*
    *wong   [la]ᵥᴄ wumwa
    1sg.FREE go.to house
    Intended: 'I went home.'

For first and second person, a bound pronoun will always be present; it expresses the subject of the clause,[46] sometimes accompanied by an optional free pronoun. For third person subjects that are co-expressed by a full NP, we find variation. When the NP is quite long (for instance, if it contains a relative clause), the VC usually shows a bound pronoun. Conversely, there are a number of examples in the data with a short NP subject where this is not the case. The sentences below give some examples of a third person singular subject expressed by a full NP, with and without a bound pronoun copy present on the verb. In (7) and (8) below, a subject expressed by a full NP headed by a directly possessed kinship term is cross-referenced by a bound pronoun on the VC. These are not instances of topicalisation (see Section 12.3.3), since the subject NP and the VC are part of a single minor prosodic unit.

(7) *tamong ipe pul la rang, ipwa "narung…"*
    [tama-ng]ɴᴘ/ₐ i=[pe   pul]ᵥᴄ=∅   la   ta-ng
    father-1sg.PERT 3sg=PFV tell=3sg.ZERO go.to CLF.POSS-1sg.PERT
    iₐ=[pwa]ᵥᴄ [naru-ng]ᴄₒₘₚₗ:Qᵤₒₜ
    3sg=say   son-1sg.PERT
    'My father spoke to me, he said, "My son…"' (KM060111_0025)

---

[46] The term *subject* refers to the grammatical subject position, which can be filled by either of the core arguments, S or A, depending on the valency of the verb. There is no formal difference between S and A. See Chapter 8 for more on grammatical relations.

(8) irê taman tarê irêro arei ngai som e?
    [irê tama-n   tarê]~NP/A~ irê=[to   arei]~VC~ [ngai som]~NP/O~ e
    3pc father-PERT 1pc.INCL 3pc=HAB say   name one.ANIM TAG
    'Our fathers used to call [it] one [specific] name, right?'
    (PK290411_1_0020)

In (9) and (10), a third person singular subject expressed by a full NP is not cross-referenced by a bound pronoun on the VC. In (9), it is expressed by a directly possessed noun and in (10) by a proper name. Dropping of the third singular subject bound pronoun seems to happen predominantly when the subject is expressed by a directly possessed noun (such as a kinship term) or a proper name. It remains a matter for further research under exactly what pragmatic circumstances the bound pronoun tends to be dropped. Across the board, the phenomenon seems to be quite marginal, however, since the great majority of subjects are cross-referenced by a bound pronoun on the VC.

(9) a tinan to apui a uro ngan
    a [tina-n]~NP/A~ [to apui]~VC~=Ø   a   u~A~=[to   ngan]~VC~=Ø~O~
    and mother-PERT HAB cook=3pl.ZERO and 3du=HAB eat=3pl.ZERO
    'And the mother used to cook [the fish] and they used to eat [them].'
    (KW290611_0019)

(10) Alup Sauka pe yaya rou China
    [Alup Sauka]~NP/A~ [pe [yaya tou]]~VC~ [China]~O~
    A.S.   PFV swim give   C.
    'Alup Sauka swam after China [name of a dog].' (MK060211_0035)

The direct object NP of a transitive verb is often not overtly expressed, especially when it is the sentence topic (see also Chapter 12).[47] Inanimate objects are never indexed on the VC, whereas animate objects are obligatorily indexed when there is no full NP referent in the clause. When the object is indexed with a bound pronoun, the latter appears as an enclitic on the last element of the VC. This means that adverbs or coverbs of manner (see Section 3.6.1.1) appear between the main verb and its object, whether this is expressed as an enclitic or as a full NP. A number of examples are given below. In (11) and (12), the main verb is followed

---

[47] There are, in fact, a number of transitive verb forms ending in -i, such as *sui* 'fry', *apui* 'cook' and *yei* 'grate, scrape'. When a 3sg bound object pronoun cliticises on these verbs, it is not easily discernible from the final vowel of the verb. However, there is an equally large number of verbs ending in other phonemes, for which the object pronoun is easily recognisable.

by an adverbial and a postverbal aspectual particle, respectively. These are in turn followed by a bound pronoun enclitic indexing the O argument. In (12), the clause *kingan puni* forms a complement clause to the main verb *pwa* 'want to' (see Section 6.5.1 below).

(11) *ipwa kingan puni*
 i_A=[pwa]$_{VC}$ [[ki$_A$-ngan pun]$_{VC}$=i$_O$]]$_{Compl:Pot}$
 3sg=want.to IRR.3sg-eat INTF=3sg
 'He wanted to really eat him/eat him completely.' (LL010711_0067)

(12) *ingan mat nêmip*
 i_A=[ngan mat nêm]$_{VC}$=ip$_O$
 3sg=eat die be.finished=3pl
 'He killed and ate them all.' [*lit.* 'he eat die finish them']
 (KW290611_0057)

In (13) and (14), the O argument is expressed by a full NP and seems to be cross-referenced by a bound object pronoun on the VC. Such a cross-reference makes sense since the O argument has human reference in both examples.

(13) *iro yuai ip lau*
 i_A=[to yuai]$_{VC}$=ip [laue]$_O$
 3sg=HAB call=3pl people
 'He usually calls the people [of his clan].' (LL300511_2_0005)

(14) *kapwa kariu keleyekip not tang teo...*
 kapwa [ka$_A$-tiu keleyek]$_{VC}$=ip [not ta-ng
 if IRR.3sg-collect turn=3pl child CLF.POSS-1sg.PERT
 te-yo]$_O$
 LIG-DEM.INT
 'If my children are handled by other people...' (NP260511_0007)

It is debatable, however, whether the third person plural pronoun in these cases has to be regarded as a bound pronoun, cliticised to the VC. As discussed in Section 5.2, free pronouns are often used within the NP (preceding the noun) as a number marking device. This is probably also the case in (13) and (14) above. There are two arguments that support such an assumption. First of all, we also find inanimate full NP O arguments that show a non-singular pronominal form, as in (15) below.

(15) *ipe ro akêp ip malet*
    i_A=[pe   to   akêp]_VC [ip malet]_O
    3sg=PFV CONT pick.up   3pl rock
    'He was picking up rocks.' (LM190611_0021)

In this case and similar ones, *ip* is better analysed as a strategy to mark non-singular number on the full NP referent. The second reason why cases like (13) and (14) probably do not contain object bound pronouns is that no examples have been found of the third person singular object bound pronoun =*i* combined with a full NP. Thus, an animate O argument is indexed on the VC *only* if the full NP referent has been elided, as is the case in e.g. (11) and (12). It is thus different from the S/A argument, which is obligatorily cross-referenced (with the apparent exception of the cases discussed above) when it is expressed as a full NP.

6.2 Aspect

Paluai has a variety of aspectual particles preceding and possibly also following the verb (an overview is given in Section 3.3.1.4). Holt (1943), quoted in Comrie (1976: 3) defines aspects as "different ways of viewing the internal temporal constituency of a situation". Aspect therefore defines the temporal flow (or lack thereof) of a given situation, a basic distinction being whether a situation is looked upon as bounded and possibly unitary, without reference to any flow of time during the situation, or non-bounded, with reference to the nature of the flow of time during the situation, but no reference to temporal bounds of the situation. The first is called perfective aspect, while the latter is called imperfective aspect. In English, this opposition can be illustrated by the pair *I ate* vs. *I was eating*. Within the imperfective, two sub-domains are usually distinguished: habitual and continuous/progressive (Bybee, Perkins, and Pagliuca 1994).

Aspect is different from tense. It is not always easy to distinguish between tense, aspect and modality; some Paluai forms may in fact show features of two of them, or of all three. Tense "relates the time of the situation referred to to some other time, usually the moment of speaking" (Comrie 1976: 2). It relates the here-and-now of the speech event to another time in the past, present or future. Aspect, on the other hand, is used to first and foremost refer to the internal composition and boundedness of a situation, rather than to foreground a temporal relation. In Paluai, aspect is not obligatorily indicated. Since there is no category of tense, this means that when aspectual particles are absent, a clause is completely unmarked with regard to its temporal reference. In these cases, it has to be inferred from the discourse context whether the clause refers to the past or the present. In addition, some aspects lend themselves more to a past interpretation and others more to a present interpretation. Any clause, however, is marked for reality status, either with *kV-* for irrealis (except for second person singular), or zero for realis; see Section 6.4 below. Because a clause referring to a future event is always marked as irrealis, absence of irrealis marking forces a past or present interpretation irrespective of aspectual marking (again, except for 2sg).

In what follows (Section 6.2.1), preverbal "core aspectual" particles will be discussed first, followed by preverbal particles indicating "secondary aspect". Postverbal aspectual particles are the topic of Section 6.2.2. The discussion will focus predominantly on realis predicates. Discussion of how the particles function in irrealis predicates will be left until after reality status has been discussed in Section 6.4.

6.2.1 Preverbal aspectual particles

6.2.1.1 Core aspect

Three forms are analysed as having core aspectual meaning. They are mutually exclusive: no predicates have been found showing a combination of them. When they occur, these forms occupy the first slot in the VC, directly after the subject bound pronoun and/or irrealis prefix.[48] The forms are given in Table 58.

Table 58: Core aspect particles.

Particle	Aspectual meaning	Lexical source
no	Imperfective	possibly *no* 'only, just'
pe	Perfective	possibly *pe* 'do, make'
an	Perfect	unknown

What the three forms have in common is that their lexical source is harder to establish than that of the secondary aspect particles discussed below, indicating that they have grammaticalised further. The remainder of this section discusses each of the three forms in turn.

---

[48] One more form that can occupy this slot and is mutually exclusive with the core aspectual particles is the modal particle *sa*, discussed in Section 6.5.2.

6.2.1.1.1 Imperfective *no*

The particle *no* is frequently used to indicate a temporal flow during an event. The event can be either a state or an action; in the latter case, the continuative/habitual particle *to* will usually also be present. Whether or not *no* is accompanied by continuative/habitual *to* may depend on the semantics of the verb and maybe also on the context. More discussion can be found below in Section 6.2.1.2.1. Two examples of the use of *no* are given below. The particle *no* sometimes shows vowel assimilation, as discussed in Section 2.2.5.1; this phenomenon has not been observed for the homophonous adverb *no*.

(16) *i reo, ngano pwa i mui le i sa?*
i te-yo nga$_A$=[no pwa]$_{VC}$ [i mui le i sa]$_{Compl:Quot}$
3sg LIG-DEM.INT 1sg=IPFV think 3sg dog or 3sg what
'That there, I think it's a dog, or what is it?' (Game1_280812_0082)

(17) *a ipnoro yiki a ipno ro yiki, a sin no lol a pulêng no bu, a ipno ro yiki*
a ip$_A$=[no to yik]$_{VC}$=i$_O$ a ip$_A$=[no to
and 3pl=IPFV CONT search.for=3sg and 3pl=IPFV CONT
yik]$_{VC}$=i$_O$ a [sin]$_S$ [no lol]$_{VC}$ a [pulêng]$_S$ [no pu]$_{VC}$
search.for=3sg and sun IPFV be.dark and dawn IPFV break
a ip$_A$=[no to yik]$_{VC}$=i$_O$
and 3pl=IPFV CONT search.for=3sg
'They were searching and searching for him, while the sun was setting and the dawn was breaking, they were (still) searching for him.' (MS250311_0043)

The particle *no* can refer to events either happening at the moment of speaking, as in (16), or happening prior to the moment of speaking, and thus in the past, as in (17). This has to be inferred from the discourse context, however, since the particle itself is tense-neutral and only indicates that the event is viewed as unbounded, and that time passes during the event. Therefore, *no* is analysed as a marker of imperfective aspect.

Because imperfective aspect refers to an event during which there is a temporal flow, it tends not to go together very well with verbs that are inherently punctual or imply reaching a limit (Comrie 1976), such as *lot* 'fall' or *angou* 'arrive'. With these verbs, the use of *no* and in particular *no to* implies iterative aspect (e.g. 'fall down repeatedly'). Complex clauses, where the use of *no* implies simultaneousness or co-occurrence of events, may be an exception to this:

(18) *no re nganingning wot a ono lot*
no te nga$_S$=[ning.ning wot]$_{VC}$ ya wo$_S$=[no lot]$_{VC}$
when 1sg=REDUP.see go.level then 2sg=IPFV fall
'Just as I was looking over [at you], you fell down [were in the process of falling down].' (field notes 10/05/2011)

With verbs that have inherent "accomplishment" in their meaning, such as 'write', use of the imperfective with past time reference implies that the action was not completed:

(19) *mino reo, ngano rayei pas*
mino te-yo nga$_A$=[no tayei]$_{VC}$ [pas]$_O$
yesterday LIG-DEM.INT 1sg=IPFV write letter
'Yesterday, I wrote on a/the letter.' [for a while; it is not completed yet] (elicitation 11/10/2012)

No can have overtones of doing something 'instead of' something else, or doing something other than planned. Two examples are given below. The translation of (20) does not suggest an imperfective meaning; however, *akêp* means in fact both 'pick up' and 'hold'. Since the speaker picked up the picture and is still holding it, imperfective meaning does apply in this case.

(20) *ngamaakêp nganngan pwên, ma ngano akêp muyou*
nga$_A$=[ma=akêp]$_{VC}$ [nganngan]$_O$ pwên ma nga$_A$=[no
1sg=NEG$_1$=pick.up food NEG$_2$ but 1sg=IPFV
akêp]$_{VC}$ [muyou]$_O$
pick.up snake
'I didn't pick up the food, but I picked up the snake instead.' (Game1_021012_0510)

(21) *ngapwa kope kou, ma pwên, ngano yaya*
nga$_A$=[pwa]$_{VC}$ [ko$_A$-pe [kou]$_O$]$_{Compl:Pot}$ ma pwên nga$_S$=[no yaya]$_{VC}$
1sg=want.to IRR.1sg-do fishing but NEG 1sg=IPFV swim
'I wanted to go fishing, but no, I went swimming [instead].' (elicitation 01/05/2011)

Combined with the irrealis marker, *no* may confer modal meaning to a clause. See Section 6.5.3.2 for more information.

6.2.1.1.2 Perfective pe

pe is analysed as a marker of perfective aspect. In many instances, a "background" situation is described accompanied by an imperfective particle, and the event that is in focus or that is foregrounded is then introduced in the discourse accompanied by perfective *pe*. A clear example is the following:

(22) ipno rok wot onga ippe yong tuktuk tou
 ip$_S$=[no tok wot]$_{VC}$ onga ip$_A$=[pe yong]$_{VC}$ [tuktuk.tou]$_O$
 3pl=IPFV sit go.level and.so 3pl=PFV hear drum.beat
 'They were sitting around and then they heard the drumbeat for calling people.' [their chief was beating the drum in order to summon them] (LL300511_1_0017)

The first situation is viewed as extending in time; the second is viewed without regard to the time passing while it is happening (this doesn't imply that it is punctual). The first situation started before the second and continued until the second had already been terminated (but not for much longer, since the subjects decided to pack their belongings and answer the summoning). This is typical for the use of perfective and imperfective (Comrie 1976; Dixon 2012: 35), in particular in narratives. The perfective (in realis predicates) refers to an event that occurred prior to the here-and-now of the speech act, in other words in the past. In contrast to the imperfective discussed above, the perfective does not lend itself well to an interpretation in which the event is happening in the present. This is because it is hard to conceptualise an event currently occurring as a bounded, unitary totality. Because Paluai does not have tense, the perfective will often be used to refer unambiguously to past events, whereas the imperfective can have both past and present interpretation based on the discourse context. The perfective can, in combination with irrealis marking, also refer to future events; for further discussion, see Section 6.5.3.1.

When used with verbs that have inherent "accomplishment" in their semantics (such as 'write'), *pe* usually indicates that the event has indeed been completed:

(23) mino reo, ngape rayei pas
 mino te-yo nga$_A$=[pe tayei]$_{VC}$ [pas]$_O$
 yesterday LIG-DEM.INT 1sg=PFV write letter
 'Yesterday, I wrote a/the letter.' [it is completed] (elicitation 11/10/2012)

In this case, too, the use of the perfective indicates that the event is viewed without respect to its duration, i.e. without respect to the time that passed while the letter was being written. Instead, the event is viewed as a bounded whole.

With stative verbs, *pe* indicates a change of state, as exemplified below. In contrast to the perfect discussed in the next section, the perfective only indicates the inception of the state and does not give information on whether the state is still ongoing. Thus, it can be used for states that are no longer ongoing at the narrative present.

(24) ippe mapwai rai re imat
      ip$_S$=[pe mapwai]$_{VC}$ ta-i   te   i$_S$=mat
      3sg=PFV know   CLF.POSS-3sg.PERT SUB   3sg=die
      'They got to know about him [*lit.* 'became knowledgeable about him'], that he died.' (MS250311_0060)

(25) ippe yop touong a ngape kaêrêt
      ip$_A$=[pe yop.tou]$_{VC}$=ong$_O$ a   nga$_S$=[pe kaêrêt]$_{VC}$
      3pl=PFV chase=1sg   and   1sg=PFV   be.afraid
      'They chased me and I got frightened.' (LL030611_0055)

6.2.1.1.3 Perfect an

The particle *an* is a marker for the perfect. It is presumably at a late stage of grammaticalisation; indications for this are that it seems completely devoid of lexical meaning and that its origins could not be traced. Moreover, *an* is phonologically more dependent than some other preverbal particles: for singular number it has partly fused with the bound pronoun proclitics, as discussed below, and its final /n/ assimilates or drops in some environments (see Section 2.2.7.1).

Although, strictly speaking, the perfect is not an aspect (Comrie 1976), it is often discussed together with aspect. The perfect "indicates the continuing present relevance of a past situation" (Comrie 1976: 52). It therefore expresses a relation between two time points, "on the one hand the time of the state resulting from a prior situation, and on the other the time of that prior situation" (Comrie 1976: 52). By virtue of its definition, the perfect is not compatible with the irrealis, since it refers to a prior event that has in fact happened (a realis). Since *an* cannot be combined with any other particles, it always directly follows the (bound) pronoun. In the singular, the proclitic has partly fused with it, producing the following surface forms:

(26) 1sg: /ŋa=an/ → [ŋan]
    2sg: /wɔ=an/ → [ɔn]
    3sg: /i=an/ → [m]

Some typical examples of the use of the perfect are provided in sentences (27) to (29). The events described in each of these examples continue to have relevance for the here-and-now of the speech act. In (27), the person referred to is actually at the place where the utterance took place at the moment of speaking. In (28), the fact that two pigs have already been eaten means that their meat cannot be shared out at the moment of speaking, with the rest of the pig meat. In (29), the fact that the subject (a bush spirit) has been deceived means there will be repercussions for the people who took part in the deception. The perfect indicates a change in the state of affairs, which is still ongoing at the moment of speaking and which, in the case of (28) and (29), also cannot be undone: the pigs cannot be un-eaten, and the spirit cannot be un-deceived.

(27) *Ponaun in si*
    [Ponaun]$_S$ i=[an si]$_{VC}$
    P.   3sg=PRF come.down
    'Ponaun (already) has arrived [so he is here now].' (ANK020995_0013)

(28) *arêan ngan yumou mino*
    arê$_A$=[an ngan]$_{VC}$ [yumou]$_O$ mino
    2pc=PRF eat   two.ANIM   yesterday
    'You have (already) eaten two [pigs] yesterday.'
    (YK290411_2_0051)

(29) *apan sukong*
    ap$_A$=[an suk]$_{VC}$=ong$_O$
    2pl=PRF deceive=1sg
    'You have deceived me.'
    (PN100411_0025)

Thus, the perfect views events as states (cf. Timberlake 2007). It refers to an event that has accomplished a change in the world that is still ongoing, and thus relevant, at the narrative present. It is often used with stative verbs. However, the changed state still has to be ongoing for the perfect to be felicitous, otherwise the perfective aspect is more compatible. The distinction between imperfective,

perfective and perfect for stative verbs is illustrated by the following elicited examples:

(30) *mino reo, niong no porok*
 mino te-yo [nia-ng]₍S₎ [no porok]₍VC₎
 yesterday LIG-DEM.INT stomach-1sg.PERT IPFV be.painful
 'Yesterday, my stomach was (being) painful.' (elicitation 11/10/2012)

(31) *mino reo, niong pe porok*
 mino te-yo [nia-ng]₍S₎ [pe porok]₍VC₎
 yesterday LIG-DEM.INT stomach-1sg.PERT PFV be.painful
 'Yesterday, my stomach became painful.' (elicitation 11/10/2012)

(32) *niong in porok*
 [nia-ng]₍S₎ i=[an porok]₍VC₎
 stomach-1sg.PERT 3sg=PRF be.painful
 'My stomach has become painful.' (elicitation 11/10/2012)

Sentence (30) indicates that some amount of time passed while the state of "being painful" was going on and gives no information about its inception and cessation. Sentence (31) gives information about the stomach's change of state from "not being painful" to "being painful", but not about the duration of the new state, and whether or not it is still ongoing. Sentence (32) gives information about the same change of state, perhaps in the not-so-distant past, but at the same time links it to the present, indicating that the new state of 'being painful' is still on-going. In the event that the speaker's stomach became painful yesterday afternoon and ceased to be painful last night, both (30) and (31), but not (32), would be accurate descriptions of what happened (although the same state of affairs would be looked at from different angles). In case the speaker's stomach became painful yesterday and is still painful today, (32) would be an accurate description. However, it is more likely that in this case the adverb *pa* 'yet, still' would be used, in combination with a verb not marked for aspect, to indicate the (overly) long duration of the stomach ache.

The perfect is also used to indicate that a given situation has occurred before another situation described in the same clause. It then functions as an anterior (Bybee et al. 1994: 54).

(33) *epkape pat ip sapon kei re epan san*
ep_A=ka-[pe   pat]_{VC} [ip sapo-n kei [te ep_A=[an
1pl.EXCL=IRR.NS-PFV plant   3pl top-PERT herb REL 1pl.EXCL=PRF
san]_{VC}]_{RC}]_O
cut
'We will plant the herbs that we have (already) cut.'
(NP220611_1_0018)

The perfect appears to be similar to the completive aspect, but actually the two are quite different. Completive aspect (indicated by the verb *nêm* 'be finished' following the main verb; see Section 6.2.2.1) can only be used with active verbs, not with stative ones. It indicates that the action has taken up a stretch of time and is now finished. *an* does not necessarily imply this: it is similar to the perfective in that it disregards the internal composition of the event. Thus, *an* can be used with all verbs, whereas *nêm* is only compatible with active verbs, and within the active verb class only with non-punctual verbs. *nêm* can be used e.g. with *ngan* 'eat', indicating completion of the 'eating' event, which necessarily takes some time, but not with e.g. *angou* 'arrive', which is punctual.[49] In addition, *nêm* does not indicate a relation between two points in time, contrary to the perfect.

The perfect can be used to indicate completion, provided the completion has present relevance. The elicited example (34) once again involves the verb 'write':

(34) *ngan tayei pas mino*
nga_A=[an tayei]_{VC} [pas]_O mino
1sg=PRF   write   letter   yesterday
'I have (already) written the/a letter yesterday.' [it is done, I need not worry about it anymore, etc.] (elicitation 11/10/2012)

When example (34) is compared to examples (19) or (23), where the imperfective and perfective were used, the pragmatic difference between perfect and the other two aspects also becomes apparent. Sentences (19) and (23) could be answers to a question such as *What did you do yesterday?*. Sentence (34), however, is most felicitous as an answer to the question *When will you write the*

---

49 *nêm* could be used felicitously with verbs such as *angou* when there is a non-singular subject, indicating that a set of iterative 'arrival' events has been completed. In this case, however, it is not used in the same sense as the perfect.

letter?. Thus, the perfect focuses on a change of state in the world, and the relevance this has for the current state of affairs.

6.2.1.2 Secondary aspect

In addition to the core aspectual particles discussed in the previous section, there are a number of preverbal particles categorised under "secondary aspect", because they can be combined with one of the core aspectual particles (but can also occur without them). They always occur in the slot immediately left of the main verb and are mutually exclusive with each other. The attested forms are listed in Table 59.

Table 59: Secondary aspect particles.

Particle	Aspectual meaning	Lexical source
to	Continuative, Habitual, Iterative	'be, stay (for a shorter duration)'
tu	Stative continuative	'be, remain (for a longer duration)'
yen	Progressive	'lie'

One feature shared by the secondary aspect particles is that they seem to be lower on the cline of grammaticalisation: they are all attested as full lexical verbs. The lexical verb counterparts probably originate from posture verbs, but *to* and *tu* have developed into existential verbs 'be' rather than posture verbs. Other Oceanic languages, e.g. Boumaa Fijian (Dixon 1988) and Loniu (Hamel 1994) also distinguish preverbal particles that have developed from posture verbs. Hamel (1994: 106) mentions the cognate forms *tɔ* 'be in or at a place', *sɔ* 'be in/on a place; stand', and *yɛ* 'be in/on a place; sit'.

6.2.1.2.1 Continuative/habitual *to*

The verb *to(k)* occurs both as a main verb and as a preverbal particle. *tok* is used when no other constituent (e.g. a locative) follows the main verb, otherwise we find *to*. The preverbal form *to* can refer to a habitual situation, described by Comrie (1976: 28) as "a situation which is characteristic of an extended period of time, so extended in fact that the situation referred to is viewed not as an incidental property of the moment but, precisely, as a characteristic feature of a whole period."

Alternatively, it can refer to a continuative event, one that is ongoing for some duration at the time referred to. Typical of continuatives/progressives is that they are projected to continue into the immediate future, but could also easily change or cease (Timberlake 2007). Sometimes, it is not really possible to decide whether a particular instance of *to* denotes habitual or continuative aspect, even in context; this indicates that they are two subtypes of the imperfective whose meanings lie very closely together. Often, imperfectives develop out of progressives (Bybee et al. 1994; Timberlake 2007). In the current analysis, *no* rather than *to* is analysed as imperfective, since it is mutually exclusive with *pe*. *to*, on the other hand, occurs in combination with both *no* and *pe*, as in the examples below:

(35) *ino ro lêp la um*
    i$_A$=[no to lêp]$_{VC}$=Ø$_O$ la   wumwa
    3sg=IPFV CONT take=3sg.ZERO go.to house
    'He was taking [it] home.' (NP210511_1_0050)

(36) *ippe ro pul kasiek*
    ip$_A$=[pe to pul=Ø$_O$ kasiek]$_{VC}$
    3pl=PFV HAB tell=3sg.ZERO not.well
    'They used to tell [it] incorrectly.' (MS250311_0083)

To is also used to indicate gnomic aspect, referring to a general truth. The example in (37) refers to all dogs and all cuscus, across all situations:

(37) *mui iro gat kamou ran ngoyai osa?*
    [mui]$_A$ i=[to   gat]$_{VC}$ [kamou]$_O$ ta-n   ngoyai osa
    dog   3sg=HAB have speech   CLF.POSS-PERT cuscus TAG
    'After all, the dog has a conflict with the cuscus.' (LL010711_0089)

When *to* follows *no*, continuative rather than habitual meaning is indicated; reference is made either to a past situation or a situation ongoing at the here-and-now of the speech event. In combination with *pe*, *to* can either have habitual meaning, as in (36) above, or continuative meaning, which yields a predicate with an iterative interpretation, as in (38) below, where the action of picking up occurs over and over again. In fact, (36) can be analysed as having iterative meaning in addition to habitual: it is likely that the story would have been told incorrectly many times over.

(38) *ipe ro akêp ip malet*
    i$_A$=[pe to   akêp]$_{VC}$ [ip malet]$_O$
    3sg=PFV CONT pick.up 3pl rock
    'He was picking up the rocks.' (LM190611_0021)

Continuative, in contrast to progressive (which is discussed in Section 6.2.1.2.3 below), is not incompatible with stative verbs, nor with active verbs representing states that can continue by inertia (such as sleeping, knowing, drifting, etc.). When a considerable amount of effort or energy is put into the action to keep it ongoing, progressive *yen* rather than *to* can be used, but *to* is not incompatible with active verbs. Thus, *to* has a more general use and is more frequent than *yen*.

Since many stative verbs are ambiguous between a stative reading 'be X' and a change-in-progress reading 'become X' (see Section 3.3.1.2), the use of the continuative with them indicates that a *change* of state, rather than a state, is ongoing. This happens for instance with *kôk* 'be hot/heat up' or *piak* 'be/become wet'. Verbs referring to a state that can only change in a more or less punctual fashion, such as *mat* 'die, be dead', are encountered in this type of construction as well:

(39) *ngaro mat tepwo-om*
 nga$_S$=[to mat]$_{VC}$ tepwo-om
 1sg=CONT die right.now
 'I am dying right now.' (elicitation 08/05/2011)

With non-stative verbs that have punctual semantics, use of *to* without any other aspectual marking yields a predicate with habitual meaning:

(40) *wut panu ro angou a la ro ningning ai*
 [wut panu]$_S$ [to angou]$_{VC}$ a [la to ning.ning]$_{VC}$ a=i
 every land HAB arrive and go.to HAB REDUP.see OBL=3sg
 'Everybody used to come here and go look at it.' (MS250311_0081)
 *'Everybody was arriving and looking at it.'

Sentence (40) cannot have a continuative or durative interpretation in the strict sense, since the verb *angou* 'arrive' is punctual: *He was arriving for several hours* is strange. It thus necessarily has a habitual interpretation. There is, however, a durative sense in the habitual, because the individuated occasions of 'arriving' take place over a longer stretch of time. This, again, shows the close connection between continuative, iterative and habitual.

6.2.1.2.2 Stative continuative *tu*
Because of its semantic and formal similarity to *to(k)*, it is hard to tell whether *tu* is in fact a different particle, with different meanings and functions, or an

allomorph of *to*. Based on the corpus, it appears that the two are different and have grammaticalised from different sources, but that they may be converging. This is supported by the fact that many speakers insist that they are "the same". Other Oceanic languages have maintained a cognate of *tu* as a posture verb, contrastive in meaning to a *to(k)* cognate. Boumaa Fijian, for instance, has *tuu* 'stand; (i.e. exist) at a place (of tall things)' and *to'a* 'sit on the heels; squat; be (i.e. exist) at a place'. As an aspectual modifier, *tuu* refers to a condition that is permanent or to something happening over an extended period, whereas *to'a* refers to something "happening continuously during *this* period of time, but not necessarily before or afterwards" (Dixon 1988: 76; italics in original). Although in Paluai *to* and *tu* are encountered only as existential verbs and aspectual particles, and no longer as posture verbs, they seem to have retained the semantic difference that relates to the permanence of the situation they refer to.

Tu is attested as a main verb, heading a predicate.[50] As a full verb, it has the more stative meaning of 'remain, be in a place/state', whereas *to(k)* has more dynamic or active overtones of 'reside, dwell; sit'. *tu* also seems to refer to longer durations of time than *to(k)*. Still, both as a particle and as a full verb, *to(k)* is also used for stative situations (as described above) and *tu* is also used for dynamic ones. The following sentences provide some examples of how *tu* is used: as a main verb in (41) and as preverbal particle in (42) and (43).

(41) *ino ru re onga ililiu la um*
 i$_S$=[no tu]$_{VC}$ te onga i$_S$=[liliu]$_{VC}$ la wumwa
 3sg=IPFV stay SUB and.so 3sg=return go.to house
 'He remained there [for a while] and then he returned home.' (WL020711_0080)

(42) *ngamaru mapwai pwên itarak ai kerin samnon*
 nga$_S$=[ma=tu mapwai]$_{VC}$ pwên i$_S$=[tarak]$_{VC}$ a=i
 1sg=NEG$_1$=STAT.CONT know NEG$_2$ 3sg=climb OBL=3sg
 kerin samnon
 year how.many
 'I don't know for how many years he climbed.' (NS220511_1_0017)

---

50 There is no counterpart *tuk* to *tu* (similar to the existence of *tok* as a counterpart to *to*). This is another indication that the two are different morphemes rather than allomorphs.

(43) *kono ru kolkol lai rea*
 [ko_S-no tu kol.kol]_{VC} la a=i te-ya
 IRR.1sg-IPFV STAT.CONT REDUP.wait go.to OBL=3sg LIG-then
 'Then, I will be waiting for that.' (MK050311_0030)

It is possible that *tu* is used more often to refer to situations in the past and the future than to situations happening in the present. However, this is a tendency at best, since example (42), for instance, refers to the here-and-now of the speech act. In all cases where *tu* may be used, it seems that *to* could be used instead. The reverse is not the case, since *tu* never has habitual meaning. In addition, *to* is overwhelmingly more frequent, with a total of 2,600 instances of *to* encountered in the corpus compared to 185 of *tu* (across all meanings and functions in both cases).

6.2.1.2.3 Progressive *yen*

Progressive *yen* has roughly the same meaning as continuative *to*, but it is not as common. *Yen* is mostly used when a considerable amount of effort or energy has to be put into the action to keep it ongoing. Examples are e.g. stirring (44) and squeezing (45). This puts *yen* in contrast with the more general continuative particle *to*, which can also be used for situations where this is not the case: knowing, sleeping, waiting, drifting, etc., of which some examples were given above. *Yen* as a secondary aspect particle shows assimilation or elision of the final /n/ in some environments (see Section 2.2.7.1), a process that is not attested for its full-verb counterpart.

(44) *wono yen yet*
 wo_A=[no yen yet]_{VC}=Ø_O
 2sg=IPFV PROG stir=3sg.ZERO
 'You keep on stirring [it].' (CA120211_2_0025)

(45) *ope yen piy me ai yapi*
 wo_A=[pe yen piy]_{VC}=Ø_O me a=i yapi
 2sg=PFV PROG squeeze=3sg.ZERO come OBL=3sg sago
 'You keep on squeezing [it] into the sago.' (CA120211_1_0036)

6.2.2 Postverbal aspectual particles

In contrast to the preverbal particles (indicating aspect or otherwise), which can occur in sequence, it seems that only one element can occur after the main verb inside the VC. In addition, as already mentioned, it is often difficult to determine the status of the element following the main verb. It could be analysed as another verb, a coverb or an adverb. This has of course implications for how the entire construction should be analysed. Typical of elements following the main verb is that they give information about the manner in which the action described by the main verb is carried out. In what follows, two elements are discussed that can occur as main verbs but also following a main verb, and that can take on aspectual meaning. There are other verb-like forms that can follow a main verb within the VC, but many of them do not occur by themselves as a main verb and are thus better analysed as adverbs. In other cases, the combination of two verbs leads to unpredictable changes in meaning; these combinations are better analysed as (semi-)lexicalised SVCs (see Chapter 9). The forms discussed here are shown in Table 60 below.

Table 60: Postverbal aspectual particles.

Element	Aspectual meaning	Lexical source
nêm	Completive	'be finished'
wot	Durative, Iterative	'go on a level, away from the DC'

6.2.2.1 Completive *nêm*

Completive aspect means that an activity is done completely and thoroughly. This implies that the activity that is described has taken up some time and, possibly, effort to complete, and therefore completive aspect cannot be used with all verbs. It is not found with stative verbs, or with punctual verbs (e.g. *angou* 'arrive'), for instance. *nêm* always follows the main verb, as in (46). In (47) and (48), *nêm* appears between the main verb and the object. If the O argument of a transitive verb is indexed by a bound pronoun, the latter cliticises to *nêm*, as in (47). *nêm* is also attested as a (stative) main verb, as in (49).

(46) *epla rok si onga ep nganngan nêm*
 eps=[la   tok.si]vc onga   eps=[ngan.ngan   nêm]vc
 1pl.EXCL=go.to sit.down and.so 1pl.EXCL=REDUP.eat be.finished
 'We went to sit down / we sat down, and we finished eating.'
 (KM060111_0073)

(47) *ingan mat nêmip*
i$_A$=[ngan mat nêm]$_{VC}$=ip$_O$
3sg=eat die be.finished=3pl
'He killed and ate them all.' [*lit.* 'he eat die finish them'] (KW290611_0057)

(48) *kala wurut nêm kanum...*
[ka$_A$-la wurut nêm]$_{VC}$ [kanum]$_O$
IRR.NS-go.to make.mounds be.finished garden
'When they finish making mounds in the garden...' (KM190211_0029)

(49) *pian te irok o, inêm*
[pian [te i$_S$=[tok]$_{VC}$]$_{RC}$ yo]$_S$ i=[nêm]$_{VC}$
good REL 3sg=stay DEM.INT 3sg=be.finished
'The goodness that was there, it is finished.' (PK290411_3_0046)

The use of *nêm* with verbs that have plural countable objects indicates that the action was done to all individuated items or persons, as in (47). Thus, *nêm* often most of all refers to a property of the object of a transitive verb, i.e. that it is completed or finished. In fact, sequences of a main verb followed by *nêm* have much in common with cause-effect/resultative SVCs in which a stative V2 describes the effect or result (to the object) of the action described by V1 (these constructions are discussed in Section 9.1.1). This is another indication that it is hard to draw the line between different types of verbal sequences.

6.2.2.2 Durative *wot*

The directional *wot* 'go on the same level, away from DC' is frequently used in postverbal position to indicate that an activity is continuing for a stretch of time. It is comparable to (and may be a calque of) Tok Pisin constructions such as *Em i wokabaut i go i go i go* 'He's walking and walking and walking'. *Wot* has grammaticalised to such an extent that it can also be used with activities that do not involve motion. A few examples are given below.

(50) *kino yen pe wot a poyen teo kipe som*
[ki$_A$-no yen pe]$_{VC}$=Ø$_O$ wot a [poye-n teyo]$_S$ [ki-pe
IRR.3SG-IPFV PROG do=3sg.ZERO go.level and residue-PERT DEM.INT
som]$_{VC}$
IRR.3SG-PFVform
'It will keep on doing [that] for a while and its residue will form.'
(CA120211_2_0029)

(51) *uno rok wot*
 u$_S$=[no   tok   wot]$_{VC}$
 3du=IPFV sit   go.level
 'They lived (together) for an extended time.' (LL10711_0011)

Wot seems to differ from other postverbal elements in one respect, though: it occurs following an overt O argument, as in (52) below.

(52) *uno ro pe mangat wot wot wot*
 u$_A$=[no   to   pe]$_{VC}$ [mangat]$_O$ wot   wot   wot
 3du=IPFV CONT do   work   go.level go.level go.level
 'They were working for a long time.' (LM190611_0005)

There are no examples in which *wot* precedes a bound object pronoun. It may therefore, in fact, not be part of the VC. Instead, constructions involving durative *wot* may be more similar to switch-subject SVCs with a directional as V2 (which are discussed in Section 9.1.5). It is likely that *wot*, rather than *la* (also indicating 'motion away from DC', see below), has grammaticalised in this way because *wot* is atelic, whereas *la* is telic and necessitates an endpoint to the action. The semantics of the verb *wot* was more in accordance with the durative meaning to begin with and, thus, it was this verb that grammaticalised along this path.

6.3 Direction and associated motion

Directionals, in particular the most common forms *la* 'go (away from DC), thither' and *me* 'come (towards DC), hither', are also found preverbally. They follow core aspect particles, but precede secondary aspect ones. The directional paradigm and the semantic distinctions made within it are discussed in Section 3.3.1.3. While preverbal instances of directionals have the same forms as their full verb counterparts, the constructions they appear in are not analysed as SVCs. Rather, the directionals are analysed as grammaticalised preverbal particles on a par with, for instance, *to* 'be' for habitual or continuative aspect (see Section 6.2.1.2.1 above). This is due to their distribution: they only occur in a slot that occupies a place between two slots reserved for particles. Preverbal directionals also cannot receive stress, in contrast to their full verbal counterparts.

When preverbal directionals occur with lexical (intransitive) motion verbs, they specify a path for the translational motion encoded by the main verb. With non-motion verbs, preverbal directionals indicate deictically anchored motion prior to the event encoded by the main verb ('go/come and do X'); in this functional slot, directionals can be considered markers of associated motion (AM) (Schokkin forthcoming). Only a prior motion sense is attested in these cases. With transfer verbs, a directional in the preverbal slot has a different function from one that occurs postverbally in a SVC: rather than indicating the motion that the O argument undergoes as a result of the transfer, it indicates motion associated with the A argument (see Section 9.1.5 for more on SVCs including a directional). Preverbal *la* with a simultaneity interpretation has lost all motion semantics, but rather indicates that an event takes place at a distance removed from the DC.

The following sections contain examples of the occurrence of the various directionals in the preverbal slot. For a detailed overview of the semantics of Paluai directionals and their grammaticalisation paths, see also Schokkin (2013, forthcoming).

6.3.1 *la* 'go to'

Example (53) indicates a sequence of events, more specifically a movement away from the DC followed by an action. Cases of preverbal *la* where there is simultaneity with the main verb have acquired another meaning, namely that the action referred to by the main verb is taking place at a distance removed from the DC (usually the speaker). Cognates of *la* with this function are found in other Oceanic languages (Lichtenberk 1991); Paluai examples are given in (54) and (55). These cases usually show a secondary aspectual particle following the directional, indicating that the event is in progress. There is no sense of motion in these clauses, in contrast to AM uses of *la* such as in (53). When no core aspectual and/or irrealis markers are present (as is usually the case), the event is understood as in progress at the here-and-now of the speech event.

(53) *ngapwa kola ning Ponaun*
    nga_A=[pwa]_VC [[ko_A-la   ning]_VC [Ponaun]_O]_Compl:Pot
    1sg=want.to   IRR.1sg-go.to see   P.
    'I want to go and see Ponaun.' (ANK020995_0008)

(54) *ila ro nu*
 i_S=[la to nu]_{VC}
 3sg=go.to CONT bathe
 'He is off bathing [right now, at a place removed from the DC].' (052b_0211)

(55) *urêro pe yong yamat te ila ro pe kolon wat*
 wurê_A=[to pe yong]_{VC} [yamat [te i_A=[la to pe]_{VC}
 1pc.EXCL=CONT PFV hear person REL 3sg=go.to CONT make
 [kolo-n]_O wat]_{RC}]_O
 voice-PERT up.high
 'We were hearing a person who was shouting [*lit.* 'making voice'] up high.' (NP210511_2_0024)

6.3.2 *me* 'come'

The following example shows a use of *me* as a marker of AM, indicating prior motion towards the DC. While the main verb is a transfer verb, preverbal *me* is not related in any way to the motion of the O inherent in the transfer, which is expressed by *la* following the main verb. *Me*, in this case, only entails the motion associated with the A of the main verb, prior to the action entailed by that verb.

(56) *ippe me rou liliu la kep*
 ip_A=[pe me tou liliu]_{VC}=Ø_O la ka-ep
 3pl=PFV come give return=3sg.ZERO go.to CLF.food-1pl.EXCL.PERT
 'They came and gave [the leftover food] back to us.' (KM060111_0087)

6.3.3 *wot* 'go horizontally, away from DC'

Preverbal *wot* means 'go ahead (and)', which is again symptomatic of a sequence of events. When *wot* follows the main verb, the predicate acquires a different, aspectual meaning: 'do for an extended period of time' (see Section 6.2.2.2 above).

(57) *eppe wot lêpi ran mwanen Parugui*
 ep_A=[pe wot lêp]_{VC}=i_O ta-n mwane-n Parugui
 1pl.EXCL=PFV go.level take=3sg CLF.POSS-PERT brother-PERT P.
 'We went ahead and took her of/from her brother Parugui.' (KM060111_0035)

6.3.4 *sot* 'go upwards'

Sentence (58) could have two interpretations: either a sequential AM use of a directional verb, or an instance where there is simultaneous motion and action, i.e. 'going up' and 'hunting' happening at the same time.

(58) *eppwa kasot yik ngoyai*
    ep$_A$=[pwa]$_{VC}$ [[ka$_A$-sot   yik]$_{VC}$ [ngoyai]$_O$]$_{Compl:Pot}$
    1pl.EXCL=want.to IRR.NS-go.up search.for cuscus
    'We were about to go uphill and hunt for / hunting for cuscus.'
    (NP210511_2_0011)

In sentence (59) below, there is no such ambiguity: the actions described by *sot* and the main verb *pe* happen simultaneously. The speaker talks about the weaving technique for a particular kind of basket, which is woven starting at the bottom and then working upwards, weaving the opening (called the 'mouth') last. Example (59) describes this process. Thus, even though there is strictly speaking no translational motion entailed in this event, the speaker prefers to use a preverbal directional in a metaphorical sense.

(59) *kosot pe kolon*
    [ko$_A$-sot   pe]$_{VC}$ [kolo-n]$_O$
    IRR.1sg-go.up make mouth-PERT
    'I will make the mouth [of the basket].' [*lit.* 'I will go up-make its mouth.'] (MK050311_0021)

6.3.5 *sa* 'come upwards'

In sentence (60), we see again a directional verb used as a marker of sequential AM.

(60) *wuipe sa lêpi*
    wui$_A$=[pe   sa   lêp]$_{VC}$=i$_O$
    1du.EXCL=PFV come.up take=3sg
    'We came up and got her.' (MK060211_0044)

6.3.6 *suwot* 'go downwards'

Sentence (61) describes the making of another kind of basket. In this case, the weaver starts with the rim and works his or her way down towards the bottom. As a result, (61) is the exact opposite of sentence (59) above. Sentence (62) illustrates a sequence of events with *suwot*.

(61) *kosuwot tik nupun*
[ko$_A$-suwot   tik]$_{VC}$ [nupu-n]$_O$
IRR.1sg-go.down weave bottom-PERT
'I will weave its bottom.' [*lit.* 'I will go down-weave its bottom.']
(AK160411_2_0023)

(62) *osuwot ilili la pulen kone areo*
wo$_S$=[suwot   ilili]$_{VC}$ la   pulen.kone   a=te-yo
2sg=go.down stand.up go.to beach   OBL=LIG-DEM.INT
'You go down and stand up on the beach there.' (LL030611_0051)

6.3.7 *si* 'come downwards'

Sentence (63) shows another purposive sequence. Sentence (64), on the other hand, indicates simultaneous actions of walking, coming down (towards the shore) and carrying the bed.

(63) *kisi wut kem a...*
[ki$_A$-si   wut]$_{VC}$ [kem]$_O$ ya
IRR.3sg-come.down fetch   salt.water then
'When she will come down to the shore to fetch sea water, then...'
(LK100411_0090)

(64) *ip silal leo ipno si wau lêp pat teo*
[ip silal te-yo]   ip=[no   si   wau   lêp]$_{VC}$
3pl spirit LIG-DEM.INT 3pl=IPFV come.down move take
[pat te-yo]$_O$
bed LIG-DEM.INT
'The spirits, they were walking down to the shore carrying the bed.'
(LM190611_0026)

6.3.8 *wen* 'move horizontally'

The three directionals indicating motion that is not deictically anchored (*sen*, *wen* and *suwen*) are relatively rare; few representative instances were found in the data of their preverbal use (none at all in the case of *sen* 'move upwards'). *Wen* indicates motion along the shore that is not deictically anchored. In sentence (65), there is a simultaneous action of moving and getting closer together, which is hard to express in English. In addition, the second instance of *wen* in this example refers to a location rather than a movement.

(65) *apkawen pir ai pulen kone re iwen telo*
ap_S=[ka-wen   pit]_{VC}   a=i   pulen.kone   [te
2pl=IRR.NS-move.level   be.close   OBL=3sg   beach   REL
i_S=[wen]_{VC}]_{RC}   te-lo
3SG=move.level   LIG-DEM.DIST
'You will gather together on the sand beach that is over there.'
(NP210511_2_0009)

6.3.9 *suwen* 'move downwards'

In (66), *suwen* is used in an imperative construction (taken from a direct quotation in a narrative). Interestingly, it is the non-deictic directional that is used in this case. When someone is told to go away, this usually implies motion away from the DC, i.e. from the speaker who utters the imperative. It is possible that, because the imperative already implies this motion from DC, it is not necessary to use a deictic directional.

(66) *ipwa, "suwen tet la paye!"*
i_A=[pwa]_{VC}   [[suwen   tet]_{VC}   la   paye]_{Compl:Quot}
3sg=say   move.down   move   go.to   down.below
'She said, "Go down!"' [i.e. get out of the house] (WL020711_0062)

6.4 Reality status

In Paluai grammar, a fundamental distinction is made between realis and irrealis clauses. Realis is formally unmarked and typically refers to actions that have been taking place or are taking place right now. Irrealis is formally marked by the prefix *kV-* (except for second person singular, which receives zero marking);

the prefix attaches to the first verb or particle in the VC. The marker for reality status is analysed as a form bound to the verb, rather than as a particle, because it is much further integrated into the grammar than the aspectual particles. Firstly, reality status is obligatorily expressed on the verb complex when the context demands this, whereas aspect appears to be always optional. In a sequence of verb complexes in coordinated clauses that refer to the same event, aspectual particles are usually left out after the first iteration, whereas the irrealis prefix occurs on each verb complex. Secondly, the *kV-* prefixes are portmanteau forms, giving information both about reality status and person/number of the S/A argument. Particles, on the other hand, are typically uninflected forms.

Irrealis typically refers to actions that have not yet taken place, or that could have taken place but didn't. Mithun (1999: 173) defines reality status as follows: "The realis portrays situations as actualised, as having occurred or actually occurring, knowable through direct perception. The irrealis portrays situations as purely within the realm of thought, knowable only to imagination."

However, in Paluai and related languages, it seems more likely that irrealis not only refers to the non-actualised but, more generally, to situations that are somehow removed from the here-and-now of the speech situation, and that are non-specific in terms of their temporality or reference to core participants. This may be the reason behind the fact that the irrealis is commonly used to refer to past habituals. For more detailed discussion, see Section 6.4.2.3.

The Paluai *kV-* prefix makes no distinction between different kinds of modality, although modal overtones can be added by combining the irrealis marker with one or more aspectual/modal particles or, alternatively, by modal adverbs. These issues will be discussed in Section 6.5.

6.4.1 Forms of the irrealis marker

The irrealis is expressed by the forms in Table 61.

Table 61: Forms of irrealis prefixes.

Form	Used with
ko-	first person singular
∅-	second person singular
ki-	third person singular
ka-	all non-singular persons (first, second and third person dual, paucal and plural)

The second person has no formal marking for irrealis in the singular, although it has in the other numbers. In addition, there is syncretism in person and number for all the non-singular forms; in other words, there is no distinction between dual, paucal and plural, as in the pronominal system.

The irrealis marker is the only verbal category (in addition to the bound pronouns) that shows a person/number distinction. This gives rise to the notion that it may consist of a *k*- formative plus a person/number morpheme (maybe an earlier version of the bound subject pronoun), both of which fused together. Synchronically, however, the markers are not further analysable.

6.4.2 Functions of the irrealis

The irrealis can have the following range of functions:
 - Marking an unrealised event or state of affairs (Section 6.4.2.1);
 - Marking immediate future (Section 6.4.2.2);
 - Marking past habitual (Section 6.4.2.3);
 - Marking a dependent clause (Section 6.4.2.4).

6.4.2.1 Unrealised event/state of affairs

In its cross-linguistically most prototypical function, the irrealis refers to an unrealised state of affairs. Since Paluai does not have tense, the irrealis is used to refer to future events, as illustrated in the following examples.

(67) *kipat sap kain te kino pat ai*
　　 [ki$_A$-pat]$_{VC}$　[sap kain　[te　[ki-no　　pat]$_{VC}$=Ø$_O$　　a=i]$_{RC}$]$_O$
　　 IRR.3sg-plant　any　kind　REL　IRR.3sg-IPFV　plant=3sg.ZERO　OBL=3sg
　　 'He will plant any kind (of thing) that he could plant in it.'
　　 (KM190211_0032)

(68) *ip lau karo wau kasông*
　　 [ip　laue]$_S$　[ka-to　　wau]$_{VC}$　kaso-ng
　　 3pl　people　IRR.NS-HAB　move　　near-1sg.PERT
　　 'The people will be walking near to me.' (LK250111_0075)

(69) *urêkabe lêpo la panu liliu*
　　 wurê$_A$=[ka-pe　　lêp]$_{VC}$=o$_O$　la　　panua　liliu
　　 1pc.EXCL=IRR.NS-PFV　take=2sg　go.to　home　again
　　 'We will take you back home again.' (NP210511_2_0058)

6.4.2.2 Immediate future

Marking immediate future is a subtype of the "marking an unrealised event" function. In Loniu, a cognate marker *k-* is analysed as indicating "potential aspect" (Hamel 1994: 111–112). It is possible that this is its most basic meaning in Paluai as well. In daily conversation, *kV-* is frequently used to refer to the immediate future. A representative example of a direct quotation from a narrative is given below.

(70) wono rok a ong kola panu menengan
 wo$_S$=[no tok]$_{VC}$ a wong$_S$ [ko-la]$_{VC}$ panu menengan
 2sg=IPFV stay and 1sg.FREE IRR.1sg-go.to land big
 'You stay, and I will go to Manus mainland.' (KW290611_0039)

When asked, consultants indicated that the use of a *kV-* prefix means that the person producing the utterance has just made up his or her mind and is about to carry out the action described by the verb, i.e. is on the verge of carrying out his or her intention (indeed, at the moment of speaking, the proposition is still only within the realm of thought). The desiderative particle *pwa* can be used to emphasise this meaning (see Section 6.5.1). Related to this is the use of the irrealis marker with inclusive pronouns for hortative meanings:

(71) ngapwa, "kay, tau kaaluk"
 nga$_A$=[pwa]$_{VC}$ [kay tau$_S$ [ka-aluk]$_{VC}$]$_{Compl:Quot}$
 1sg=say okay 1du.INCL IRR.NS-paddle
 'I said, "Okay, let's paddle."' (MK060211_0039)

However, the use of *kV-* suggests that its meaning is broader than this. It is therefore analysed as a marker of irrealis, rather than as a mere reference to the immediate future.

6.4.2.3 Past habitual

A less prototypical use of the irrealis is its reference to past habitual events. In Paluai, the irrealis is very often encountered in past habitual contexts, a feature it appears to share with other Admiralties languages (Cleary-Kemp 2015: 58). The use of the non-singular irrealis prefix *ka-*, either without a bound subject pronoun or with the third person plural form *ip*, as in (74), is typical for impersonal, generic clauses for which it is not very relevant who exactly the agent is. This construction is quite often used in instructional texts, in which hypothetical agents carry out certain actions. Some examples from this text genre are given below.

(72) *nirasip pun kala alilêt*
 panurasip pun [ka_S-la]_{VC} alilêt
 first INTF IRR.NS-go.to forest
 'First of all, one would go to the bush.' [text about how to prepare coconut oil] (CA120211_2_0004)

(73) *kayei lai sip purukei a kirok*
 [ka_A-yei]_{VC}=Ø_O la a=i sip purukei a
 IRR.NS-scrape=3sg.ZERO go.to OBL=3sg one.INANIM bowl and
 [ki_S-tok]_{VC}
 IRR.3sg-stay
 'One would grate [it] into a bowl and it would stay there.' [text about how to fry sago] (CA120211_1_0019)

(74) *ipkalêp tut ip sapon kei re nangin o*
 ip_A=[ka-lêp tut]_{VC} [ip sapo-n kei [te Ø
 3pl=IRR.NS-take completely 3pl top-PERT herbs REL 3sg.ZERO
 nangin]_{RC} yo]_O
 scent DEM.INT
 'They would take all the herbs that have a nice scent.' [text about how to plant yams] (NP220611_1_0011)

This use of the irrealis is found not only in instructions, but also in other narratives. Two examples are given below; the first one describes the customs of Paluai people long ago, the second details the behaviour of a father and mother towards their daughter.

(75) *ipkaliliu la lopwan ip panuan te pulêng teo kipe masai*
 ip_S=[ka-liliu la]_{VC} lopwa-n ip panuan te [pulêng
 3pl=IRR.NS-return go.to place-PERT 3pl before SUB dawn
 te-yo]_S [ki-pe masai]_{VC}
 LIG-DEM.INT IRR.3sg-PFV be.clear
 'They would go back to their houses before the dawn would break.' (LK250111_0027)

(76) *taman a tinan teo, ukape pul la rai la remenin telo*
　　　[tama-n　　a　　tina-n　　te-yo]$_A$　　u=[ka-pe　　pul]$_{VC}$=Ø
　　　father-PERT and mother-PERT LIG-DEM.INT 3du=IRR.NS-PFV tell=3sg.ZERO
　　　la　　ta-i　　　　la　　temenin te-lo
　　　go.to CLF.POSS-3sg go.to like　　LIG-DEM.INT
　　　'Her father and mother, they would speak to her as follows.'
　　　(LK100411_0013)

The use of the irrealis for past habitual events makes sense if the irrealis is analysed as an expression of temporal non-specificity. This was done, for instance, by Baker and Travis (1998) for Mohawk. Past habituals refer to a set of iterated events, but no single event is referred to directly. This non-specificity or atemporality is expressed by using the irrealis. In addition, in Paluai procedural texts, underspecification of the S/A argument seems to be a further characteristic of the past habitual use of the irrealis.

6.4.2.4 Marker of a dependent clause

Irrealis (without any additional aspectual marking) is often encountered in dependent clauses, marking either the protasis of a conditional construction or a Potential complement clause. The current section focuses exclusively on the use of the irrealis in main clauses. A more detailed discussion of its use in conditional clauses can be found in Section 11.1.3.5, and the use of irrealis in complement clauses with the desiderative modal *pwa* 'want to' is discussed in Section 6.5.1.

6.4.3 Dependencies between reality status and other verbal categories

If the irrealis prefix is present, a subject bound pronoun is in principle optional; see for instance examples (73) above and (77) below. This is undoubtedly due to the fact that the irrealis marker contains person and number information. A bound pronoun is still needed, however, to specify for dual, paucal or plural within the non-singular number category, since the irrealis prefix does not make this distinction. See example (76) for use of a dual pronoun, and (78) for the use of a paucal one.

(77) *koyektou lai sin*
　　　[ko$_A$-yek.tou]$_{VC}$=Ø$_O$　　la　　a=i　　sin
　　　IRR.1sg-put=3pl.ZERO go.to OBL=3sg sun
　　　'I will put [them] into the sun.' (AK160411_2_0012)

(78) *urêkabe lêpo la panu liliu*
 wurê_A=[ka-pe   lêp]_{VC}=o_O la   panua liliu
 1pc.EXCL=IRR.NS-PFV take=2sg go.to home again
 'We will take you back home again.' (NP210511_2_0058)

The irrealis marker cannot be used in combination with the perfect, discussed in Section 6.2.1.1.3 above. The perfect is by default incompatible with the irrealis, because its use indicates that a certain action has already happened and has relevance for the here-and-now of the speech situation. The resulting predicate can therefore never have irreality status. In addition, the perfect generally singles out a specific relevant event, which is not compatible with the additional sense of non-specificity characteristic of the irrealis.

An irrealis clause cannot be negated in the usual way with the preverbal clitic *ma=* and the postverbal negator *pwên*. Negated irrealis clauses have *kV-* attached to the modal particle *sa*, with the clause followed by *pwên*. "Regular" negation is discussed in Section 10.2.2, whereas negation of *sa* clauses is discussed in Section 6.5.2.2 below.

The meaning difference between a main clause that has only irrealis marking, on the one hand, and a main clause that is marked for irrealis and aspect, on the other, is not always clear-cut.[51] Irrealis-only clauses, as already mentioned above, often refer to the immediate future, whereas clauses additionally marked as perfective or imperfective can seemingly also refer to a more distant future. What is more important to note, however, is that the irrealis is atemporal, whereas aspect is not fully so. In principle, a reality status distinction does not make a statement about the temporality of the proposition. As explained above, it revolves around a distinction between a real, actualised world and a hypothetical, non-actualised world (or, in other words, a world of specific events and one of non-specificity). Temporality then becomes irrelevant. However, because in daily reality people cannot place themselves outside of time, an irrealis situation has in many pragmatic circumstances to be interpreted as unrealised with relevance to a moment in relation to the here-and-now of the communicative situation. Thus, temporality once again becomes relevant on a pragmatic plane. When there are aspectual particles present, they can give more information about the temporal boundedness or unboundedness of the event; see Section 6.5 below for discussion.

As mentioned before, irrealis is not formally expressed in the second person singular. This leads to the remarkable situation that the perfective particle *pe*

---

51 Dependent clauses show irrealis marking, but never any aspect marking, when they form the protasis of a conditional construction. These matters will be discussed in full in Section 11.1.3.5.

(discussed in Section 6.2.1.1.2 above), which for all other persons can only refer to a past event when it is not combined with an irrealis prefix, can only refer to a future event when it is used for second person singular (necessarily without a prefix). For all other persons, *pe* will be combined with an irrealis prefix when referring to a future event. The same is true for other aspectual particles, although to a lesser degree, since the imperfective particle *no*, for instance, can also refer to an event happening at the here-and-now of the speech event. It seems that the perfect is used instead of the perfective to refer unambiguously to the past with second person singular predicates, as illustrated in example (79).

(79) *on asuek pau rao?*
    wo$_{iA}$=[an asuek]$_{VC}$ [pau]$_O$ [ta-o$_i$]$_E$
    2sg=PRF   rub   coconut.oil   CLF.POSS-2sg.PERT
    'Have you rubbed coconut oil onto yourself?' (field notes 21/03/2011)

Because an irrealis reading is potentially present in all predicates with a second person singular subject, except those marked for perfect or perfective, these predicates are often ambiguous. Disambiguation between an irrealis or non-irrealis reading is only possible based on the discourse context. The following example, from an instructional text about frying sago, probably shows an instance of the irrealis, because it describes a hypothetical situation. In addition, the complement clause of *ning*, which has a third person singular subject, is marked as irrealis.

(80) *wono ro sui lak a ope ning de kipe la mwat*
    wo$_A$=[no to   sui]$_{VC}$=Ø$_O$ lak a   wo$_A$=[pe ning]$_{VC}$ [te
    2sg=IPFV CONT fry=3sg.ZERO go and 2sg=PFV see   COMP
    [ki$_S$-pe   la   mwat]$_{VC}$$_{Compl:O}$
    IRR.3SG-PFV go.to be.cooked
    'You will be frying [it] and you will see that it will become cooked.'
    (CA120211_1_0027)

The use of the second person singular, as in the example above, may be a relatively new phenomenon due to influence of Tok Pisin. A more "traditional" way of expression would be with an impersonal non-singular irrealis clause, as in example (73) above.

    The example below, on the other hand, has a realis interpretation and refers to an event that is currently happening. It may be the case that the continuative particle *to* has been added to reinforce this interpretation.

(81) *woro ning naêmwan le pwên?*
wo_A=[to   ning]_VC [naêmwa-n]_O le   pwên
2sg=CONT see   back-PERT   or   NEG
'Can you see [*lit.* 'are you seeing'] his back or not?'
(Game2_280812_0052)

6.5 Modality

Modality deals with alternatives, with different possible worlds. The alternatives "are sorted out and evaluated by some sort of authority, often the speaker, or if not the speaker, some other participant or even another situation" (Timberlake 2007: 315). Because modality deals with possible, unrealised worlds and not with the actual world, it is tied in with reality status, discussed in Section 6.4 above. Whereas for the real world there is just one alternative (the state of affairs as it is now), for possible worlds there are a great number of alternatives, and much can be said about their likelihood or probability, their necessity or inevitability, their desirability or expediency, and so forth (cf. Elliot 2000). The authority over these possible worlds can lie with the speaker, for instance when uttering a command or a wish. It can also lie with another participant who, for instance, may or may not be able or willing to carry out a certain action. Finally, it can lie with another situation, for instance in the case of a conditional, when one situation is a prerequisite for another situation to occur. Timberlake (2007: 316) distinguishes three realms of modality: epistemology, obligation, and contingency. Epistemology "has to do with knowledge about events and the world". Uncertainty of knowledge (whether a possible world is or is not true) can lead to a sentence being marked as irrealis. The second realm of modality is described as "directive" or "jussive", meaning that "the responsibility for the state of the world is transferred from one authority to another" (Timberlake 2007: 316). Obligation is one part of this, but also desiderative, directive, purposive, permission, ability, etc. The authority can rest with an individual, but it can also be impersonal and generalised. In the third realm of modality, causation and contingency, "[r]esponsibility for one situation in the world is assigned to another situation" (Timberlake 2007: 321). Since causal/contingent and conditional constructions are usually multiclausal, a fuller discussion of them can be found in Chapter 11.

By itself, irrealis in Paluai is neutral with respect to all of the above. When a predicate is marked as irrealis, this just marks it as non-actualised and/or non-specific. Thus, the assertion from Timberlake (2007) above about uncertainty of knowledge that leads to a sentence being marked as irrealis is only partially true. Irrealis, in Paluai as well, can indeed be used to mark information as less than certain, but there are no meaning shades within it as is for instance the case with English modals, where the use of *may* versus *might* indicates a difference in degree of certainty, or *must* in its epistemic sense refers to knowledge that has been inferred. Whether or not a modal meaning is also intended when a Paluai irrealis form is used, and which meaning is intended, depends on the context and sometimes in part on the aspectual particles used. Particular combinations of the irrealis prefix with aspectual particles may have modal overtones. In addition, there are several modal and/or epistemic adverbs that can modify a clause; these are discussed in Section 3.6.1.5. Thus, although reality status cannot be separated entirely from modality, in the language under discussion they do form clearly distinctive realms.

In addition, there are two preverbal particles expressing modality. *pwa*, which takes an irrealis complement clause, expresses desire, intention and imminent action; *sa*, which occupies the same functional slot as the core aspect particles, does double duty as a marker of an apprehensive stance (with positive polarity) or ability/permission (only with negative polarity).

6.5.1 Desiderative/intentional *pwa*

In addition to its use as a full lexical verb 'say, think', *pwa* is used as a particle expressing desire, intention and imminent action. The possible world in which the action is carried out is regarded as favourable by the speaker, and thus *pwa* can be regarded as an attitudinal modal operator (Timberlake 2007: 329) belonging to the realm of jussive modality. When used as a modal marker, *pwa* occurs in a realis predicate and takes a Potential complement clause (there is no dependent clause marker *te* to mark the complement clause as with other verbs; cf. Section 11.1.2). The S/A argument of the complement clause is almost always coreferential with the A argument of *pwa* (but see below for a counterexample), and the complement clause is always irrealis. It does not show any core aspectual marking and can be regarded as a pure unrealised state of affairs. Some examples are given below.

Importantly, for *pwa* to have desiderative meaning, its complement clause has to be marked as irrealis, also when the action it describes took place in the past, as in (85). When no irrealis marking is present in a complement clause of *pwa*, the sentence can only be understood literally, with *pwa* meaning 'say' or 'think' and the complement clause representing a direct quotation. However, if in sentence (82) there is no irrealis prefix, the reason for this is that the subject is second person singular, which has no overt marking for irrealis (see Section 6.4 above). Instead, the bound subject pronoun has to be repeated. In fast speech, the second instance of the 2sg pronoun often fuses with *pwa*, rendering a surface form [pɔ] ~ [βɔ].

(82) *opwa ola pe sa?*
wo$_i$=pwa   [wo$_i$=la   pe   sa]$_{Compl:Pot}$
2sg=want.to 2sg=go.to do what
'What do you intend to do?' / 'What are you going to do?' [*lit.* 'You say you go do what?'] (052b_0296)

(83) *eppwa karet*
ep$_i$=pwa   [ka$_i$-tet]$_{Compl:Pot}$
1pl.EXCL=want.to IRR.NS-move
'We want to go.' / 'We are going to / about to go.'
(052b_0091)

(84) *ngapwa kope sê nayei*
nga$_i$=pwa   [ko$_i$-pe   sê   nayei]$_{Compl:Pot}$
1sg=want.to IRR.1sg-do DIM lie
'I am going to / about to tell a little lie.' (PK290411_1_0038)

(85) *ipwa kiyokat*
i$_i$=pwa   [ki$_i$-yokat=Ø]$_{Compl:Pot}$
3sg=want.to IRR.3sg-carry=3sg.ZERO
'He wanted to / was going to carry [it].' (KW290611_0042)

Speakers indicate that *pwa* refers to a (mental) decision, just made, that the action is going to be carried out. This can possibly explain its metaphorical extension as a marker of imminent action: when a decision has been made to carry out an action, often this will happen in the immediate future. *pwa* can usually be translated with 'be about to' as well as with 'want/intend to'; in the former sense, it can also refer to subjects that have no agency or volition. The examples given above all have subjects with human referents and thus are ambiguous between a desiderative/intentional and an 'imminent action' meaning. When the subject of a *pwa* construction has inanimate reference, however, it can only have the latter meaning, as in (86):

(86) *maran teo ipwa kilot*
mata-n   te-yo   i$_i$=pwa   [ki$_i$-lot]$_{COMP}$
eye-PERT LIG-DEM.INT 3sg=want.to IRR.3sg-fall
'The lid [*lit.* 'eye'], it is about to fall off.' (Game1_280812_0067)

Sometimes, there is no coreference between the subject of the main clause and the subject of the *pwa* Potential complement clause. In (87) below, the S argument of the complement clause is not coreferential with the A argument of *pwa*. However, since *pwa* can also mean 'say', this construction could also be analysed as an example of indirect quotation: 'Kileai$_i$ said that he$_j$ should become his$_i$ son.' There is no grammatical means of distinguishing between the two interpretations. It is not an example of direct quotation, in which case *naru-* 'son' would carry first person pertensive marking.

(87) *Kileai ipwa ikila narun*
Kileai$_i$ i=pwa   [i$_j$=ki-la   naru-n$_i$]$_{\text{Compl:Pot}}$
K.   3sg=want.to 3sg=IRR.3sg-go.to son-PERT
'Kileai wanted him to become his son.' / 'Kileai wished him to become his son.' (OL20011_0045)

6.5.2 The particle *sa*

Sa is used as a modal particle referring either to apprehensive stance, with positive polarity, or to ability and permission, with negative polarity. Lichtenberk (1995) discusses a similar and maybe cognate particle *ada* in the Oceanic language To'aba'ita, which he analyses as the grammaticalisation of a lexical verb meaning 'see, look (at)'. The Paluai verb *tai* 'observe, watch (upwards)' may be the lexical source for *sa* but, apart from the similar development in a number of other Oceanic languages, all spoken on the Solomon Islands and closely related to To'aba'ita (Lichtenberk 1995: 303), there is no evidence supporting this hypothesis. First, the use of *sa* as a marker of an apprehensive stance will be discussed, followed by its use to indicate (absence of) ability and permission.

6.5.2.1 Apprehensive *sa*
Apprehensive *sa* gives information about the stance of a speaker towards a possible state of affairs. This possible world has not yet come about but it might, and this would be unfavourable to the speaker. Thus, *sa* (in this sense) can be regarded as an attitudinal modal operator. *Sa* can be used in a main clause, as a warning:

(88) *osa lot!*
wo=[sa   lot]$_{\text{VC}}$
2sg=MOD fall
'(Look out,) you may fall!'

Alternatively, it can be used in a dependent clause. In this sense, it is similar in meaning to English *lest*. The matrix clause in these cases can be marked as either realis or irrealis, and the dependent clause is marked as realis. These constructions are basically conditional; they express the idea that 'if/when X (the protasis or condition) is not met, then Y (the apodosis or consequence) will very likely come to pass, and this will be bad'. More about dependency relations between clauses can be found in Chapter 11. These dependent clause constructions with *sa* in a realis predicate are very strong assertions: they are almost certain to come true if the condition is not met. In (89), the condition is 'I cannot tell him', where 'him' refers to the speaker's brother. If this condition is not met, i.e. if the speaker *were* to tell him, this would lead to the undesired situation that the brother would beat his wife.

(89) *ngasa pul la rai pwên, te isa yeki*
 nga$_S$=[sa pul]$_{VC}$=Ø$_O$ la ta-i p wên te i$_A$=[sa
 1sg=MOD tell=3sg.ZERO go.to CLF.POSS-3sg NEG SUB 3sg=MOD
 yek]$_{VC}$=i$_O$
 hit=3sg
 'I cannot tell him, lest he beat her.' (WL020711_0056)

In (90), the condition is 'I cannot go near the fish'. If this condition is not met, i.e. when the speaker *were* to go near the fish, this would lead to the undesirable situation that the fish would rot.

(90) *kapwa karapot nik, napunan kowau la kason nik, te isa poyak*
 kapwa [ka$_A$-tapot]$_{VC}$ [nik]$_O$ napunan [ko$_S$-wau la]$_{VC}$ kaso-n
 if   IRR.NS-smoke fish NEG.IMP IRR.1sg-move go.to near-PERT
 nik te i$_S$=[sa   poyak]$_{VC}$
 fish SUB 3sg=MOD be.rotten
 'If they would smoke fish, it would be forbidden for me to go near the fish, lest it rots.' (LL030611_0026)

Sa may also be used in a coordinated construction of two main clauses. These can then be marked either for realis or irrealis. Such a coordinated construction can also be regarded as a conditional/contingent modal construction, but with the opposite effect to the one described above: 'if X is met, then Y may happen (and this is not good)'. In addition, the contingency relation between the two clauses appears to be less strong: if X comes to pass, then Y may happen but it is not as nearly certain as in the abovementioned dependent clause constructions.

In (91), the 'reason' X is 'I turn into something out in the open'.[52] If this happens, it is possible that it leads to consequence Y 'They find me', and this is undesirable to the speaker.

(91) *ngasa ro pei rare masayen, a ukasa kam kel a ukasa pwak tang*
nga$_S$=[sa to pei]$_{VC}$ [ta=te-yo masayen]$_{CC}$ ya
1sg=MOD CONT appear SPEC.COLL=LIG-DEM.INT outside then
u$_A$=[ka-sa kam]$_{VC}$ [kele]$_O$ a u$_S$=[ka-sa pwak]$_{VC}$
3du=IRR.NS-MOD catch canoe and 3du=IRR.NS-MOD meet
ta-ng
CLF.POSS-1sg.PERT
'I may turn into something out in the open. Then they may catch a canoe and find me.' (WL020711_0087)

In (92), the 'cause' X is 'They go there'. If this happens, it is possible that it leads to consequence Y 'Spirits take them, and people kill them', which is undesirable.

(92) *ipkasa lak, silalen sa lêpip, le ip yamat kasa la pemarip*
ip$_S$=[ka-sa lak]$_{VC}$ [silale-n]$_A$ [sa lêp]$_{VC}$=ip$_O$ le [ip yamat]$_A$
3pl=IRR.NS-MOD go spirit-PERT MOD take=3pl or 3pl person
[ka-sa la pe-mat]$_{VC}$=ip$_O$
IRR.NS-MOD go.to CAUS-die=3pl
'If they were to go there, bush spirits may take them, or people might kill them.' (MS250311_0046)

Sa has affinities with the irrealis, since predicates with *sa* are also less than fully actualised. They refer to a situation that is possible, but not yet realised. The situation they refer to still belongs to the realm of thought, and the speaker is apprehensive that it might come true. Because of this "subjective" stance, *sa* is categorised as a modal particle and not as a marker of the irrealis, but even when it is used in realis predicates as shown above (that is, predicates that are not marked with *kV-*), it still refers to a possible world that may, but has not yet, come about. Therefore, these predicates do not strictly speaking classify as realis, since they are not actualised.

---

[52] The example is taken from a story where a little boy encounters a bush spirit and decides he wants to be turned into a fish, because he is treated badly by his brother's wife. He then wants to become a deep-sea fish, because this will make it harder for the brother and his wife to find him.

6.5.2.2 Negated *sa* as deontic modal operator

When *sa* is encountered in a negated clause it refers to ability and/or permission (which falls under deontic modality) or, rather, to the absence thereof, i.e. prohibition. Clausal negation is discussed in more detail in Section 10.2.2. Ordinarily, it is marked by two elements: *ma=*, which occupies the first slot in the VC following the subject bound pronoun, and *pwên*, which follows the last element that is in the scope of the negation. When a clause with *sa* is negated, it does not receive the marker *ma=*, but it is only marked by *pwên*, in its usual position. These *sa* clauses indicate that the subject is not able, or not allowed, to carry out the action represented by the predicate, as in examples (93) to (95).

(93) *ip lau kasa ro ning muyom pwên*
    [ip lau]$_A$ [ka-sa   to   ning]$_{VC}$ [muya-m]$_O$ pwên
    3pl people IRR.NS-MOD CONT see   skin-2sg.PERT NEG
    'The people will not be able to see your skin.' (LK250111_0073)

(94) *ngasa ruk pal tamong pwên*
    nga$_A$=[sa tuk pal]$_{VC}$ [tama-ng]$_O$ pwên
    1sg=MOD beat break   father-1sg.PERT NEG
    'I cannot do the *tuk pal* ceremony for my father.' (PK290411_1_0041)

(95) *osa lêpi pwên*
    wo$_A$=[sa lêp]$_{VC}$=i$_O$ pwên
    2sg=MOD take=3sg   NEG
    'You cannot take [i.e. adopt] him [I won't allow it].' (LM240611_0046)

In instances such as these, *sa* functions as a deontic modal operator (Timberlake 2007: 329). The authority can vary. When ability is negated, the participant designated to carry out the action is either (physically) unable to engage in it, as in (93), or may be prevented from engaging in it due to external circumstances, as in (94). When permission is denied, it is either the speaker who acts as an authority and prohibits an action of the addressee, as in (95), or an external authority may be involved, as in (96):

(96) *osa yuai naluai pwên*
    wo$_A$=[sa yuai]$_{VC}$ [naluai]$_O$ pwên
    2sg=MOD call   garden.food   NEG
    'You cannot/should not call out for a tuber to plant.'
    (NP220611_1_0024)

Ability and permission are closely related semantically and are in addition linked to cultural notions. The fact that people are not allowed to make noise when planting yams, what sentence ((96)) refers to, is related to certain taboos around gardening. In this case, one can say that the authority is based on cultural and societal norms. In (97), on the other hand, the modal operator is more a reflection of the personal opinion of the speaker.

(97) *kasa yektou la net pwên*
 [ka$_A$-sa yek.tou]$_{VC}$=Ø la net pwên
 IRR.NS-MOD put=3sg.ZERO go.to sea NEG
 'We should not give [it] [i.e. our tradition] away.'
 (OBK040311_0126)

Expressing prohibition with *sa* may be less forceful than using the prohibitive marker *napunan* (*lit.* 'it is forbidden') and may therefore be preferred in some situations for social and politeness reasons (cf. Lichtenberk 1995 on the relation between irrealis and polite imperatives, and Chapter 10 for more on Paluai imperative mood in general).

6.5.2.3 Negated *sa* as a marker of negative polarity in the future

In order to express negative polarity with future reference, *sa* is prefixed with the irrealis marker *kV-*. One remarkable feature of predicates marked as irrealis is that they cannot be negated with the preverbal negator *ma=* (combined with *pwên*). Only when *sa* is present can a clause modified by *kV-* obtain negative polarity. Examples are sentence (97) above, and (98) below.

(98) *ikisa ningong ai kunawayen pwên*
 i$_A$=[ki-sa ning an-sê]$_{VC}$=ong$_O$ a=i
 3sg=IRR.3sg-MOD see PART-DIM=1sg OBL=3sg
 kunawaye-n pwên
 life-PERT NEG
 'She should/ought not be able to see me again for the rest of her life.'
 (WL020711_0078)

This raises an interesting question: since only *sa*, which has deontic modal overtones, can be used to negate irrealis predicates, there is no means to express negative prediction. A counterpart of *will not*, to come up with an expression like 'She will not see me again' comparable to (98) above, is not found. The question is whether this is just some idiosyncrasy of the grammar, or whether it is connected to what people believe can be inferred about the future. When the subject is not able to carry out a certain action, then it necessarily follows that this action will not be carried out. But maybe there are inhibitions in some cultures against predicting the future with such certainty, and therefore people resort to a less definitive way of referring to an event that may not happen (cf. Burridge 2002).

In (94) above, there is another factor worth considering. This utterance is from a public speech; a rhetoric element thus probably plays a role. The speaker has decided not to carry out a certain mortuary ceremony for his father, on the grounds that it has already been done in the past, as he claims. Instead of saying that he has made up his mind and it won't happen, he hedges this by claiming that he is not able to do it, as if circumstances prevent him from doing it. Thus, the responsibility for not doing the ceremony has seemingly shifted away from the speaker. In this way, it may be less likely that people will (publicly) condemn him for not organising the ceremony.

6.5.3 Irrealis and aspectual particles with modal overtones

Preverbal aspectual or directional particles prefixed by the *kV-* irrealis marker may get additional modal overtones. This seems to be the case for perfective *pe*, imperfective *no* and continuative/habitual *to*. Directional *la* is also attested with modal overtones in irrealis predicates, but only as a main copular verb.

6.5.3.1 Irrealis and perfective *pe*

Perfective *pe* (see Section 6.2.1.1.2) is very often used in combination with an irrealis prefix. It places the event further into the future, in contrast to a "plain" irrealis, which usually refers to the immediate future. As with the perfective used for past time reference, it indicates that the future event is seen as a bounded unity, without regard to the time that will pass during the event. It appears that *kV-pe* is used predominantly for events with future reference, and not for those that have past habitual reference. Some examples are given below.

(99)   *urêkabe lêpo la panu liliu*
       wurê$_A$=[ka-pe   lêp]$_{VC}$=o$_O$ la   panua liliu
       1pc.EXCL=IRR.NS-PFV take=2sg   go.to home   again
       'We will take you back home again.' (NP210511_2_0058)

(100) *ope la lopwan te lumlum kipe nengo*
 wo$_S$=[pe la]$_{VC}$ [lopwa-n [te [lumlum]$_A$ [ki-pe
 2sg=PFV go.to place-PERT REL moss 3sg.IRR-PFV
 neng]$_{VC}$=o$_O$]$_{RC}$]$_{CC}$
 climb=2sg
 'You will turn into a place where moss will grow on you.' [*lit.* 'climb you'] (LK250111_0073)

6.5.3.2 Irrealis and imperfective *no*

With irrealis marking, imperfective *no* (see Section 6.2.1.1.1) is used in the same way as in realis predicates: it refers to an event viewed as unbounded with respect to the time that passes while the event unfolds. With irrealis marking, that event is located in the future rather than in the past or present. Often, a secondary aspect particle *to* or *tu* will be present in addition to the imperfective *no*, as in (101); alternatively, a verb will be reduplicated to indicate iterative action, as in (102).

(101) *kono ru kolkol ai rea*
 [ko$_S$-no tu kol.kol]$_{VC}$ a=i te-ya
 IRR.1sg-IPFV STAT.CONT REDUP.wait OBL=3sg LIG-then
 'Then, I will be waiting for that.' (MK050311_0030)

(102) *kino pungpung nupun le kino pungpung youn teo*
 [ki$_A$-no pung.pung]$_{VC}$ [nupu-n]$_O$ le
 IRR.3sg-IPFV REDUP.smell bottom-PERT or
 [ki$_A$-no pung.pung]$_{VC}$ [you-n te-yo]$_O$
 IRR.3sg-IPFV REDUP.smell tail-PERT LIG-DEM.INT
 'He will be sniffing at his bottom, or he will be sniffing at his tail.' (LL300511_1_0084)

No can also get modal overtones, which seem to pertain to ability and possibility, more specifically the fact that a particular event could potentially happen in the future. The examples below can have an additional habitual interpretation, which is further indication that the realms of habituality and temporal non-specificity (as encoded by the irrealis) have a close connection in Paluai.

(103) *ngagat porok te ong kono yipek môsôkei*
     nga$_A$=gat [porok [te wong$_A$ [ko-no   yipek]$_{VC}$ [môsôkei]$_O$]$_{RC}$]$_O$
     1sg=have strength REL 1sg.FREE IRR.1sg-IPFV blow   conch.shell
     'I have strength that enables me to blow the conch shell.'
     (LM260511_2_0019)

(104) *kipat sap kain te kino pat ai*
     [ki$_A$-pat]$_{VC}$ [sap kain [te [ki$_A$-no   pat]$_{VC}$=Ø$_O$ a=i]$_{RC}$]$_O$
     IRR.3sg-plant any kind REL IRR.3sg-IPFV plant=3sg.zero OBL=3sg
     'He will plant whichever kind [of plants] he can plant in there.'
     (KM190211_0038)

(105) *pian te igat naluai re ipkano ret lak, kano la yil, a kano si a kano apui ngan*
     pian te i$_A$=gat [naluai]$_O$ te ip$_S$=[ka-no   tet   lak]$_{VC}$
     good SUB 3sg=have garden.food SUB 3pl=IRR.NS-IPFV move go
     [ka$_A$-no   la   yil]$_{VC}$=Ø$_O$ a   [ka$_S$-no   si]$_{VC}$ a
     IRR.NS-IPFV go.to dig=3pl.ZERO and IRR.NS-IPFV come.down and
     [ka$_A$-no   apui=Ø$_O$ ngan]$_{VC}$=Ø
     IRR.NS-IPFV cook=3sg.ZERO eat=3sg.ZERO
     'It is good that there is garden food, for they could go and dig [it] up,
     and they could come down, cook [it] and eat [it].' (WL020611_0067)

There are modal overtones in these examples, but the use of *kV-no* is not essentially modal, as shown in examples (101) and (102). However, when a durative meaning is intended, there will usually be some other indication of it, such as a continuative particle or reduplication of the verb. The use of the modal particle *sa* for deontic modality (ability and permission) was discussed in Section 6.5.2.2. *Sa* can only refer to deontic modality with negative polarity. The use of *no*, as in the examples above, may be similar to that of *sa*, but with positive polarity. *kV-no*, however, lacks the permissive dimension and only indicates that a certain possible world may come to pass, without reference to an authority. It rather indicates one of a number of alternatives.

6.5.3.3 Irrealis and continuative/habitual *to*

The particle *to* (see Section 6.2.1.2.1) can be combined with irrealis in order to indicate continuative and habitual aspect with future reference:

(106) *ip lau karo wau kasông*
     [ip laue]$_S$ [ka-to   wau]$_{VC}$ kaso-ng

3pl people IRR.NS-HAB move near-1sg.PERT
'The people will be walking near to me.' (LK250111_0075)

It seems *to* can also indicate habitual obligation. Some examples are given below. Example (107) is from a text explaining the customary ceremonies around a death, and what each party ought to do. Sentence (108) is from a text about the time that a girl would get her first period. In earlier days, the custom was that she would be hidden inside the house because men were not allowed to see her. In these cases, *kV-to* appears to function as a deontic modal operator; the authority could be characterised as "cultural norms".

(107) *ip punpot pulek, ipkaro pe kui*
     [ip pun.pot]$_A$ pulêek ip=[ka-to   pe]$_{VC}$ [kui]$_O$
     3pl relatives.of.mother too   3pl=IRR.NS-HAB make pot
     'The side of the mother too, they should/must cook.' (SP190311_0032)

(108) *kiro roktoai lalon kalal lai*
     [ki$_S$-to   tok.to.ai]$_{VC}$ lalo-n   kalal ta-i
     IRR.3sg-HAB stay   inside-PERT wall   CLF.POSS-3sg.PERT
     'She must/should stay inside her fenced-off area.'
     (NK290311_2_0014)

It is possible that the deontic meaning is acquired from context alone, since in both cases it is clear from the narrative that the speaker refers to a prescriptive state of affairs. This is often the case in narratives where the procedure for a particular traditional custom is described. However, an indication that *to* has indeed deontic modal overtones in these cases is the fact that it does not occur with a modal sense combined with the first person singular prefix *ko-*.

6.5.3.4 Irrealis and the directional *la*

The directional *la* (see Section 6.3.1) is quite often used in combination with the third person singular irrealis prefix *ki-*, sometimes with perfective *pe* also present. *la* functions as a main copular verb in these cases and introduces an adjective or a NP constituent (see Section 7.7.2 for more on *la* as a copula). Consider examples (109) to (111):

(109) *kanen tasom nêm kila yamyaman...*
     [kane-n   ta=som   nêm]$_S$ [ki-la]$_{VC}$
     meat-PERT SPEC.COLL=one.ANIM be.finished IRR.3sg-go.to

[yamyaman]_CC
red
'Its whole body is like as if it's red...' (Game1_021012_0532)

(110) i o irok tepwo, ikipe la nganngan taip kurun not
[i   yo   i=tok   te-pwo]_S   i=[ki-pe
3sg   DEM.INT   3sg=sit   LIG-DEM.PROX   3sg=IRR.3SG-PFV
la]_VC [nganngan ta-ip   kutun not]_CC
go.to   food   CLF.POSS-3pl.PERT small child
'This here, it must be food for small children.' (Game1_021012_0622)

(111) naman kipe la remenin teo
naman   [ki_S-pe   la]_VC [temenin te-yo]_CC
perhaps   IRR.3SG-PFV   go.to   like   LIG-DEM.INT
'Perhaps it will be like this...' (Game1_021012_0393)

It seems that these instances of the irrealis plus copula *la* are used when a speaker is guessing, inventing a story, or trying to make an assertion based on inference or visual clues, as in (110). The main reason for using the irrealis in these cases seems to be that the speaker is unsure about the truth value of the information and cannot vouch for it. Thus, in these cases, use of the irrealis ties in with epistemic modality. Many of these instances come from a recording of Game 1 of the *Man and Tree* game (Levinson et al. 1992). In this recording, the game was played by two male speakers who slightly bent the rules: one of the players had to guess what was on the photograph that the other was holding. This led to a high incidence of *kipe la* phrases. In addition, the players came up with scenarios that would fit what was shown in each photograph; these, too, would be introduced by something like (111).

6.6 Structural properties of the verb complex

Minimally, a VC consists of just a bare verb form. This is the case in the imperative mood with a second person singular subject (for more discussion, see Section 10.1.2.1):

(112) toksi!
tok.si
sit.down
'Sit down!'

In the declarative and interrogative mood, person/number information of the subject is obligatorily expressed. The subject is indexed only once per verb complex, by a bound pronoun and/or, in the case of the irrealis, by the *kV-* prefix. The object is indexed on the verb complex when it is animate and has no full NP reference, again only once per VC.

When main clauses are coordinated, as in (113), the subject has to be indexed on each verb complex, even if it is identical. In this sense, Paluai differs from languages such as English, where the subject of the second clause can be elided under these circumstances: *I got afraid and Ø fell into the sea*. In Paluai, however, TAM particles can be left out when coordinated clauses share the same subject. It is a striking feature of Paluai narratives that clauses are often repeated at least once and that only the first clause shows the TAM particles, which are elided in subsequent clauses. However, the irrealis prefix has to be present in each instantiation. In example (113), the bound subject pronoun is repeated in the coordinated clause, but the perfective particle is not. More on clause coordination can be found in Chapter 11.

(113) *wurêpe suwen suk. wurêsuwen suk a...*
 wurê$_S$=[pe suwen]$_{VC}$ suk wurê$_S$=[suwen]$_{VC}$ suk a
 1pc.EXCL=PFV move.down shore 1pc.EXCL=move.down shore and
 'We went down to the shore. We went down to the shore and...'
 (MK060211_0006)

The verb complex of a declarative clause can be schematically represented as in (114). There are three preverbal TAM slots: 1) core aspect and *sa*, 2) directional, and 3) secondary aspect, which are preceded by a subject bound pronoun and/or irrealis prefix. There appears to be only one postverbal TAM/adverbial slot. Slot 1 may also be filled with the first element of the discontinuous predicative negation *ma=*, to the exclusion of all core aspectual particles and *sa* (see Section 10.2.2 for more information).

(114) (ProSubj)=IRR°-(CoreAsp) (DIR) (SecAsp) VERB** (PV)(=ProObj)†
 kV- pe to nêm
 no tu (wot)‡
 an* yen Manner Adv
 sa

° The subject bound pronoun is obligatory for realis clauses and 2sg irrealis clauses, but not for other persons in the irrealis. Marking for irrealis is obligatory when the reference demands so, but formally unexpressed for 2sg.

*When *an* is present, no other preverbal matter is allowed.
**The main verb (i.e. the head of the VC) can consist of a SVC.
†The bound object pronoun only occurs when the full NP O argument is elided and refers to an animate being.
‡In transitive clauses, *wot* occurs after the bound object pronoun and thus may not be part of the VC.

Table 62: Overview of possible TAM combinations and meanings.

Slot				Meaning
IRR	1	2	3	
kV-	–	-	–	Immediate future, imminent action
-	pe	-	–	Perfective (past reference)
-	pe	DIR	–	Perfective path or AM (past reference)
-	pe	la	to	Iterative (past reference, remoteness from DC)
-	pe	-	to	Iterative (past reference)
kV-	pe	-	-	Perfective (future reference)
kV-	pe	DIR	-	Path or AM (future reference)
kV-	pe	la	to	Iterative (future reference, remoteness from DC)
kV-	pe	-	to	Iterative (future reference)
-	no	-	-	Imperfective (past or present reference)
-	no	la	-	Imperfective (past or present reference, remoteness from DC) [*rare*]
-	no	la	to	Imperfective + continuative (past or present reference, remoteness from DC) [*rare*]
-	no	-	to	Imperfective + continuative (past or present reference)
kV-	no	-	-	Imperfective (future reference) or Modal
kV-	no	la	-	Imperfective (future reference, remoteness from DC) [*rare*]
kV-	no	la	to	[not attested in the corpus]
kV-	no	-	to	Imperfective + continuative (future reference)
-	an	-	-	Perfect
-	sa	-	-	Apprehensive modality (present/future reference)
kV-	sa	-	-	Apprehensive modality (future reference)
kV-	sa	la	-	Apprehensive modality with remoteness from DC (future reference)
-	–	DIR	-	Path or AM
-	–	la	to	Remoteness from DC
kV-	–	DIR	-	Path or AM (future reference)
kV-	–	la	to	Remoteness from DC (future reference)
-	–	-	to	Habitual (past/present reference)
kV-	–	-	to	Habitual (future reference) or Modal

Table 62 shows the various combinations that are possible for the three preverbal slots with realis and irrealis marking, plus an indication of their meanings. It has to be kept in mind that the instances of irrealis referred to in this table occur in main clauses. In dependent clauses, the irrealis has other functions. In addition, this table refers to clauses with positive polarity. Many TAM distinctions for positive clauses are neutralised under negation (see Section 10.2.2.4). Other forms, such as the apprehensive modal particle *sa*, take on other meanings under negation. The directional slot will be filled with *la* when 'remoteness from DC' is indicated, but it can take other directionals for path or AM meanings. For slot 3, secondary aspect, only the form *to* is represented in the table, but it can be replaced with *tu* or *yen*, except in the case of habitual meanings.

7 Predicates II: non-verbal and copula predicates

A non-verbal predicate does not contain a verb but takes an element from another word class as its head. Nouns, adjectives, numerals and several interrogative words can all function as head of a non-verbal predicate. In addition, the possessive classifiers *ta-* or *ka-*, and the preposition *pari* 'from, belonging to', are encountered in non-verbal predicates.[53] In contrast to verbs, members of other word classes cannot take bound pronouns when functioning as the head of a predicate, nor can non-verbal predicates be marked by TAM particles. Non-verbal predicates are always intransitive, unlike verbal predicates, which can be either transitive or intransitive. Non-verbal predicates cannot form a command. However, they are negated by the same operation as verbal predicates (see Section 10.2.2.1).

Paluai non-verbal predicates have a zero copula. The existential/posture verb *to(k)* 'be; stay, remain' sometimes shows up in what appears to be a copular use, but it may not be a genuine copula in all cases. This will be discussed in Section 7.7.1. On the other hand, the directional verb *la* 'go to' developed into a change-of-state marking semi-copula (Hengeveld 1992), 'become'. In what follows, the formal features of several types of non-verbal predicates will be discussed, followed by the meanings they can express. Next, a discussion will follow about the presence or absence of a copula in Paluai. Because comparative constructions are always non-verbal, they, too, will be discussed in this chapter.

7.1 Predicates with a noun as head

The following sentences are examples of clauses containing a nominal predicate, i.e. a predicate with a noun as head. Verbless clauses containing a nominal predicate usually express a relation of identity or class membership between the subject and the nominal predicate, as illustrated in examples (1) and (2), and examples (3) and (4), respectively.

---

53 A pronominal suffix indicating person and number needs to be added to *ta-* and *ka-*, and *pari* needs to be followed by a NP. The possessive forms are also often accompanied by a NP. Thus, these forms do not make up a non-verbal predicate on their own; see Sections 7.4 and 7.5.

(1) *wong yamtan*
 [wong]₍ₛ₎ [yamta-n]₍NVPRED₎
 1sg.FREE owner-PERT
 'I am its owner.' (LM260511_1_0063)

(2) *wong maêwen*
 [wong]₍ₛ₎ [maêwe-n]₍NVPRED₎
 1sg.FREE grandchild-PERT
 'I am his grandchild.' (KW290611_0064)

(3) *ma rapo, i reo i pein silal*
 ma tapo [i te-yo]₍ₛ₎ i [pein silal]₍NVPRED₎
 but actually 3sg LIG-DEM.INT 3sg woman pirit
 'But in fact, she was a spirit woman.' (PN100411_0021)

(4) *not teo i mwen*
 [not te-yo]₍ₛ₎ i [mwen]₍NVPRED₎
 child LIG-DEM.INT 3sg man
 'The child was a boy.' (OL200111_0042)

The subject of a verbless clause usually has a human referent, and the head noun is usually a personal noun, a kinship term or another noun that typically refers to a human being. When the S argument is a full NP containing a longer phrase and/or ending with a demonstrative, there is usually a pronoun copy directly preceding the non-verbal predicate, similar to what happens in the case of verbal predicates. The only difference consists in the fact that non-verbal predicates are always preceded by a free pronoun and never by a bound one.

Verbless clauses can be used in a deictic way, together with a deictic demonstrative, to point at something. Examples (5) and (6) would be appropriate answers to the question 'What's this/that?'. In this case, there is no pronoun copy preceding the nominal predicate.

(5) *i repwo pou*
 [i te-pwo]₍ₛ₎ [pou]₍NVPRED₎
 3sg LIG-DEM.PROX pig
 'This is a pig.' (052b_0138)

(6) *i relo mui*
     [i   te-lo]$_S$ [mui]$_{NVPRED}$
     3sg LIG-DEM.DIST dog
     'That [over there] is a dog.' (052b_0142)

7.2 Predicates with an adjective as head

Property-concept predicates (Stassen 1997) in Paluai show a split between verbal encoding, with a stative verb that can take a bound subject pronoun and TAM marking, and non-verbal encoding, by means of an adjective that acts as predicate head and takes neither. Stative verbs were discussed in Section 3.3.1.2. Examples (7) to (9) show verbless clauses with a predicate that has an adjective as its head.

(7) *taim te wong pa namwi...*
     taim te [wong]$_S$ pa [namwi]$_{NVPRED}$
     when SUB 1sg.FREE yet small
     'When I was still small (a little child)...' (SY100411_0005)

(8) *pusok sê reo i somwai*
     [pusok sê te-yo]$_S$ i [somwai]$_{NVPRED}$
     island DIM LIG-DEM.INT 3sg great
     'That little island is great.' (LL030611_0008)

(9) *maran sip namwi a maran sip menengan*
     [mata-n sip]$_S$ [namwi]$_{NVPRED}$ a [mata-n sip]$_S$
     eye-PERT one.INANIM small   and eye-PERT one.INANIM
     [menengan]$_{NVPRED}$
     big
     'One eye is small and one eye is big.' (Game1_021012_0460)

Verbless clauses containing a predicate headed by an adjective usually express an attributive relation: the adjective describes an attribute or property of the S argument. When the subject is represented by a full NP, it could be argued that a verbless clause is not in fact a clause but a NP, with a modifier following the head noun. There is evidence that this is not the case, though. First of all, this type of construction does not have to contain a full NP, but can have a free pronoun as its subject; cf. example (7). Secondly, elements that modify the adjectival

228 — 7 Predicates II: non-verbal and copula predicates

predicate can occur between the subject NP and the predicate, just as in the case of a clause that does contain a verb. Example (7), for instance, contains a temporal adverb placed between the S argument and the predicate.

7.3 Predicates headed by a numeral

As mentioned in Section 4.5.3.2, numerals can form the head of a non-verbal predicate. The examples shown there are repeated below.

(10) *urê ramwen urê ngonomou, a irê rapein, irê ngunan*
[wurê ta=mwen]$_S$ wurê [ngonomou]$_{NVPRED}$ a [irê
1pc.EXCL SPEC.COLL=man 1pc.EXCL six.ANIM and 3pc
ta=pein]$_S$ irê [ngunan]$_{NVPRED}$
SPEC.COLL=woman 3pc five
'We men, we are six, and them women, they are five.'
(OL201210_0042)

(11) *pou re kope yiuek kup tan tamong nganoyumou*
[pou [te ko$_A$-pe yiu-ek= $_{O1}$ [kup
pig REL IRR.1sg-PFV pull.in-APPL=3pl.ZERO pigs.lined.up
ta-n tama-ng]$_{O2}$]$_{RC}$]$_S$ [nganoyumou]$_{NVPRED}$
CLF.POSS-PERT father-1sg.PERT eight.ANIM
'The pigs with which I will pull the rope for my father [as part of a traditional ceremony] are [i.e. number] eight.' (YK290411_2_0049)

(12) *Ngat i no nantasom o*
[Ngat]$_S$ i [no nan=ta=som yo]$_{NVPRED}$
N. 3sg only PART=SPEC.COLL=one.ANIM DEM.INT
'Ngat is the only one [i.e. an only child].' (LM240611_0045)

A verbless clause headed by a numeral instantiates an attributive relation, in much the same way as a clause containing an adjectival predicate.

7.4 Predicates containing *ta-* or *ka-*

Apart from the recent Tok Pisin loan *gat* 'have', there is no verb expressing possession in Paluai. Instead, possession is usually expressed within the NP (see Section 5.5) and can be direct or indirect. A directly possessed noun can head a non-verbal predicate, as illustrated in sentences (1) and (2) above. Indirect possession is expressed by means of the classifier *ta-*, which always receives a suffix indicating person/number of the Possessor (see Section 4.2.3.1). *Ta-* usually follows the head noun (the Possessee) within the NP, but it is occasionally found introducing a non-verbal predicate, as shown below in (13)–(15); if the Possessor is also expressed as a full NP, this follows *ta-*.

(13) *pwapwa repwo i ran Nulik*
    [pwapwae te-pwo]$_S$ i   [ta-n   Nulik]$_{NVPRED}$
    story   LIG-DEM.PROX 3sg CLF.POSS-PERT N.
    'This story is about Nulik.' (WL020611_0002)

(14) *môsôkei i ran tupung ta Ngat Porambei*
    [môsôkei]$_S$ i   [ta-n   tupu-ng
    conch   3sg CLF.POSS-PERT grandparent-1sg.PERT
    ta=Ngat Porambei]$_{NVPRED}$
    SPEC.COLL=N.P.
    '[Blowing] the conch shell belongs to my grandfather, Ngat Porambei.' (LM260511_2_0002)

(15) *i repwo i raip pein*
    [i   te-pwo]$_S$ i   [ta-ip   pein]$_{NVPRED}$
    3sg LIG-DEM.PROX 3sg CLF.POSS-3pl.PERT woman
    'This here is for women.' (Game1_021012_0133)

Ka-, the possessive classifier for edible objects, can be used in the same way as *ta-* in a non-verbal predicate:

(16) *i repwo kom*
    [i   te-pwo]$_S$ [ka-m]$_{NVPRED}$
    3sg LIG-DEM.INT   CLF.food-2sg.PERT
    'This is yours.' [food, to eat]

A verbless clause with *ta-* or *ka-* instantiates an attributive relation between the S argument and the entity referred to by the suffix on the classifier. In most cases, these clauses do not refer to possession in the strict sense, i.e. ownership, but rather indicate that there is a close connection between the two entities. Sentence (15), for example, is about a necklace shown in a picture. The speaker comments that this necklace is meant to be worn by women, rather than men.

7.5 Predicates containing *pari*

The form *pari* 'from, belonging to' can introduce a prepositional phrase that modifies a noun within a NP (see Sections 4.4.2 and 5.6). It is also frequently encountered in non-verbal predicates. The following sentences provide examples. *pari* is followed directly by a NP in case the latter is headed by a local noun, otherwise the preposition *a=* (cliticised to the 3sg pronoun *i*) needs to be present. In (17) and (18), *pari* refers to origin. In (19), it refers to both origin and ownership, and in (20) it refers to content.

(17) *wong pari Lipan*
      [wong]$_S$ [pari   Lipan]$_{NVPRED}$
      1sg.FREE belonging.to L.
      'I am from Lipan.' (KM060111_0003)

(18) *ipan pwa, "naman pein teo i pari Lou"*
      ip$_A$=an pwa [naman [pein   te-yo]$_S$ i   [pari
      3pl=PRF think perhaps woman LIG-DEM.INT 3sg belonging.to
      Lou]$_{NVPRED}$]$_{Compl:O}$
      Lou
      'They had thought, "Perhaps this woman is from Lou."' [i.e. that she was a living person, but she turned out to be a spirit] (PN100411_0020)

(19) *pwapwa repwo i pari Marako*
      [pwapwa te-pwo]$_S$ i   [pari   Marako]$_{NVPRED}$
      story   LIG-DEM.PROX 3sg belonging.to M.
      'This story belongs to Marako.' (WL020611_0060)

(20) *pwapwa repwo i pari ai mwamwanget*
      [pwapwae te-pwo]$_S$ i   [pari   a=i
      story   LIG-DEM.PROX 3sg belonging.to OBL=3sg
      mwamwanget]$_{NVPRED}$
      be.lazy
      'This story is about indolence.' (WL020611_0059)

Verbless clauses with *pari* are rather similar to those with *ta-* described above. They also instantiate an attributive relation between the S argument and the entity referred to by the NP following *pari*. *Pari* mostly refers to origin, but it can also be ambiguous between origin and ownership, as in (19), and its use has been metaphorically extended to include content/meaning and purpose. *Pari* can only introduce a NP referring to an inanimate entity, which explains why *ta-* is used in (13) and (14), where the NP refers to a human being.

7.6 Predicates headed by an interrogative word

Non-verbal predicates in a verbless clause can be headed by several interrogative forms all of which query identity or attribution. An overview is given in Table 63.

Table 63: Question markers that can function as heads of a non-verbal predicate.

Form	Translation	Questions	Word class
sa	what	Identity of object	Noun
sê	who	Identity of person	Noun
samnon	how many	Quantity	Numeral
samai-	what of	Direct possession relation (part-whole or kinship)	Noun
tenepa	how	Quality, Attribute	Adjective

The expected answer to a question with any of these interrogative words would also be a verbless clause. Each of them is related to a word class that heads a non-verbal predicate. For more on interrogative clauses, see Section 10.1.1. Examples of non-verbal predicates headed by one of the interrogative words listed in the table are given below.

(21) *ngayom sa?*
 [ngaya-m]$_S$ [sa]$_{NVPRED}$
 name-2sg.PERT what
 'What's your name?'

(22) *i reo i sê?*
 [i te-yo]$_S$ i [sê]$_{NVPRED}$
 3sg LIG-DEM.INT 3sg who
 'Who is this?'

(23) *kerin tao samnon?*
      [kerin ta-o]~S~   [samnon]~NVPRED~
      year   CLF.POSS-2sg.PERT how.many
      'How old are you?' [*lit.* 'Your years are how many?']

(24) *Pokut i samaim?*
      [Pokut]~S~ i   [samai-m]~NVPRED~
      P.   3sg what.of-2sg.PERT
      'How is Pokut related to you?' (052b_0015)

(25) *tare kanopa kaywun le i reneba?*
      [ta=te-yo]~S~   kanopa [kaywun]~NVPRED~ le i   [tenepa]~NVPRED~
      SPEC.COLL=LIG-DEM.INT like   white   or 3sg how
      'This [thing], it is like white or how [does it look]?' (Game2_280812_0142)

7.7 Potential copula clauses

As mentioned above, there are predicates headed by the posture/existential verb *to* 'be; stay, remain' or the directional *la* 'go to'. They could be analysed as property-concept predicates (Stassen 1997) with a verb that has a copular function. For this reason, they will be discussed here, even though these clauses are strictly speaking verbal.

7.7.1 Clauses with *to* 'be'

The following types of constructions are encountered with *to*:
– *to* followed by an adjective;
– *to* followed by the possessive classifier *ta-* plus pronoun suffix.

Examples of *to* followed by an adjective are given in (26) and (27). *To* followed by *ta-* is exemplified in (28) and (29).

(26) *kola sayek net are kiro mwanen*
      ko~A~-la   sayek   [net]~O~   a=te   ki~S~-to   [mwanenen]~CC~
      IRR.1sg-go.to rub.with salt.water OBL=SUB IRR.1sg-be straight
      'I will rub it with salt water so that it will be straight.' (AK160411_1_0007)

(27) *i sê kiro yauron teo*
 i [sê$_S$ ki-to [yauron]$_{CC}$ te-yo]
 3sg who IRR.3sg-be short LIG-DEM.INT
 'It is one that will be short.' (KW290311_0029)

(28) *Korup iro rang*
 [Korup]$_S$ i=to [ta-ng]$_{CC}$
 K. 3sg=be CLF.POSS-1sg.PERT
 'Korup is mine.' [i.e. he belongs to my lineage] (ANK020995_0010)

(29) *pwapwaen iro rang*
 [pwapwae-n]$_S$ i=to [ta-ng]$_{CC}$
 story-PERT 3sg=be CLF.POSS-1sg.PERT
 'The story is mine/belongs to me.' (YK290411_1_0030)

The combination of *to* with an adjective is encountered very infrequently, and only with irrealis predicates, which refer to hypothetical or future events or states (see Section 6.4). The irrealis marker *kV-* can only attach to a verb and not to a member of another word class, which is probably the reason why *to* is attested as a copula in these cases. Interestingly, *to* is also not encountered in the corpus with predicates headed by a noun, which encode a semantic relation of identity or class membership. This type of relation presumably has more time-stability and is thus in less need of TAM marking.

In cases like (28) and (29), *to* is most likely not a copula. *Ta-* is not only attested with a possessive sense, but also as a locative (for instance, it marks the Goal argument of a ditransitive clause in case that argument is animate; see Sections 4.4.3 and 8.1). This use of *to* plus the possessive classifier can be interpreted as an abstraction from the locative sense of *ta-*. Another indication for this may be that the form *ka-*, which functions as a possessive classifier for edible objects (see Section 7.4 above), is not encountered in combination with *to*.

When combined with *to*, *ta-* has comitative overtones; it could also be translated as 'with'. It is not entirely clear why in some cases *to* is present; as we have seen in Section 7.4, *ta-* can also be part of a non-verbal predicate by itself. The semantics of the subject NP may play a role. In (28), it refers to a human being and in (29) to a story; the latter are seen as the properties of clans and not individuals. It may not be possible to use *ta-* directly with these kinds of entities.

7.7.2 Clauses with *la* 'go'

More often than *to*, it is the directional *la* 'go to', which grammaticalised into a change-of-state marker 'become', that will be used to express an attributive relation. The resulting clause can be analysed as a copula predicate; examples are given below. In fact, auxiliaries with meanings such as 'become' are better analysed as semi-copulas, because they do add an element of meaning to the predicate. Leaving them out changes the meaning of the clause (Hengeveld 1992: 35). The copula complement can be either an adjective, as in (30) and (31), or a noun, as in (32) and (33).

(30) *muyan kipe la pian*
 [muya-n]$_S$ ki-pe la [pian]$_{CC}$
 skin-PERT IRR.3SG-PFV go.to good
 'His skin will become nice.' (NP260511_0021)

(31) *wong kola paiwon*
 [wong]$_S$ ko-la [paiwon]$_{CC}$
 1sg.FREE IRR.1sg-go.to strong
 'I will become strong.' (SY100411_0013)

(32) *uwot wum, are ila poyep*
 u$_S$=wot wumwa a=te i$_S$=la [poyep]$_{CC}$
 3du=go.level house OBL=SUB 3sg=go.to afternoon
 'They went home, for it had become afternoon.' (MS250311_0041)

(33) *Kileai ipwa ikila narun*
 Kileai i$_S$=pwa i$_S$=ki-la [naru-n]$_{CC}$
 K. 3sg=want.to 3sg=IRR.3sg-go.to son-PERT
 'Kileai wanted him to become his son.' / 'Kileai wished for him to become his son.' (OL20011_0045)

In addition, a copula-like *la* can follow a main verb in a sequence. This construction may have originated as a SVC (see Chapter 9), but has probably led to the reanalysis of *la* as a periphrastic marker of adverbial phrases derived from adjectives (see also Section 3.6.2.2). The *la* phrase now functions as a postverbal manner adverbial modifier, indicating 'in the state of'. Whether or not the construction in its entirety has change-of-state semantics depends on the semantics of the main verb. Compare for instance examples (34) and (35). Clearly, in

(34), a change of state is indicated, whereas (35) refers to an unchanging state of affairs. This is probably due to the difference in lexical semantics between *pei* 'appear' and *tok* 'be, stay'. The development of *la* as an adverbial marker also probably gave rise to constructions such as in (36), with *la pwên* as an adverbial negator (see Section 10.2.3.1). In addition, when following a main verb, *la* can have a noun as complement, as in (37).

(34) *muyan kipe pei la pian*
 [muya-n]$_S$ ki-pe pei [la pian]$_{Adv}$
 skin-PERT IRR.3sg-PFV appear go.to good
 'His skin will become nice.' (NP260511_0023)

(35) *uro rok la pian palsi*
 u$_S$=to tok [la pian]$_{Adv}$ palosi
 3du=HAB stay go.to good past
 'They used to live (together) well in the past.' (LL010711_0092)

(36) *iyik lêp tinan la pwên*
 yi$_A$=yik lêp [tina-n]$_O$ [la pwên]$_{Adv}$
 3sg=search.for take mother-PERT go.to NEG
 'He searched in vain for his mother.' (KW290611_0054)

(37) *samin teo, koripêl la kalomwen*
 [samin teyo]$_{TopO}$ ko$_A$-tipêl=Ø [la kalomwen]$_{Adv}$
 end.of.rope DEM.INT IRR.1sg-braid=3pl.ZERO go.to handle
 'The ends of the twines I will braid into the handles [of the basket].' (AK160411_2_0024)

7.8 Comparative constructions

Comparative constructions are formed by verbless clauses with an adjectival predicate as their head, modified by a phrase introduced with the general prepositional clitic *a=*. This Oblique constituent contains the Standard of comparison. The element being compared is referred to by the S argument and the Parameter of comparison is formed by the adjective that heads the predicate. It appears that only adjectives, and not (stative) verbs, can be Parameter of comparison. The adjective is often modified by the adverb of degree *pun*, which functions as a general intensifier and can in this case be translated as 'much'.

The adverb of degree *paran* 'very' is not encountered with comparative constructions. A number of examples are given below.

(38) *naman i pian pun ai aronan toktoai ran musau pari tepwo*
    naman   [i]$_S$ [pian pun]$_{PARAMETER}$ a=i   [arona-n toktoai
    perhaps   3sg   good   INTF   OBL=3sg way-PERT live
    ta-n   musau   pari   tepwo]$_{STANDARD}$
    CLF.POSS-PERT foreign.place belonging.to LIG-DEM.PROX
    'Perhaps it [the way of life of our ancestors] is much better than the Westernised lifestyle of today.' (BK040311_0043)

(39) *i menengan ai i rayamyaman*
    [i]$_S$ [menengan]$_{PARAMETER}$ a=i   [i   ta=yamyaman]$_{STANDARD}$
    3sg   big   OBL=3sg 3sg SPEC.COLL=red
    'It [the yellow one] is bigger than the red one.' (Game2_021012_0261)

When the head adjective is both preceded and followed by the intensifier *pun*, the construction can be analysed as a superlative. In this case, there is no Standard of comparison. This construction is rare, however; moreover, the only representative example found in the data is a copula construction with *la* and it is possible that the first instance of *pun* modifies *la* rather than *pian*.

(40) *naman kape la pun pian pun*
    naman   ka$_S$-pe   la   [pun [pian pun]$_{CC}$]
    perhaps   IRR.NS-PFV   go.to   INTF   good   INTF
    'Perhaps, this will be the best for us.' [*lit.* 'we will be the best'] (BK040311_0036)

Comparative and in particular superlative constructions are not at all common in spontaneous speech, although they can easily be elicited. Normally, speakers have a preference to compare two entities in coordinated verbless clauses with two gradable adjectives that form a pair of antonyms, for example *menengan* 'big' and *namwi* 'small':

(41) *i rakaywun i menengan, ma i re i rayamyaman kayan telo i namwi*
    [i   ta=kaywun]$_S$ i   [menengan]$_{NVPRED}$ ma i te   [i
    3sg   SPEC.COLL=white 3sg   big   but   3sg
    ta=yamyaman   kayan   te-lo]$_S$ i   [namwi]$_{NVPRED}$
    SPEC.COLL=red   black   LIG-DEM.DIST 3sg small
    'The white one is big, but the dark red one is small.' (Game1_021012_0581)

Comparatives of equality are usually formed by either the verb *(pe) ngonomek* 'be corresponding, be the same' or by a non-verbal predicate headed by *no sip* 'the same (*lit.* 'only one').

(42) *taywêp ino pe ngonomek*
 [ta=yuêp]$_S$ i=no pe ngomomek
 SPEC.COLL=two.INANIM 3sg=IPFV do be.corresponding
 'The two are just the same.' (Game2_021012_0156)

(43) *menenganan taywêp nêm no sip*
 [menengan-an ta=yuêp nêm]$_S$ [no sip]$_{NVPRED}$
 big-NOM SPEC.COLL=two.INAMIM be.finished only one.INANIM
 'They are of the same size [*lit.* 'the bigness of the two is only one'].'
 (Game2_280812_0201)

8 Grammatical relations and valency

This chapter discusses intra-clausal grammatical relations in the context of verb valency. First, a number of definitions related to grammatical functions and valency will be discussed. After that, it will be shown how different grammatical relations are formally represented in Paluai. Next, some strategies to either increase or reduce verb valency will be discussed. It will be shown that Paluai has lost many of the morphological means for altering verb valency that are attested for Proto-Oceanic; however, SVCs (discussed in more detail in Chapter 9) have partly taken over this task.

8.1 Expression of core and peripheral arguments

8.1.1 Core arguments S, A and O

Most verbs are subcategorised for either one or two arguments; these are called core arguments. A verb with one core argument is intransitive. When a verb is subcategorised for two core arguments, it is transitive. Throughout this work, the core argument of an intransitive verb, the intransitive subject, is represented by S, whereas the two core arguments of a transitive verb are called transitive subject (expressed by A) and object (O), following Dixon (2010b). There are very few genuine three-place verbs in Paluai; likely candidates are discussed in Section 8.1.2.2 below.

Paluai has nominative-accusative alignment: both S and A precede the verb complex when expressed by full NPs or free pronouns, and they are usually cross-referenced by means of a bound pronoun attached to the verb complex as a proclitic.[54] Cases in which the subject is not cross-referenced on the verb are discussed in Section 6.1. The O argument, in contrast, immediately follows the verb complex. It is indexed by an enclitic bound pronoun to the VC only when it is animate and not expressed by a full NP; thus inanimate objects not expressed by a full NP have zero reference. Examples (1) and (2) show unmarked constituent order in Paluai: SV for intransitive, and AVO for transitive clauses.

---

54 The term "verb complex" is used to refer to the conglomerate of the main verb, its preceding TAM and/or directional particles and its postverbal adverbials (if present), as discussed in Chapter 6.

https://doi.org/10.1515/9783110675177-008

(1) *kamoun pe ret*
 [kamou-n]ₛ pe tet
 speech-PERT PFV spread
 'Word of it has spread.' (KM060111_0045)

(2) *Namwai wutantek mumura rai*
 [Namwai]ₐ wut.antek [mumura ta-i]ₒ
 N. bail.out vomit CLF.POSS-3sg.PERT
 'Namwai bailed out his vomit.' (MK060211_0024)

The subject, which can be either a S or an A argument, always has to be overtly expressed, either by a full NP or free pronoun, a bound pronoun and/or irrealis prefix, or both a free and a bound form. Thus, there are no declarative predicates whose subject is completely unexpressed. For objects, the situation is different. As mentioned before, O arguments retrievable from the discourse context will be elided. In these cases, they are indexed on the verb complex only when the referent is animate. This makes it difficult to establish from texts whether a particular verb is transitive (i.e. subcategorised for an O argument that was elided) or ambitransitive. This issue will be discussed in Section 8.2.

8.1.2 The E argument

Both transitive and intransitive verbs may be subcategorised for an additional "extended" argument, referred to by E, following Dixon (2010a, 2010b). The E argument is, in principle, obligatorily expressed and thus can also be regarded as a core argument rather than an optional (peripheral) argument; it has to be pointed out, though, that the distinction between core and peripheral arguments "is never a hard and fast one" (Dixon 2010a: 101). The E argument is never directly indexed on the verb by a bound pronoun, in contrast to (animate) O arguments. While O arguments are always referred to by a bare NP without any flagging of their grammatical function, E arguments take additional Oblique marking: the preposition *a=* for an inanimate referent, and *ta-* for an animate one. An important distinction that sets the E argument apart from peripheral arguments, specifically with regard to verbs of transfer, is that genuine E arguments are the only ones that can occur without a preceding directional.

The E argument is found with a number of intransitive verbs, as discussed below, yielding an extended intransitive clause with S and E arguments. It is also found with some transitive verbs, yielding an extended transitive (or ditransitive) clause with A, O and E arguments.

Table 64: Verbs of emotion belonging to the extended intransitive subclass.

Form	Meaning
kaêrêt	be afraid (of)
kolu- sosol	mourn, be sad (about) (*lit.* 'inside mourns')
maloa- wop	take fright (of) (*lit.* 'spirit flies')
mwamwanget	be tired (of), fed up (with)
mwamwasêk	be ashamed (of)
nayet	be happy (about)
nia- palak	be angry (with) (*lit.* 'stomach is bad')
nopnop	be jealous (of)
nunuau	be energetic, keen (on)
pilel	laugh (about)
tou put	regret, have regrets (about)
wayêt	be sorry, feel sadness (for)
yangyang	like, love
teng	cry (about)

8.1.2.1 Extended intransitive verbs

An overview of intransitive verbs in Paluai that subcategorise for an E argument is provided in Tables 64 and 65. The E argument is not always overtly expressed, however, and thus may not, strictly speaking, classify as a core argument. The reason why these verbs are to be set apart from other intransitive verbs (that can take peripheral arguments) is that the Oblique marker can take both the form of the preposition *a=* and the locative/possessive *ta-*. *a=* is cliticised to the 3sg pronoun *i*, yielding the surface form *ai*, while *ta-* is always suffixed by a form from the indirect possession paradigm (see Section 4.2.3.1). With regular peripheral arguments, such as locatives or instrumentals, the Oblique marker can only be *a=*. Some of the verbs in this class have frozen applicative counterparts, in which the

Table 65: Other verbs belonging to the extended intransitive subclass.

Form	Meaning
inap (Tok Pisin loan)	be sufficient (for)
mapwai	know (about)
masai	be clear (about)
mêpmêp	dream (about)
pangai	think (about/of)
pwak	meet (with), encounter; find; be related to
tenten	speak to an ancestor; pray (to)

E argument is expressed instead as O: an example is *tengsek* 'cry for, mourn' originally derived from *teng* 'cry' (see Section 8.3.2.3).

Furthermore, there seems to be a semantic basis to the classification. The verbs in question tend to express states rather than actions, there is low volition and control on the part of the subject argument, and the argument marked as Oblique is not highly affected. They can quite neatly be divided into two semantic subclasses. The majority, shown in Table 64, refer to emotional or physical sensations. They take an Experiencer subject and a Stimulus E argument, which is expressed by an Oblique constituent marked by *a=* for inanimates and *ta-* for animates. When there is no overtly expressed E argument, there is nonetheless an entity that has to be regarded as the stimulus or cause of the sensation. In the words of a consultant: "you cannot just be afraid; you have to be afraid of *something*." Examples of the verb *kaêrêt* are given in (3) and (4). The E argument in (3) is animate; in (4), it is inanimate.[55]

(3) *ngaru kaêrêt tan muyou*
     nga$_S$=tu   kaêrêt   ta-n   [muyou]$_E$
     1sg=STAT.CONT be.afraid CLF.POSS-PERT snake
     'I was afraid of the snake.' (Game1_021012_0562)

(4) *ipto kaêrêt ai aronan kauwat taip*
     ip$_S$=to   kaêrêt   a=i   [arona-n   kauwat   ta-ip]$_E$
     3pl=HAB be.afraid OBL=3sg way-PERT tradespartner CLF.POSS-3pl.PERT
     'They used to be afraid of the ways of their tradespartners.' (MS250311_0046)

Some of the other verbs in this subclass (shown in Table 65) are verbs of cognition. They take a "Cogitator" subject, with the thing or person that she has knowledge about expressed as an Oblique.[56] The status of *mêpmêp* is slightly uncertain, since its only occurrence in the corpus is with an applicative suffix (see Section 8.3.2 below). *pwak* has many meanings. It can refer to finding a person or an object one has been looking for, to meeting a person (in both of these cases, 'encounter' would be an apt translation), but it can also refer to being related to a certain thing or person, or being stuck (said of a door, for instance). In any case, once again we see low control and volition of the subject. The same goes for *inap* 'be sufficient'. The verb *tenten* 'pray', on the other

---

[55] The constituent that forms the E argument is headed by the inanimate abstract noun *arona-*; the entity represented by the argument is therefore considered to be inanimate.
[56] The term "Cogitator" is from Dixon (2010a, 2010b).

hand, is probably more similar to the extended transitive verbs discussed below. It is used in situations similar to those where verbs of SPEAKING are used, but it is not subcategorised for a Theme (the speech transferred from one person to the other).

There is variation with *mapwai*, as it can seemingly take either an overt O or an E argument to express the semantic role of Theme related to the 'knowing' situation. Below, two examples of the verb *mapwai* are given with an animate and inanimate E argument, respectively, followed by an example where *mapwai* takes an O argument.

(5)   *ila ro mapwai rurê a urêro mapwai rai*
       i$_S$=la   to   mapwai   ta-urê$_E$ a
       3sg=go.to   CONT   know   CLF.POSS-1pc.EXCL.PERT   and
       wurê$_S$=to   mapwai   ta-i$_E$
       1pc.EXCL=CONT   know   CLF.POSS-3sg.PERT
       'She knows about us and we know about her.' (OL201210_0153)

(6)   *epmaru mapwai liliu ai pwên*
       ep$_S$=ma=tu   mapwai   liliu   a=i$_E$ pwên
       1pl.EXCL=NEG$_1$=STAT.CONT   know   again   OBL=3sg   NEG$_2$
       'We do not know about it anymore.' (NP190511_2_0032)

(7)   *ngasa mapwai puyamat tang pwên*
       nga$_A$=sa   mapwai   [pun.yamat   ta-ng]$_O$ pwên
       1sg=MOD   know   genealogy   CLF.POSS-1sg.PERT   NEG
       'I would not be able to know my genealogy.' (OBK040311_0078)

It appears that *mapwai* can take two types of objects: a "close" one, marked as O, or a "remote" one, marked as E. This is reminiscent of the English verb 'know', which can also take either a direct object or a prepositional phrase starting with the preposition *about*. In Paluai, there seems to be a restriction on the O argument of *mapwai* that is not observed with the O arguments of other verbs: the O argument has to be expressed by a full NP and cannot be pronominal. Pronominal arguments to *mapwai* are always Obliques, i.e. introduced with *a=* or *ta-*.

The distinction between Oblique arguments with *a=* and those with *ta-* is purely semantically based: it exclusively depends on animacy. Inanimate referents are marked with *a=*, whereas animate referents are marked with *ta-*. It could be claimed that it is logical for a locative or instrumental Oblique not to be marked with *ta-*, since this type of Oblique does not generally refer to animate beings; there is thus no real reason to regard the subclass of verbs under consideration as special. This is a valid point. However, there are a number of counter-claims that could be made in favour of the current analysis. The first one is that these verbs all represent transitive situations, in which two arguments are involved and one is affected by the other, but in contrast to many other transitive situations, expressed by regular transitive verbs, one of the arguments is not expressed simply as O, but gets a different marking. There is hence a formal difference between these and regular transitive clauses. Secondly, and related to this, the fact that one of the arguments is marked differently from a "regular" O suggests that the relationship between the subject and the other argument somehow deviates from the prototypical transitive situation. This is indeed the case: as mentioned above, these verbs characteristically involve low control and volition on the part of the subject, and the second argument (the one that gets Oblique marking) is not really affected. This is different from a typical transitive situation (e.g. with the verb 'eat'), where the A argument is in control of the action and the O argument (in the semantic role of Patient) is highly affected.

The affectedness/non-affectedness dimension of the non-subject argument may interrelate with the animacy dimension. After all, the higher the non-subject argument is on the animacy scale and the less affected it is, the more likely it is that the subject argument is less in control and more affected, and vice versa. This does not mean that animate referents cannot function as Patients. They surely can, but when they do, they will be expressed as regular O arguments and be cross-referenced on the verb with a bound pronoun.

A third observation in support of the fact that the second arguments of the verbs discussed above are not regular peripheral arguments, even though they are not obligatorily overt, is that, also when absent, they are somehow central to the event described by the verb and will be understood from the context. In this way they contrast with "regular" peripheral arguments, which can be added to any clause, regardless of its number of core arguments, and can usually be replaced with adverbial phrases.

8.1.2.2 Extended transitive or ditransitive verbs

A number of transitive verbs are subcategorised for an E argument; they are shown in Table 66. These verbs take an A and an O argument, and the additional E argument is most often overtly expressed. One important consideration in favour of treating these constituents as core arguments of the verb, is that they are the only ones that can occur without additional marking by a directional, even when there is a clear instance of transfer. Constructions that involve a directional are regarded as SVCs and discussed in more detail in Section 9.1.5.

Table 66: Verbs belonging to the extended transitive subclass.

Form	Meaning
apek	hit, shoot
asuek	rub on
ayek	keep, withhold
lêp	take, receive
touek	show, teach
yuêt	to ask

A number of interesting observations can be made about this small group of verbs. Firstly, more than half end in *-ek*, and may thus be frozen applicatives. Apart from the AFFECT verbs *apek* and *asuek*, extended transitive verbs belong to the classes Dixon (2010a: 104) identifies as GIVING and SPEAKING, typically referring to an interpersonal event in which an object or speech is transferred from one person to another. Secondly, apart from *touek*, it appears that the third argument these verbs are subcategorised for is referring to a transfer Source, rather than a Goal: note that *lêp* 'receive, take' is in this class but not *tou* 'give', and *yuêt* 'ask' but not *pul* 'tell'. *tou* and *pul* are without exception found with a directional introducing the Goal argument and are thus not analysed as three-place verbs. The abovementioned verbs can certainly occur with a directional introducing the third argument; crucially, however, this is not necessary.

Comparing the various verbs with respect to the participant they select as the third argument brings to light a number of different patterns, which will be discussed in turn. Firstly the focus will be on *apek* 'hit, shoot' and *asuek* 'rub'. They fall into the semantic category of AFFECT verbs, which typically take an Agent, a Target and a Manip argument (Dixon 2010a: 104), referring to the person doing the action, the thing or person being affected by the action, and the implement the action is carried out with, respectively. What makes *apek* and *asuek* special among Paluai AFFECT verbs is that they take the Manip, the implement with which the action is carried out, as their O argument (which can be elided), and the Target as an obligatory E argument. The Target can be either animate or inanimate; the Manip argument is typically inanimate. There are several other verbs of affect, but they all take the Target as O argument and the Manip as optional Oblique instrumental.

apek and *asuek* are frozen applicatives, and are similar to productively formed applicatives in that the Manip is the semantic role taking the O marking. The difference between them is that the Target semantic role gets Oblique marking, whereas it is expressed as a "regular" O with productive applicatives, yielding a double object construction (see Section 8.3.2 below). Example (8) shows and contrasts the use of *apek* with another AFFECT verb, *sok* 'hit'; an example of *asuek* is given in (9).

(8) *ilêp pailou rai reo ma iapek tan parian deo, ino sok mat parian ai pailou reo*
    i$_A$=lêp   [pailou ta-i   te-yo]$_O$ ma   i$_A$=apek=Ø$_O$
    3sg=take   spear   CLF.POSS-3sg.PERT   LIG-DEM.INT   and   3sg=hit=3sg.ZERO
    ta-n   [paria-n   te-yo]$_E$ i$_A$=no   sok mat   [paria-n]$_O$
    CLF.POSS-PERT   wife-PERT   LIG-DEM.INT   3sg=IPFV   hit   die   wife-pert
    a=i   [pailou te-yo]$_{INSTR}$
    OBL=3sg   spear   LIG-DEM.INT
    'He took his spear and hit [it] at his wife, he killed his wife with the spear.' (WL020711_0135)

(9) *on asuek pau rao?*
    wo$_A$=an asuek   [pau]$_O$ ta-o$_E$
    2sg=PRF   rub   coconut.oil   CLF.POSS-2sg.PERT
    'Have you rubbed coconut oil onto yourself?' (field notes 21/03/2011)

Lêp 'take, receive' takes an A with Agent role, an O with Theme role and an E with Source role, usually referring to a human being. However, the verb is mostly encountered without a specified Source argument. When there is an explicit source, it is not usually introduced by a directional. An example with an animate O referent is given in (10).

(10) *eppe wot lêbi ran mwanen Parugui*
     ep$_A$=pe   wot   lêp=i$_O$ ta-n
     1pl.EXCL=PFV go.horizontally   take=3sg   CLF.POSS-PERT
     [mwane-n   Parugui]$_E$
     brother-PERT   P.
     'We went and took her from her brother Parugui.' (KM060111_0035)

Touek 'show, teach' takes an A with Agent role, but participates in two different patterns with regard to its other roles, i.e. Theme (that which is shown/taught) and Recipient (the person to whom it is shown/taught). In the first pattern, the Theme is marked as O and the Recipient receives Oblique marking; a directional introduces the argument referring to the recipient. An example is given in (11).

(11) *ippe rouek patpat la rep*
 ip$_A$=pe touek [patpat]$_O$ la ta-ep$_{AG}$
 3pl=PFV show bed go.to CLF.POSS-1pl.EXCL.PERT
 'They showed the customary marriage bed to us.' (KM060111_0068)

In the second pattern, the Theme is marked as E, with *a=*, while the Recipient is the O and indexed on the verb. There is no directional introducing the E argument. It is possible that the alternation in patterning corresponds to a semantic alternation between 'show' and 'teach'. An example of the second pattern is given in (12).

(12) *wong korouekap ai peinan net*
 wong$_A$ ko-touek=ap$_O$ a=i [peinan net]$_E$
 1sg.FREE 1sg.IRR-teach=2pl OBL=3sg making.of sea
 'I will teach you fishing.' (PN100411_0023)

The final verb to be discussed is *yuêt* 'ask', where the A argument represents the Asker and the O argument the Asked (generally a human being). The Question, i.e. what the person is asked about, is represented by an E argument, which is however not required.

(13) *ngoyai ipul la ran mui a iyuêri ai pangpangai rai*
 [ngoyai]$_A$ i=pul=Ø$_O$ la ta-n [mui] a
 cuscus 3sg=tell=3sg.ZERO go.to CLF.POSS-PERT dog and
 i$_A$=yuêt=i$_O$ a=i [pangpangai ta-i]$_E$
 3sg=ask=3sg OBL=3sg thought CLF.POSS-3sg.PERT
 'The cuscus talked to the dog and asked him about his opinion.' (LL010711_0012)

8.1.3 Peripheral arguments

In addition to core arguments, a predicate may have one or more peripheral arguments. This type of argument (which is also often referred to as an "adjunct" in the literature) is not usually seen as obligatory and can refer, among others, to location, time, instrument or reason. In Paluai, peripheral arguments are typically marked by the preposition *a=*; this type of marking is referred to throughout by means of the term Oblique. Section 4.4.1 gives examples of the semantic range of constituents introduced by *a=*. Examples of locative constituents consisting of local nouns appear in (14) and (15); examples of spatial nouns, in (16)–(18). When a locative constituent is formed by a local noun (see Section 3.2.1) or a directly possessed body part or spatial noun, it is not marked by *a=* and thus formally indistinguishable from a core argument. When there is an overtly expressed O argument, either an enclitic as in (16) or a full NP as in (18), it always directly follows the verb complex and precedes the locative constituent. Thus, in most clauses, particularly with transitive verbs, constituent order serves to make a distinction between core and peripheral arguments. But sometimes, e.g. in (15), it is not possible to determine the syntactic status of a constituent based on formal criteria alone, and the semantics and behaviour of both the verb and the noun heading the constituent need to be taken into account. Since a constituent following *lot* 'fall' needs to take the *a=* marker in other cases, it is concluded that *lot* is an intransitive verb and takes an optional locative constituent, not an O argument, and that the place name *Pityilu* is a local noun.

(14) *wuikala au nayek ansê alilêt*
 wui$_S$=ka-la   wau   nayek an-sê   [alilêt]$_{LOC}$
 1du.EXCL=IRR.NS-go.to move about PART-DIM forest
 'We would go and walk around a bit in the bush.' (KM050995_0003)

(15) *ila lot Pityilu*
 i$_S$=la   lot [Pityilu]$_{LOC}$
 3sg=go.to fall P.
 'He fell from the sky on Pityilu [small island to the north of Manus mainland].' (KW290611_0046)

(16) *pang nu ru nganui lalon kanum areo*
 pang$_A$ no   tu   ngan=ui$_O$ [lalo-n   kanum]$_{LOC}$
 rain   IPFV STAT.CONT eat=1du.EXCL inside-PERT garden
 a=te-yo
 OBL=LIG-DEM.INT
 'We were drenched by rain [*lit.* 'rain ate us'] inside the garden there.' (KM050995_0026)

(17) *ilaro ilili maranu pou reo*
 i$_S$=la   to   ilili   [mata-n   u   pou te-yo]$_{LOC}$
 3sg=go.to CONT stand.up in.front-PERT 3du pig LIG-DEM.INT
 'He is standing in front of the two pigs.' (Game2_021012_0245)

(18) *sap pulen sê iro wot to pe nangin naêmwan um*
 [sap pulen sê]_A i=to wot to pe [nangin]_O
 which thing DIM 3sg=CONT go.level CONT make smell
 [naêmwa-n wumwa]_LOC
 backside-PERT house
 '[He went to see] what kind of thing was causing a smell behind the house.' (LL300511_1_0051)

Instrumental constituents, on the other hand, are always marked by *a=* regardless of the semantics of the head noun. The difference between a locative and an instrumental Oblique formed by the body part noun *mina-* 'hand' is shown in (19) and (20). In (20), the O argument is topicalised and thus fronted (for more on this, see Section 12.3.3).

(19) *kei ilaro minan*
 [kei]_S i=la to [mina-n]_LOC
 tree 3sg=go.to be hand-PERT
 'The stick is in his hand.' (Game2_280812_0039)

(20) *kei rai reo ilaro nêktou ai minan i raalmaru*
 [kei ta-i te-yo]_O [i]_A=la to nêk.tou=Ø_O
 tree CLF.POSS-3sg.PERT LIG-DEM.INT 3sg=go.to CONT hold=3sg.ZERO
 a=i [mina-n i ta=almaru]_INSTR
 OBL=3sg hand-PERT 3sg SPEC.COLL=right
 'His stick, he is holding [it] with his right hand.' (Game2_280812_0019)

Peripheral arguments may be added to a clause regardless of the verb's valency, and so can potentially occur with intransitive, transitive, extended intransitive and extended transitive clauses. However, they do occur most often with intransitive clauses. When the main verb involves motion, a locative peripheral argument will often be introduced by a directional verb serialised with the main verb. This is discussed in more detail in Section 9.1.5. Peripheral arguments, in particular with instrumental semantics, may be promoted to O with an applicative operation. Applicatives are discussed in more detail in Section 8.3.2 below.

8.2 Valency

In this work, transitivity is seen as a feature of the clause, which consists of a predicate and its accompanying core and peripheral arguments. Valency, on the other hand, is regarded as a feature of verb roots. Whether a clause is syntactically transitive or intransitive is determined by the valency of the verb that fills the predicate. In Paluai, most verbs are strictly subcategorised for either one or two arguments, and thus can be said to have a valency of one or two, respectively. The language has a small number of ditransitives (with a valency of three), as discussed in Section 8.1.2.2.

Valency can be changed by morphological derivations, which will be discussed in the next section, or by means of a SVC, which will be discussed in Section 9.1.3. It is, however, rather tricky to establish base valency for verbs. As mentioned in Section 8.1, the O argument is usually elided when it is inanimate and can be retrieved from the discourse context; it is also not indexed on the verb complex, in contrast to animate O arguments. Thus, whether or not a verb is ambitransitive can only be established with certainty for those verbs for which it makes sense to have an animate O argument. If these are strictly transitive, it would be ungrammatical not to index an O argument with animate reference not expressed by a full NP. This turns out to be the case for all verbs encountered for which it makes sense to have an animate O. The O is, in these cases, always indexed on the verb complex, and thus these verbs can be considered strictly transitive. Examples are the verbs *ngan* 'eat', *tapôn* 'hide (tr.)', *pul* 'tell' and *pwapwasek* 'speak about'. The latter two verbs are interesting, because their O argument is usually the (inanimate) message, either represented by the noun *kamou* 'speech, words' or by direct quotation. A direct quotation is introduced by another verb of cognition/communication, *pwa* 'say, think', which can only have a direct quote as its object (in the form of a complement clause – see Section 11.1.2.3). *pwa* appears in a separate clause coordinated with the *pul* clause, and the S/A arguments of both verbs are coreferential. Direct quotations are therefore introduced as in (21):

(21) *aso-ong ino pul la rang a ipwa "si ret la pulen kone areo"*
 [asoa-ng]$_{S/A}$ i=no   pul la   [ta-ng]$_{AG}$ a
 husband-1sg.PERT 3sg=IPFV tell go.to CLF.POSS-1sg.PERT and
 i$_A$=pwa [si   tet   la   pulen.kone a=te-yo]$_{Compl:O}$
 3sg=say come.down move go.to beach   OBL=LIG-DEM.INT
 'My husband spoke to me and he said, "Come down to the beach there."'
 (LL030611_0045)

At first sight, *pul* appears to be an intransitive verb that usually takes an Oblique constituent introduced by a directional referring to the person spoken to (Receiver); the semantic role of Message is represented by the O argument of a separate verb *pwa*. However, *pul* can in fact take an O argument; what is more, if this O argument refers to an animate being, it has to be cross-referenced.

(22) *ngapwa kopul kamoun kun*
 nga$_A$=pwa [ko$_A$-pul [kamou-n kun]$_O$]$_{Compl:Pot}$
 1sg=want.to IRR.1sg-tell speech-PERT basket
 'I'm going to tell a story about baskets.' (MK050311_0001)

(23) *woning pou i re i nian teo, te rau puli reo?*
 wo$_A$=ning [pou$_i$ i [te i$_S$ [nia-n]$_{NVPRED}$]$_{RC}$ te-yo [te
 2sg=see pig 3sg REL 3sg stomach-PERT LIG-DEM.DIST REL
 tau$_A$ pul=i$_{Oi}$]$_{RC}$ te-yo]$_O$
 1du.INCL tell=3sg LIG-DEM.INT
 'Do you see the pig that is pregnant, that we were talking about?'
 (Game4_280812_0192-193)

In (23), the O argument of the verb *pul* is *pou* 'pig', an animate referent represented as a full NP in the preceding (matrix) clause. *Pul* occurs in a relative clause; the common argument is the O (see Section 11.1.1 for more on relative clauses). Since the head noun has an animate referent, this has to be indexed on the verb in the relative clause by a bound pronoun. This goes to show that *pul* is in fact strictly transitive.

There are, however, a number of verbs that may be ambitransitive. First of all, some verbs occur as transitives when used by themselves, but as intransitives in a SVC. These verbs are usually S = A ambitransitives. One example is *neng*, which means 'climb (tr.)' when used by itself, but 'step (intr.)' when used in a SVC. These cases will be discussed in Chapter 9. There is only one potential ambitransitive of the S = O type: *siei* 'tear'. Its transitive and intransitive use are shown below. However, in its intransitive variant *siei* only occurs with *pulêng* 'dawn' as its subject, and is thus rather restricted in its use.

(24) *wosiei tiap tang*
 wo$_A$=siei [tiap ta-ng]$_O$
 2sg=tear sarong CLF.POSS-1sg.PERT
 'You tore my sarong.' (elicitation 14/09/2012)

(25) *pulêng tu siei rea*
 [pulêng]ₛ tu siei te-ya
 dawn STAT.CONT tear LIG-then
 'The dawn was breaking then.' (LM190611_0047)

8.3 Valency-changing derivations

Although Paluai has a number of valency-changing derivations, these are quite limited compared to what is available for many other Oceanic languages and the reconstructed protolanguage (cf. Evans 2003). The transitivising suffix *-i for instance, which has reflexes in a great number of Oceanic languages, is not attested as a productive morpheme in present-day Paluai. However, a reflex of the transitivising suffix *-akin[i] is found, as well as a potential reflex of the causative prefix *pa[ka]-. With regard to valency-reducing devices, there is a mechanism of productive reduplication that derives an intransitive verb from a transitive one. There are also potential reflexes of the POc detransitivising prefixes *ma- and *ta-, but these are not productive.

The job of increasing valency has been taken over largely by SVCs. Since this is an important topic that needs discussion in its own right, valency-increasing verb serialisation will not be discussed here but in the next chapter, which is devoted to SVCs. In what follows, morphological operations to increase valency are discussed first; valency-reducing operations are next.

8.3.1 Causative *pe-*

Only intransitive stative verbs can be transitivised by means of a causative derivation. This is one of the criteria that has been used to distinguish stative verbs as a separate word class (see Section 3.3.1.2). When a causative operation is applied, the S argument of the intransitive verb moves into the O slot, and an additional A argument is introduced, the Causer. A causative is formed by means of the form *pe-*, prefixed to the verb. There are two possible sources for this form: it could be an instance of the full verb *pe* 'make, do', resulting in a periphrastic causative, or it could be a reflex of the abovementioned causative prefix *pa[ka]-. Evans (2003) notes that the POc prefix only derived causatives from stative verbs, which is an indication that *pe-* is a reflex of it.[57] Evidence in support of the prefix

---

[57] In fact, POc probably had two variants: *pa-, which derived causatives from Actor subject verbs, and *paka-, which derived causative from Undergoer subject verbs. In many Oceanic

status of *pe-* is that the derived form is regarded as a whole and can, for instance, be nominalised. A causative *pemat* 'kill' can be formed from the stative verb *mat* 'die, be dead', which is nominalised as *pemat-an (yamat)* 'murder' (lit. 'CAUS-die-NOM (person)'). In this regard, causatives differ from light verb constructions such as *pe kui* 'cook', lit. 'make pot' (see Section 3.3.1.6). With the latter, the nominalising suffix immediately follows *pe* (plus, sometimes, an *-i* formative): *pe-i-nan kui* 'cooking' (lit. make-NOM pot'). In the case of causatives, *pe-* is a prefix, whereas in the case of light verb constructions, *pe* is a full verb.

The following examples show the use of the causative. The additional A, expressing the Causer role of a causative, does not have to refer to an animate being with a high degree of volition and control, but can also refer to a weather phenomenon or other natural force, as in (27), (29) and (30).

(26) a. *imat*
   i$_S$=mat
   3sg=die
   'He is dead.'
   b. *ippe pemari*
   ip$_A$=pe   pe-mat=i$_O$
   3sg=PFV   CAUS-die=3sg
   'They killed him.' (KW290611_0046)

(27) a. *poko reo in ket*
   [poko   te-yo]$_S$ i=an   ket
   water.container LIG-DEM.INT 3sg=PRF be.full
   'The water container is full (has filled).'
   b. *pang kisi peket poko*
   [pang]$_A$ kisi   pe-ket   [poko]$_O$
   rain   IRR.3sg-come.down CAUS-be.full water.container
   'The rain will come down and fill up the water containers.'
   (LM260511_1_0050)

(28) a. *kun tang inali*
   [kun   ta-ng]$_S$ i=nali
   basket CLF.POSS-1sg.PERT 3sg=be.lost
   'My basket is lost.'

---

languages, intervocalic [k] has been lost, so the two prefixes became formally the same (cf. Evans 2003). The Paluai prefix would be a reflex of the latter, since as we will see below it only derives causatives from Undergoer subject (stative) verbs.

b. *ngan penali kun tang*
nga_A=an pe-nali [kun ta-ng]_O
1sg=PRF CAUS-be.lost basket CLF.POSS-1sg.PERT
'I have lost my basket [*lit.* 'made my basket be lost'].'
(elicitation 17/05/2011)

(29) a. *nganngan imwat*
[nganngan]_S i=mwat
food 3sg=be.cooked
'The food is done (cooked).'
b. *yep teo iro pemwat nganngan*
[yep te-yo]_A i=to pe-mwat [nganngan]_O
fire LIG-DEM.INT 3sg=HAB CAUS-be.cooked food
'The fire usually causes the food to be done.' (elicitation 12/09/2012)

(30) a. *ponat in kôk*
[ponat]_S i=an kôk
soil 3sg=PRF be.hot
'The soil has heated up.'
b. *sin kipe pekôk ponat*
[sin]_A ki-pe pe-kôk [ponat]_O
sun IRR.3SG-PFV CAUS-be.hot soil
'The sun will heat up the soil.' (elicitation 12/09/2012)

The reason why only stative verbs can enter into causative constructions is that they take an Undergoer subject. The distinction between Actor and Undergoer subjects is well established within Oceanic linguistics and was first made by Pawley (1973). It can be defined as follows (Evans 2003: 25):

> Verbs in Oceanic languages can be divided into two groups on the basis of the macro-role of the S argument of their intransitive form. [...] Actor and Undergoer, as used here, do not refer directly to semantic roles, but rather represent the interface between semantic roles and the morphosyntax. That is, they are a conglomeration of semantic roles into two categories, each category behaving differently in terms of morphosyntax.

Thus, the morphosyntactic behaviour of stative verbs differs from that of active verbs because their subjects have a different macrorole. This is borne out for Paluai stative verbs: they can enter into a causative operation that active verbs cannot enter into. Only the S argument of a stative verb can be demoted to an O argument through a causative construction, because it is underlyingly an Undergoer, a semantic role that is consistent with the O argument position.

There is a way to form a periphrastic construction with active verbs that is similar to a causative construction. (31) shows what it looks like:

(31) a. *ngapei a ipe ilili*
nga$_A$=pe=i$_O$ a i$_S$=pe ilili
1sg=make=3sg and 3sg=PFV stand.up
'I made him stand up.' [*lit.* 'I made him and he stood up.'] (elicitation 12/09/2012)
b. **ngapeililii*
*nga$_A$=pe-ilili=i$_O$
1sg=CAUS-stand.up=3sg
Intended: 'I made him stand up.'

This type of construction was only produced during elicitation sessions and not in spontaneous speech.[58] The S of the active verb is not demoted to O, but stays as the verb's S in an additional clause. The "causative" notion is expressed by the transitive dummy verb *pe* 'make, do', which takes an object enclitic coreferential with the subject of the active verb in the second clause. There are two reasons why this construction is not a real causative: firstly, it does not form one predicate but is made up of two coordinated clauses, and secondly, the S of the intransitive verb does not become the O of the transitive verb. The reason why verbs such as *ilili* cannot enter into a causative is that their semantics does not allow this. They have subjects with an underlying Actor macrorole, which is not in accordance with the O argument position.

It has to be noted that many causatives of stative verbs have lexical counterparts that would be preferred by speakers in everyday language use. For instance, instead of the causative *pekôk* 'heat up, make hot' (from the stative verb *kôk* 'be hot'), speakers prefer to use more specific transitive verbs such as *apin* 'heat up (food)', *nan* 'heat up (leaves)', *tun* 'boil (e.g. water)' etc. Overuse of the causative is seen as a sign of "childish" language use. Therefore, many instances of causatives were only found under elicitation, and not in spontaneous language use. None of these are ungrammatical; they are merely dispreferred.

There are a number of stative verbs for which it is infelicitous to form a causative with *pe-*, but that satisfy other criteria for stativity. An example is *nanet* 'ripen, be ripe', which can be modified by the adverb of degree *paran* 'very' and has a derived adjective *(n)antenen* 'ripe'. Both facts are consistent

---

[58] It may thus very well be a result of translation from Tok Pisin or perhaps English and is therefore analysed as having marginal status in the grammar.

with the criteria for stativity. However, a derived causative *penanet 'make ripe' was rejected by native speakers. This causative would necessarily have an inanimate causer such as *sin* 'sun', which appears not to be a problem for other causatives; compare for instance (29) and (30) above, both of which show an inanimate causer. There are two possible explanations. Either, verbs that cannot take a *pe-* causative do not have Undergoer subjects, but Actor subjects. Or, alternatively, in cases such as *nanet* a causative relation is deemed too tentative by speakers. It is possible that a cause-effect relation has to be rather direct for the causative construction to be felicitous. Thus, a fire or the sun heating something up would be fine, because there is a direct and perceptible relation between the two events of heating and becoming hot. However, the sun causing fruit to ripen is seen as too distant a relation to be represented by a causative construction, because it is not perceptible.

Because they satisfy other criteria that distinguish them as stative verbs, the verbs in this class for which a causative cannot be derived are still analysed as such, and not as active verbs with Actor subjects. The impossibility of the causative derivation is based on semantic, rather than syntactic grounds. When it is not possible to derive a morphological causative, this is because the relation between the causing event and the resulting event is deemed too obscure.

8.3.2 Applicative -(C)ek

Applicatives and quasi-applicatives (cf. Dixon 2012: 299) come in various flavours. In what follows, applicatives productively derived from verbs are discussed first, followed by frozen quasi-applicatives. Productive applicatives can mainly be classified as Instrumental, but there appear to be a handful of examples with other semantic functions. More general discussion on types of applicatives encountered cross-linguistically can be found, for instance, in Mithun (2001) or Dixon (2012).

8.3.2.1 The productive applicative

8.3.2.1.1 Form and function of the productive applicative

The *-ek* suffix on transitive verbs can productively derive an instrumental applicative.[59] An instrumental Oblique constituent of a verb (normally a peripheral

---

[59] The frozen applicative forms discussed below may have retained thematic consonants, hence the use of *-(C)ek* in the section title. The productive applicative, on the other hand, is never attested with a thematic consonant.

argument marked with *a=*) will be promoted to O position and subsequently elided, since it is retrievable from the discourse context. The original O is not demoted to Oblique status, because it does not receive marking with *a=* or *ta-*, but always follows the promoted constituent. Based on currently available data, these constructions are best analysed as double object constructions. The only difference between the two objects is that one of them directly follows the verbal form (and its suffix) and is elided. It therefore behaves more like a prototypical O argument than the second one.[60]

The applicative is typically encountered in one particular discourse/information structure context. It is used as an anaphorical device to refer back to an item mentioned just before, usually in the previous clause. This use of a reflex of the POc form *akin[i] as a "trace element" is attested for at least two other Oceanic languages: Bauan Fijian and Tongan (Evans 2003: 149). A few examples of the use of the applicative are given below. Examples (32)–(34) show cases where an instrument is mentioned (often introduced for the first time) in one clause and its use is described in a coordinated clause. In all three examples, the A argument of the first and the second clause are coreferential with each other, as are the O argument of the first and O1 argument of the second clause.

(32) *ope lêp suep a ope yilek ponat*
 wo$_A$=pe lêp [suep]$_O$ a wo$_A$=pe yil-ek=Ø$_{O1}$ [ponat]$_{O2}$
 2sg=PFV take hoe and 2sg=PFV dig-APPL=3sg.ZERO soil
 'You will take a hoe and you will dig the ground with it.' [*lit.* 'dig-with [it] the ground'] (KS030611_1_0015)

(33) *kope lêp samel a ong kobe ayitek lalon*
 ko$_A$-pe lêp [samel]$_O$ a wong$_A$ ko-pe
 IRR.1sg-PFV take knife and 1sg.FREE IRR.1sg-PFV
 ayit-ek=Ø$_{O1}$ [lalo-n]$_{O2}$
 separate-APPL=3sg.ZERO inside-PERT
 'I will take a knife and I will separate the inside layer of the bark with it.' (MK050311_0012)

---

60 It is not entirely clear, based on current data, to what extent the applicative construction is a "true" double object construction. There may be other subtle ways in which the two O arguments behave differently. This requires further testing.

8.3 Valency-changing derivations — 257

(34) *ipe lêp nipen kopup sê re onga ipe ro sanek parun ngoyai reo*
 i$_A$=pe   lêp   [nipen   kopup   sê
 3sg=PFV take part.of.round bamboo DIM
 te-yo]$_O$ onga   i$_A$=pe   to   san-ek=Ø$_{O1}$
 LIG-DEM.INT and.so 3sg=PFV CONT cut-APPL=3sg.ZERO
 [patu-n   ngoyai]$_{O2}$
 head-PERT cuscus
 'He took the sliver of bamboo and he cut the hair of the cuscus with it.'
 (LL010711_0020)

In example (35), the applicative is used in a relative clause (see Section 11.1.1 for more on relative clauses). Here, the O1 of the relative clause is coreferential with the subject of the matrix clause. Interestingly, it is not indexed on the verb, even though it has an animate referent. This may be one way in which applied objects are distinct from base objects.

(35) *pou re kope yiuek kup tan tamong nganoyumou*
 [pou]$_S$ [te   ko$_A$-pe   yiu-ek=Ø$_{O1}$
 pig   REL IRR.1sg-PFV pull.in-APPL=3sg.ZERO
 [kup   ta-n   tama-ng]$_{O2}$]$_{RC}$ nganoyumou
 rope.with.pigs CLF.POSS-PERT father-1sg.PERT   eight.ANIM
 'The pigs with which I am going to pull the rope of my father are eight in number.' (YK290411_2_0046)

That the applicative suffix *-ek* is productive is shown by its use with loans from Tok Pisin:

(36) *kope lêp yon a kope wasimek pelet tepwo*
 ko$_A$-pe   lêp   [yanu]$_O$ a   ko$_A$-pe   wasim-ek=Ø$_{O1}$
 IRR.1sg-PFV take water   and IRR.1sg-PFV wash-APPL=3sg.ZERO
 [pelet te-pwo]$_{O2}$
 plate LIG-DEM.PROX
 'I'm going to get water and wash these plates with it.'
 (field notes 28/10/2012)

The construction exemplified above is by far the commonest way in which two consecutive clauses containing an instrumental are expressed. The applicative is rarely encountered promoting peripheral arguments with different semantics, even in contexts where there is coreferentiality; one example is given in the next section. It is unclear why this is the case. Based on the amount of frozen applicative morphology found, as discussed below, it appears that the applicative operation was once much more productive. It ceased to be that way for all contexts apart from when there are two arguments with instrumental semantics that are coreferent with each other.

8.3.2.1.2 Semantics of the productive applicative

The main semantic function of the productive applicative is Instrumental: it promotes an instrumental Oblique to O function. There are only a handful of examples where an applicative appears to have another semantic function. One is shown in (37):

(37) ippe lêp si masayen a ipno rabuiek la nupun ip
 ip_A=pe lêp=$\emptyset_O$ si masayen a ip_A=no
 3pl=PFV take=3pl.ZERO come.down outside and 3pl=IPFV
 tabui-ek=$\emptyset_{O1}$ la [nupu-n ip]$_{LOC}$
 shoot-APPL=3pl.ZERO go.to bottom-PERT 3pl
 'They came outside taking [their tails] and they put them onto their bottoms.' [*lit.* they shoot-on [them] to their bottoms'] (LL300511_1_0064)

This rather strange example comes from a children's story about a tribe of dogs, where the chief dog one day decides that all his subjects have to take off their tails and put them into a specially built storehouse. However, the house catches fire and the dogs hastily go in to rescue their tails. They put them randomly back onto their bottoms and thus end up with the wrong tails. The NP *youn ip* 'their tails' is the elided O argument from the first sentence. The AFFECT verb *tapui* 'shoot' (normally) takes the Target as its O (here, *nupun ip* 'their bottoms') and the Manip as an optional Oblique (here, *youn ip* 'their tails'). However, in this case the Manip is promoted to O by an applicative operation (and elided), and the Target is expressed by a locative Oblique introduced by the directional *la* (the locative NP is not marked by *a*= because *nupu-* is a directly possessed body part noun).

An interesting feature of the applicative suffix is its use to mark reciprocal constructions, perhaps in combination with a reflex of the POc reciprocal prefix **paRi-*. This use is attested for a few other Oceanic languages as well (Evans 2003: 142). More information about reflexive and reciprocal constructions can be found in Section 8.3.4 below. The "reciprocal applicative" is attested for both transitive and extended intransitive verbs; an example of each is given in (38) and (39).

(38) *upe mwangeku*
 u$_A$=pe-mwang-ek=u$_O$
 3du=RECIP-watch-APPL=3du
 'The two of them looked at each other.' (LL010711_0011)

(39) *wui Maiau pe yangyangek tui*
 [wui Maiau]$_A$ pe-yangyang-ek [ta-ui]$_E$
 1du.EXCL M. RECIP-love-APPL CLF.POSS-1du.EXCL.PERT
 'Maiau and I fell in love with each other.' (KM060111_0009)

Thus, it can be concluded that productive applicatives are mainly to be characterised as Instrumental; only a few have Locative meaning. They further developed into a marker of reciprocal constructions.

8.3.2.2 Frozen applicatives with three core arguments

The verbs *apek* 'hit, shoot', *asuek* 'rub' and *touek* 'show; teach' were discussed above in Section 8.1.2.2, where some examples can also be found. They differ from productive Instrumental applicatives (but not from the example of Locative applicative found), because the third argument is always marked as an Oblique, with *a=* or *ta-*. It can be said that these verbs form more typical applicatives than the productive ones. It is likely that the applicative operation was more productive in the past, and had more uses than only the Instrumental one discussed above in Section 8.3.2.1. This is also evident from the large number of other frozen forms, discussed below.

8.3.2.3 Other frozen applicatives

A large number of verb forms have lexicalised with an applicative suffix included. Either, they have a form which ends in *-(C)ek* and a form which doesn't, with the difference in meaning between the two not always predictable. Or, alternatively, the verb only comes in a form ending in *-(C)ek* and the non-applicative counterpart has become obsolete.

A number of the forms ending in *-(C)ek* can more appropriately be classified as adverbs. They differ from verbal *-(C)ek* forms because they often have a nominal counterpart, and are not attested as the head of a predicate, but only following an independent verb. These forms are discussed briefly in Section 3.2.4.4. However, the boundary between verbal and adverbial forms is fuzzy. For instance, there are clearly verbal forms that have a nominal counterpart, such as the transitive *moleyek* 'decorate', which is related to *molea-* 'colour'. On the other hand, there are forms that are predominantly attested following a main verb, but are also occasionally found heading a predicate, such as *ngonomek* 'correspond to, (be) corresponding'. These forms usually do not have a nominal counterpart.

Most of the forms that are unambiguously verbs are transitive, but there are a few intransitive forms. Many of these seem to be inherently fully or partly reduplicated, and none of them have an unambiguous counterpart without the suffix. They are shown in Table 67.

Table 67: Intransitive frozen applicative verbs.

Form	Meaning	Optional Oblique
kumkulumuek	rinse the mouth	–
(mot)moreyek	joke	–
ngulnguluek	hang	Locative, where S hangs onto
parek	be hidden	Locative, where S is hidden
peruruek	be ignorant	Stimulus, what S is ignorant about
terepelek	run	Locative, where S is running to

As can be seen, most of the intransitive verbs have an optional Oblique argument. Thus, the presence of the *-(C)ek* suffix is hard to explain, because one would expect these verbs to take the Oblique argument as an O argument instead, due to the presence of the suffix. One possibility is that *-ek* refers to a reflexive meaning (cf. use of the applicative for reciprocals discussed above). The semantics of several of the verbs, such as 'hide oneself' (*parek*), is in line with this. However, reflexives are usually formed by adding the object bound pronoun (making the clause ambiguous between a reflexive and a non-reflexive reading), and a reflexive reading seems a bit far-fetched for some of the other verbs. These forms appear to be left-overs of a process that may once have been productive but that is obscure in the present-day language.

The remainder of the frozen quasi-applicatives are transitive; some have a non-applicative counterpart but many do not. Several have a thematic consonant: in addition to *-ek* we also find *-sek*, *-tek*, *-ngek*, *-nek* and *-lek*. These thematic consonants may have a semantic basis. This suggestion has been made for a number of other Oceanic languages, such as Bauan Fijian and Manam, but is controversial (cf. Evans 2003: 213). Below, the semantics of each of the possible varieties of the suffix will be discussed. Some items in the tables do not have a non-suffixed counterpart, but are included because of formal similarity with the other items. However, it is not possible to say with certainty that these forms are quasi-applicatives, unless a non-suffixed counterpart were to be found in the data.

8.3.2.3.1 Frozen applicatives with no thematic consonant

Table 68 lists some forms that are attested with the *-ek* suffix, without a thematic consonant. The verbs that were chosen to make the semantic extensions more palpable have an applicative and a non-applicative version, but with a slightly unpredictable change in meaning.[61]

Table 68: Some frozen applicative forms with *-ek*.

Form	Related form	Valency	Arguments
iliek 'stretch'	*ilili* 'stand' (intr)	Tr	A 'stretcher' and O 'thing stretched'
keleyek 'turning around'	*kel* 'mix' (tr)	N/A (Adv)	–
koloek 'wait for, expect'	*kol* 'wait for' (tr)	Tr	A 'waiter' and O 'thing waited for'
lomêek 'plant (yams)'	*lom* 'be ready for harvesting' (intr)	Tr	A 'planter and O 'thing (or garden) planted' (Obl: place where O is planted)
memeyek 'spoil, ruin'	*memeyeng* 'make trouble' (intr)	Tr	A 'spoiler' and O 'thing spoiled'
sisiek 'sweep out/away'	*sisi* 'sweep (tr)'	Tr	A 'sweeper' and O 'thing swept' (Obl: where O is swept into)
weiek 'coil'	*wei* 'wipe with circular motion' (tr)	Tr	A 'coiler' and O 'thing coiled' (Obl: thing O is coiled on)

With *iliek* and *lomêek*, the *-ek* suffix seems to have a causative-like function, deriving a transitive verb from an intransitive one. With *keleyek*, however, more or less the reverse seems to be the case: *-ek* derives a non-verbal form from a transitive verb. With the other verbs, *-ek* derives another transitive verb from a transitive verb, altering its meaning. It seems to mainly intensify the meaning of the verb it is derived from.

8.3.2.3.2 Frozen applicatives with thematic consonant /s/

In Table 69, examples of applicatives ending in *-sek* are given. Many of the forms with *-sek* have a counterpart without it; particularly in the case of emotive and

---

[61] The suffixed verb forms have sometimes retained a final vowel that has been lost in the non-suffixed form. This is a process similar to that for the suffixed and non-suffixed nouns described in Chapter 3.

Table 69: Some frozen applicative forms with -sek.

Form	Related form	Valency	Arguments
apsek 'sprinkle'	apek 'hit' (tr)	Tr	A 'sprinkler' and O 'thing which is sprinkled' (Obl: place where O is put)
lilisek 'forget about, ignore'	–	Adv	–
nesek 'reveal'	–	Tr	A 'revealer' and O 'thing revealed'
pilelsek 'laugh at/about'	pilel 'laugh' (intr)	Tr	A 'laugher' and O 'thing/person laughed at'
piysek 'squeeze out'	piy 'squeeze' (tr)	Tr	A 'squeezer' and O 'thing squeezed out' (Obl: place where O is squeezed into)
pwapwasek 'speak about'	pwa 'say' (tr)	Tr	A 'speaker' and O 'thing spoken about' (Obl: person spoken to)
tengsek 'cry about, mourn'	teng 'cry' (intr)	Tr	A 'mourner' and O 'thing/person mourned'
wayêsek 'worry about'	wayêt 'worry' (intr)	Tr	A 'worrier' and O 'person worried about' (Obl: reason for worry)
yokosek 'mix'	yok 'grab together' (tr)	Tr	A 'mixer' and O 'thing mixed' (Obl: thing O is mixed into)

cognitive verbs (laugh, cry, speak), the relation and meaning difference between the two is transparent. The intransitive verbs *pilel* 'laugh', *teng* 'cry' and *wayêt* 'worry' can take an optional Oblique expressing the Stimulus argument (see Section 8.1.2.1). The Stimulus, however, can be expressed as an O argument as well, by means of the counterpart ending in *-sek*. This implies furthermore that the action is more controlled and volitional on the part of the subject; compare e.g. the slight meaning difference between *teng* 'cry (about)' and *tengsek* 'mourn'. With other verbs, presence of *-sek* seems to denote repetition or intensification of the action described by the verb.

The historical source of the /s/ in these forms is most likely a root-final consonant, which was lost in the non-suffixed counterpart when applicativisation was still much more productive. This consonant was subsequently reanalysed as part of the suffix. The historical development from POc *tangis 'weep' to its Paluai reflexes *teng* and *tengsek* illustrates this. This development is attested in other Oceanic languages, such as Fijian (Arms 1973). It is likely that use of the reanalysed suffix including the consonant was subsequently extended to include other verbs similar in meaning to e.g. *teng* and that it thus became a thematic suffix.

8.3.2.3.3 Frozen applicatives with thematic consonant /t/

Table 70 lists examples of applicatives ending in *-tek*. Between vowels, /t/ is realised as [r]. *-tek* seems to indicate inward motion and objects that are brought into contact for an extended amount of time, although some of its uses appear to be quite idiosyncratic.

Table 70: Some frozen applicative forms with *-tek*.

Form	Related form	Valency	Arguments
langtek 'speak ill of a deceased person'	*lang* 'lift up' (tr)	unclear	unclear
lirek 'explain; sort out'	–	Tr	A 'sorter' and O 'thing sorted out'
porek 'attach (to s.t. strong)'	*porok* 'strength' (N)	Tr	A 'attacher' and O 'thing attached' (Obl: thing O is attached to)
sarek₁ 'wear (clothes)'	–	Tr	A 'wearer' and O 'thing worn'
sarek₂ 'put into, insert'	–	Tr	A 'inserter' and O 'thing inserted' (Obl: where O is inserted into)
yektek 'arrange'	*yek* 'hit' (tr)	Tr	A 'arranger' and O 'thing being arranged'

8.3.2.3.4 Frozen applicatives with thematic consonant /ŋ/

This consonant is only attested in three verb forms, shown in Table 71. They seem to denote outward motion.

Table 71: Frozen applicative forms with *-ngek*.

Form	Related form	Valency	Arguments
angek 'spread out'	–	Tr	A 'spreader' and O 'thing spread out' (Obl: place where O is put)
sulngek 'send (a person)'	–	Tr	A 'sender' and O 'person sent' (Obl: what O is sent for/where O is sent)
tamngek 'invite, send for'	–	Tr	A 'inviter' and O 'person invited' (Obl: reason for invitation)

8.3.2.3.5 Frozen applicatives with thematic consonant /n/

The sequence *-nek* is attested in the forms shown in Table 72. They seem to denote downward motion, with some idiosyncratic meanings.

Table 72: Frozen applicative forms with *-nek*.

Form	Related form	Valency	Arguments
monek 'spy on'	–	Tr	A 'spy' and O 'person spied on'
pitnek '(suddenly) think of s.b.'	*pit* 'be close' (intr)	unclear	unclear
samnek 'chew; guzzle down'	–	Tr	A 'chewer' and O 'thing chewed'
soknek 'dispense with (liquid)'	*sok* 'hit with thrown implement' (tr)	Tr	A 'dispenser' and O 'thing dispensed' (Obl: where O is put)
tanek 'create, design'	–	Tr	A 'creator' and O 'thing created'

8.3.2.3.6 Frozen applicatives with thematic consonant /l/

The sequence *-lek* is attested in the forms shown in Table 73. They seem to denote upward motion, although there are also some idiosyncratic meanings.

Table 73: Frozen applicative forms with *-lek*.

Form	Related form	Valency	Arguments
pipilek 'defame, slander'	*pipia* 'rubbish' (N)	Tr	A 'defamer' and O 'person defamed'
sapolek 'throw over (with arc)'	*sapo-* 'top' (N)	Tr	A 'thrower' and O 'thing thrown' (Obl: place where O is thrown)
sulek 'raise'	*sulu* 'tighten'	Tr	A 'raiser' and O 'thing raised'
wolek 'ordain, bless'	–	Tr	unclear

8.3.2.4 Applicatives: overview and conclusions

Evans (2003: 199) identifies five different role-marking functions of *akin[i] in POc, depending on the semantics of the verb. These are shown in Table 74.

It seems that these role-marking functions can partly be recognised in the Paluai reflex of POc. The concomitant function of the suffix may have led to grammaticalisation into a marker of reciprocal constructions, but as such, a concomitant function of *-(C)ek* is not found in present-day Paluai. The suffix *-sek* with a thematic consonant /s/, although no longer productive, is clearly reserved for emotive verbs and speech/cognition verbs, to indicate the role of stimulus and content, respectively. The applicative suffix is not attested as a marker of the

Table 74: Participant role-marking functions of *akin[i].

Type of verb	Role marked by *akin[i]
motion verbs	Concomitant
psychological / emotional states	cause / stimulus
speech and cognition	content
excretion / secretion	product
process-action verbs	instrument, benefactive

product of verbs of excretion and secretion. Finally, the productive applicative is frequently attested marking an Instrument of a process-action verb (typical transitive verbs such as 'cut', 'dig', 'wash', etc. can be regarded as process-actions verbs). Applicative as a marker of Benefactive, however, is not attested, except possibly in a few instances of the verb *touek* 'show; teach' (see Section 8.1.2.2). This semantic role is typically marked by a SVC and *ta-/ka-*; these constructions are discussed in Chapter 9.

8.3.3 Valency-reducing operations

8.3.3.1 Reduplication

As mentioned before, reduplication can function as a valency-reducing operation. Full or partial reduplication of a base transitive verb makes it intransitive, with the A argument of the transitive construction becoming the S of the intransitive construction. This operation is discussed in Section 3.3.2.1.2.1; examples given there are repeated here.

(40) *epngan yapi*
 ep$_A$=ngan [yapi]$_O$
 1pl.EXCL=eat sago
 'We ate sago.'

(41) *epnganngan nêm*
 ep$_S$=ngan.ngan   nêm
 1pl.EXCL=REDUP.eat be.finished
 'We finished eating.' (KM060111_0075)

(42) *ipkalomêek suei le mwayen, kalolomêek nêm...*
ip_A=ka-lomêek [suei le mwayen]_O ka_S-lo.lomêek
3pl=IRR.NS-plant short.yam or yam IRR.NS-REDUP.plant
nêm
be.finished
'They will plant short or long yam. (When) they finish planting...'
(KM190211_0035)

8.3.3.2 Fossilised prefixes *ma-* and *ta-*

Evans (2003: 267) discusses two POc prefixes, *ma- and *ta-, which are defined as "stative verb derivatives". *ta- furthermore indicated that the state came about spontaneously. In present-day Paluai, many stative verbs (and some active intransitive verbs) start with *ma-* or *ta-* and were thus likely derived from a prefixed POc form. A number still have a transitive counterpart attested. Tables 75 and 76 show some examples of these verbs.

Table 75: Derived intransitive verbs with a *ma-* prefix.

Form	Valency	Related form
makap 'be thin'	Stat intr	–
maleu 'be quick'	Stat intr	–
maling 'be lost'	Stat intr	–
malo 'disappear'	Act intr	–
mamat 'be awake'	Stat intr	*mat* 'be dead, unconscious' (stat)
mapông 'be cracked, popped open (of fruit)'	Stat intr	*pông*$_1$ 'wake s.b. up by poking gently' *pông*$_2$ 'popping sound'
mapwai 'know'	Act intr	*pwa* 'say, think' (tr)
mari 'be asleep'	Stat intr	–
masai 'be clear'	Stat intr	–
maut 'sink'	Stat intr	–
mayai 'be quick'	Stat intr	–
mayeng 'be cracked, be split'	Stat intr	*yeng* 'slice' (tr)

The *ta-* and *ma-* prefixes likely functioned as intransitivising devices, yielding what can be characterised as an anticausative derivation. The O argument of the transitive verb became the S argument of the derived intransitive verb, and the A argument was deleted. Because the resulting intransitive verb has an Undergoer subject, it functions as a stative verb. Examples comparing *sil* 'peel off' and its

Table 76: Derived intransitive verbs with a *ta-* prefix.

Form	Valency	Related form
takau '(go) directly'	Act intr	–
tangalau 'be tall (humans)'	Stat intr	*alau* 'grow' (intr)
tapal 'be broken'	Stat intr	*pal* 'break' (stat)
tapoy 'be loose, loosened'	Stat intr	–
tapwak 'be peeled off'	Stat intr	*pwak* 'be stuck' (stat)
tarak 'climb, ascend'	Act intr	*tak* 'rise, point upwards' (stat)
tasil 'be peeled off'	Stat intr	*sil* 'split, divide' (tr)

derivative *tasil* 'peeled off' are given in (43) and (44). However, the process has probably long ago ceased to be productive, as is also evident from a lot of irregularity in the present-day forms. *pal* 'break', for instance, is currently only attested as a stative verb and not as a transitive one.

(43) *ong kope sil antek kunuan*
wong$_A$ ko-pe   sil   antek   [kunua-n]$_O$
1sg.FREE IRR.1sg-PFV split put.away bark-PERT
'I'm going to take off the [outer layer of the] bark.' (MK050311_0006)

(44) *asilon lopwan ilili rai relo in tasil*
[asilo-n   lopwa-n   ilili   ta-i   te-lo]$_S$
side-PERT place-PERT stand.up CLF.POSS-3sg.PERT LIG-DEM.DIST
i=an   tasil
3sg=PRF   peeled.off
'The [paint at the] side of the thing he is standing on has peeled off.'
(Game4_280812_0345)

Not for all verbs starting with *ma-* or *ta-* does it make sense to have a transitive counterpart (e.g. *mari* 'be asleep') and thus it is quite likely that there never has been such a verb. For others, a transitive counterpart probably has existed but has become obsolete. The tables show that *ma-* is more regular and transparent than *ta-*. There are two verbs starting with *ma-* (*malo* 'disappear' and *mapwai* 'know') which are not stative, i.e. they have Actor subjects. Even more active intransitive and transitive verbs starting with *ta-* are attested.

8.3.4 Reflexive and reciprocal constructions

8.3.4.1 Reflexives

Several transitive verbs can take an O argument that is coreferential with the A argument, yielding a reflexive construction. Reflexive constructions do not receive any special marking on either the verb or any of its arguments; thus, reflexive constructions look the same as regular transitive clauses. That A and O are coreferential is evident only from the semantics of the verb and the discourse context. Out of context, certain utterances with third person reference may be ambiguous between a reflexive and a non-reflexive reading; example (45) could indicate either *He$_i$ looked at himself$_i$* or *He$_i$ looked at him$_j$*. Since reflexive constructions usually have an animate A argument, the coreferential O argument is indexed by a bound pronoun enclitic on the verb. If one of the arguments is stated as a full NP, it is always the A argument, as in (46) and (47).

(45) *imaru riyi lai sesap pwên*
    i$_i$=ma=tu   tiy=i$_i$ la   a=i   sesap   pwên
    3sg=NEG1=STAT.CONT observe=3sg go.to OBL=3sg something NEG$_2$
    'He didn't look at himself with anything.' (LL010711_0043)

(46) *muyou reo iweieki la ro patan yoy*
    [muyou te-yo]$_A$ i$_i$=weiek=i$_i$ la   to   pata-n   yoy
    snake   LIG-DEM.INT 3sg=coil=3sg go.to be top-PERT stone
    'The snake has coiled itself on top of the stone.' (Game1_280812_0008)

(47) *pulei iro asueki rang*
    [pulei]$_A$ i$_i$=to   asuek=i$_i$ ta-ng
    cat   3sg=CONT rub=3sg CLF.POSS-1sg.PERT
    'The cat is rubbing itself against me.' (field notes 30/04/2011)

(48) *ngano rayenong pun lalon net teo*
    nga$_i$=no tayen=ong$_i$ pun lalo-n   net te-yo
    1sg=IPFV submerge=1sg INTF inside-PERT sea LIG-DEM.INT
    'I submerged myself fully into the sea.' (LL030611_0056)

As is evident from these examples, the controller does not have to be human, but it seems it has to be at least animate. There appear to be no inherently reflexive verbs, i.e. verbs that always have to occur in a reflexive construction.

The verb *tiy* 'observe', rather than *ning* 'see', may be preferred for the action of looking at oneself, as in (45).

Occasionally, a reflexive construction is found with a coreferential Oblique, rather than an O argument:

(49) *on asuek pau rao?*
   wo$_{iA}$=an asuek [pau]$_O$ ta-o$_{iE}$
   2sg=PRF rub coconut.oil CLF.POSS-2sg.PERT
   'Have you rubbed coconut oil onto yourself?' (field notes 21/03/2011)

Sometimes, a reflexive construction appears to be used in order to satisfy the requirements of the verb's argument structure, as in examples (50) and (51) below. Because *top* 'drop' and *yêk* 'feel' are transitive verbs, there needs to be an O argument. Since the O argument is coreferential with the A argument of the verb, it is animate and has to be indexed on the verb complex. Because the Activity complement clause in (51) functions as an Oblique to the verb (it is marked by the preposition *a=*; see also Section 11.1.2), a dummy O argument has to be present.

(50) *ipe ropi la net a ipe yaya*
   i$_i$=pe top=i$_i$ la net a i=pe yaya
   3sg=PFV drop=3sg go.to sea and 3sg=PFV swim
   'He let himself fall [lit. 'dropped himself'] into the sea and swam.'
   (KW290611_0055)

(51) *iyêki are sông iraii*
   i$_i$=yêk=i$_i$ a=te sông i=tai=i$_i$
   3sg=feel=3sg OBL=COMP hunger 3sg=take.possession.of=3sg
   'He felt himself getting hungry.' [lit. 'hunger taking possession of him.']
   (WL020611_0018)

8.3.4.2 Reciprocals

Reciprocal constructions also include coreference of participants, but in a slightly different way: a participant X acts in a certain way towards Y and Y simultaneously acts in the same way towards X. This, at least, is the case for two-participant reciprocals, which are the only reciprocals attested in the data. There is no reciprocal pronoun or other form used specifically to mark reciprocal constructions. However, as mentioned in Section 8.3.2.1, it seems that the applicative suffix is used to indicate a reciprocal meaning, perhaps in combination with a reflex *pe-* of the POc reciprocal prefix **paRi-* (this prefix is not attested without the applicative suffix). The "reciprocal applicative" is attested for both transitive and extended intransitive verbs; an example of each is given below. Two-participant reciprocals are indicated by dual pronouns coreferential with each other. Again, if one of the arguments is stated (partly) as a full NP, it is always the A argument, as in (53).[62]

(52) *upemwangeku*
 u$_i$=pe-mwang-ek=u$_i$
 3du=RECIP-watch-APPL=3du
 'The two of them looked at each other.' (LL010711_0011)

(53) *wui Maiau peyangyangek tui*
 [wui Maiau]$_i$ pe-yangyang-ek ta-ui$_i$
 1du.EXCL M. RECIP-love-APPL CLF.POSS-1du.EXCL.PERT
 'Maiau and I fell in love with each other.' (KM060111_0009)

Example (54) below is a slight variation on the ones given above, since here it is a Possessor that is coreferential with the proclitic referring to the A argument. Again, there is a reciprocal prefix and applicative suffix on the verb.

(54) *upeyek keleyekek maran u la sip*
 u$_i$=pe-yek keleyek-ek mata-n u$_i$ la sip
 2du=RECIP-hit turn-APPL face-PERT 3du go.to one.INANIM
 'They have turned facing each other.' (Game4_280812_0121)

Reciprocal constructions are quite rare in spontaneous speech. Sometimes, a reduplicated form of a transitive verb is used to refer to a reciprocal action. In (55) below, nominalisation of the transitive verb *kam* 'hug' is shown.

(55) *taukaro pe kamkam*
 tau$_S$=ka-to pe kam.kam
 1du.INCL=IRR.NS-HAB make REDUP.hug
 'The two of us should hug (each other).' [*lit*. 'The two of us should do hugging.'] (PK290411_3_0025)

---

[62] The use of the dual pronoun together with a full NP as in (53) is typical for Paluai and other Oceanic languages. See Section 5.9 and Lichtenberk (2000) on "inclusory pronominals" for a more detailed discussion.

In addition, there is a verb *ngonomek* with a very general meaning 'be corresponding, be the same' that could be said to have inherently reciprocal semantics.

(56) *umape ngonomek pwên*
      u$_S$=ma=pe   ngonomek   pwên
      3du=NEG$_1$=PFV   be.corresponding   NEG$_2$
      'They do not resemble each other.' (Game3_280812_0009)

9 Serial verb constructions

The Paluai corpus contains a large number of sequences that appear to form a single predicate consisting of more than one verbal element. Not all of these are analysed as serial verb constructions (SVC). This chapter discusses in detail the criteria a construction needs to fulfil in order to be analysed as a SVC, and which subtypes of SVCs can be distinguished. SVCs can be contrasted with the morphological process of compounding, on the one hand, and with the syntactic operations of coordination and subordination, on the other. Diachronically, it is possible that preverbal particles, adverbial forms or prepositions have developed out of full verbs that were once part of SVCs.

It is widely agreed that a sequence of verbs must comply with the following criteria to be regarded as a SVC (based on Aikhenvald (2006) and references therein):
1. It must form a single predicate, without any overt marking of coordination, subordination or syntactic dependency;
2. It must describe what is conceptualised as a single event;
3. It must have intonation properties that are the same as those of a monoverbal clause;
4. It must have one tense, aspect and polarity value;
5. Each component must be able to appear on its own.

For Paluai SVCs, all of the criteria apply. An important distinction with regard to SVCs is made between symmetrical and asymmetrical SVCs (Aikhenvald 2006). Symmetrical SVCs obtain all their elements from the open, unrestricted class of lexical verbs, whereas asymmetrical SVCs recruit one element from a semantically and/or grammatically restricted set of verbs. The other element is typically a lexical verb. Symmetrical SVCs (as entities) are prone to lexicalisation, whereas the components from restricted subclasses in asymmetrical SVCs are prone to losing verbal status and grammaticalise into adpositions, adverbs, or case markers.

Another important distinction relates to the arguments that the serialised verbs in a SVC can share. Because Paluai does not have SVCs of more than two verbs, the elements of a SVC will be referred to as V1 and V2. The following structural types of SVCs can then be distinguished (Crowley 2002):
1. Same-subject serialisation: V1 and V2 have the same subject.
2. Switch-subject (or switch-function) serialisation: the object of V1 is the subject of V2.
3. Inclusory serialisation: the subject of V2 consists of both the subject and object of V1.

4. Multiple-object serialisation: "there may be same-subject or switch-subject relationships between the subjects of serialised verbs, each of which is transitive, and each of which has its own object" (Crowley 2002: 41).
5. Ambient (or event-argument (cf. Aikhenvald 2006)) serialisation: V2 makes a qualification about the manner in which an action is performed, without V1 and V2 sharing any arguments.

In Paluai, only types (1) and (2) are found, which is in line with general typological tendencies for SVCs (Aikhenvald 2006: 55). In what follows, the discussion will turn to asymmetrical SVCs first, followed by symmetrical SVCs. It is doubtful whether symmetrical serialisation is a productive process in the present-day language. Symmetrical SVCs are contiguous (i.e. no elements are allowed to occur between V1 and V2), but this is also the case for some types of asymmetrical SVCs.

9.1 Asymmetrical SVCs

Asymmetrical SVCs have in common that one of their components is picked from a restricted subset of verbs. The other verb is generally chosen from the unrestricted class of lexical verbs. In Paluai, asymmetrical SVCs are more common than symmetrical ones, which is in line with cross-linguistic typology. Aikhenvald (2006: 56) states that if a language has limited serialisation, it generally only has asymmetrical SVCs. The restricted members of asymmetrical SVCs are prone to grammaticalisation: they often develop into adpositions, case markers, TAM markers, etc.

Asymmetrical SVCs in any given language can usually be divided into a number of different types based on their semantics and/or grammatical functions. What exactly is considered an asymmetrical SVC is a matter for discussion, with quite far-reaching implications for Paluai. As discussed in Chapter 6, preverbal secondary aspect particles and directionals have full verb counterparts that can head a predicate on their own. Verbal inflectional morphology, if present, only affects the first element in the VC, regardless of whether this is a full verb or a particle; it does not affect any other elements. Verbal inflection is thus not a useful criterion to distinguish full verbs and particles, and every construction containing a particle could alternatively be analysed as a SVC. Since these are very common, this alternative analysis would greatly increase the frequency of asymmetrical SVCs, which are already very frequent even under a conservative approach. In fact, there would hardly be any predicate (save non-verbal ones) not containing a SVC. Most likely, there is a grammaticalisation cline of verb to

particle, with many forms sitting somewhere along the continuum. Preverbal particles are discussed in detail in Chapter 6, together with arguments supporting their distinctive status compared to full verbs. The following semantic and grammatical types of asymmetrical SVCs can be distinguished:
1. Cause-effect/resultative
2. Adverbial
3. Valency-increasing
4. Posture verb as V2
5. Directional as V2
    a. Specifying Goal of ditransitive clause
    b. Specifying Oblique of intransitive clause

Asymmetrical SVCs are usually productive or semi-productive, but semi-lexicalised combinations are also attested. In addition, the restricted members of types (1) and (2) are prone to develop into adverbs (i.e. forms that do not occur as predicate heads) and the restricted members of types (4) and (5) may develop into preposition-like items and lose verbal status. In what follows, asymmetrical SVCs will be discussed in turn.

9.1.1 Cause-effect/resultative SVCs

In cause-effect/resultative SVCs, V2 represents either an effect caused by the action denoted by V1, or it describes a result of the action denoted by V1. In the majority of cases, V1 is a transitive AFFECT verb; see Section 3.3.1.1. V2 is always a stative verb, and most often a BREAKING verb; see Section 3.3.1.2. Which combination of verbs is most appropriate depends on the action and the material involved. Thus, although these SVCs seem to be rather productive, their occurrence is bound by semantic constraints. A number of examples are given in Table 77.

These SVCs are of the switch-subject type: the O argument of V1 is the S argument of V2. Since stative verbs have Undergoer subjects (see Section 8.3.1), they can lend themselves to this type of construction. The resulting SVC is always contiguous: no elements can occur between V1 and V2. The O argument of the SVC always follows it. Examples (1) to (4) show how the cause-effect/resultative type of SVC is used. Note that in sentence (1), the original O argument of *pêl* 'roll' is suppressed. It is ousted by the S argument of *pwon* 'be covered', which takes the place of the O argument of the SVC. The original O of *pêl*, 'stones', is deleted and has to be inferred from the context.

Table 77: Some cause-effect/resultative SVCs.

Verb combination	Meaning	Meaning V1	Meaning V2
arei mat	bite dead	bite	die
neng pui	crush with foot (fruit etc.)	step on	be soft
pêl pwon	cover by rolling s.t. on	roll	be covered
sok mut	tear (cloth etc.)	hit with implement	tear, snap (of rope or cloth)
sok pal	break into halves (by slamming)	hit with implement	burst, break, crack (in half)

(1) *ipkapêl pwon yep teo*
 ip$_A$=ka-[pêl pwon]$_{SVC}$ [yep te-yo]$_O$
 3pl=IRR.NS-roll be.covered fire LIG-DEM.INT
 'They will cover up the fire by rolling stones onto it.' [*lit.* 'roll-cover the fire'] (MS250311_0065)

(2) *kowom mut poyengom teo*
 ko$_A$-[wom mut]$_{SVC}$ [poyenga-m te-yo]$_O$
 IRR.1sg-chop break throat-2sg.PERT LIG-DEM.INT
 'I will chop off your head.' [*lit.* 'chop-break your throat'] (NS220511_1_0027)

(3) *Komou reo isok sau Parulabei sopol*
 [Komou te-yo]$_A$ i=[sok sau]$_{SVC}$ [Parulabei sopol]$_O$
 K. LIG-DEM.INT 3sg=hit split P. one.half.round
 'Komou cut off one half of Parulabei.' [*lit.* 'hit-cut one half'] (LK250111_0069)

(4) *kino arei mari*
 ki$_A$-no [arei mat]$_{SVC}$=i$_O$
 IRR.3sg-IPFV bite die=3sg
 'He may bite him to death.' (LL010711_0090)

Cause-effect SVCs are closely related to causative constructions (see Section 8.3.1). In fact, the causative construction could have developed out of a SVC that has *pe* 'make' as its V1. It seems that the causative construction can be used as a more 'general' means to describe the accomplishment and causation of a result, whereas a cause-effect SVC is more specific as to exactly how this result was achieved. Thus, the causative *pemat* 'kill' mainly refers to the act of 'killing', whereas *arei mat* 'bite dead' and *neng mat* 'kill by standing on' specify the manner of killing. *pemut* 'tear, rip' mainly refers to the act of 'breaking', whereas *san mut* 'cut', *wom mut* 'chop' or *neng mut* 'tear by standing on' refer to the specific action that caused the material to separate.

9.1.1.1 Cause-effect SVCs with intransitive V1

There is a subtype of cause-effect SVCs in which V1 and V2 have no arguments in common. This subtype appears to be quite rare, but some examples are nonetheless given in (5) and (6) below, with the stative verb *pwon* 'be covered'.

(5) *iro yen pwon sal*
  i$_A$=to   [yen pwon]$_{SVC}$ [sal]$_O$
  3sg=CONT lie   be.covered road
  'He [the cat] is lying in the way.' [*lit.* 'lie-cover the road'] (field notes 01/11/2012)

(6) *leut kipe ret pwono*
  [leut]$_A$ ki-pe   [tet   pwon]$_{SVC}$=o$_O$
  weed   IRR.3sg-PFV grow be.covered=2sg
  'Weeds will grow so as to cover you.' [*lit.* 'grow-cover you'] (LK250111_0073)

Both examples describe a single-participant event ('lying' and 'growing', respectively), the result of which leads to another participant being blocked or covered. The SVC as a whole is transitive, just as in the case of regular cause-effect SVCs. The difference is that V1 and V2 do not share arguments, since V1 is intransitive and thus does not have an O argument. The S of V1 becomes the A argument of the SVC, and the S argument of V2 becomes the O argument.

9.1.2 Adverbial SVCs

In adverbial SVCs, V2 provides a description of the action indicated by V1 rather than an indication of its effect or result. Adverbial SVCs are very similar to verb-adverb sequences, and it is quite likely that the latter have developed out of the former. The difference between them is that with adverbial SVCs, the second element can still function as an independent verb.

In an adverbial SVC, V1 can be either transitive or intransitive; V2 is always stative.[63] Adverbial SVCs are of the same-subject type: the S or A argument of V1 corresponds to the S argument of V2. Some examples of forms often encountered in V2 position are *maleu* 'hurry' and *parek* 'be hidden'; they are illustrated in (7) and (8). Adverbial SVCs are always contiguous.

(7) *ong koret parek monok tai*
wong$_S$ ko-[tet   parek]$_{SVC}$ monok ta-i
1sg.FREE IRR.1sg-move be.hidden behind CLF.POSS-3sg.PERT
'I will go stealthily after her.' (KS030611_1_0029)

(8) *ipwa kingan maleu*
i$_A$=pwa   [ki$_A$-[ngan maleu]$_{SVC}$=Ø$_O$]$_{Compl:Pot}$
3sg=want.to IRR.3sg-eat hurry=3sg.ZERO
'He wanted to eat [it] quickly.' (Game1_021012_0404)

Some stative verbs, such as *pit* 'be close', may yield either a cause-effect/resultative SVC or an adverbial SVC depending on the transitivity of V1. In (9), where V1 is intransitive, the S of V1 and V2 is the same. When *pit* is used with a transitive V1, however, its S argument will correspond to the O argument of V1, for example in *tou pit* 'bring together', *soyek pit* 'heap together' or *yuai pit* 'call together'. These constructions look the same as other cause-effect/resultative SVCs, as shown for instance in example (1).

(9) *onga ippe reng pit*
onga   ip$_S$=pe   [teng pit]$_{SVC}$
and.so 3pl=PFV cry   be.close
'And so, they came together to cry.' (NP210511_2_0079)

Sentence (10) shows that stative verbs that are frequently used as independent verbs, such as *lot* 'fall', can also be felicitously used in adverbial SVCs. The sentence was used to describe the first attempts at walking of little kittens.

---

63 This is why *liliu* 'return; again' is not discussed here, but in Section 9.2.1. *liliu* is attested both as an independent (active) verb and as an adverb, and thus occurs in SVCs as well as in verb-adverb sequences.

(10) *ipto wauwau lot*
 ip_S=to [wau.wau lot]_SVC
 3pl=CONT REDUP.move fall
 'They are walking about in a falling manner.' (field notes 19/10/2012)

Eventually, when it ceases to be used on its own, the second element of an adverbial SVC is likely to develop into an adverb. This can happen regardless of semantics, as is shown, for example, by the form *panak* 'steal'. This form may have been verbal at some point, but in the present-day language it only occurs following a lexical verb such as *neng* 'climb', *lêp* 'take' or *pe* 'do' to refer to the act of stealing (again, sequences with *panak* are not classified as SVCs, since *panak* does not occur independently).

(11) *ipe la lêp panak narunpein som Us*
 i_A=pe la lêp panak [narun.pein som]_O [Us]_SOURCE
 3sg=PFV go.to take steal young.woman one.ANIM Hus Island
 'He went and stole a young woman from Hus Island.' (KW290611_0034)

9.1.3 Valency-increasing SVCs with *tou* or *lêp*

Paluai has two types of asymmetrical SVCs that increase the valency of an intransitive verb, making it transitive.[64] The original S argument becomes the A of the construction, and a new O argument is added. These constructions are formed with either *tou* 'give' or *lêp* 'take' as V2, and have the semantics described in (12):

(12) with V2 *tou* 'V1 after X', 'V1 while following X'
 with V2 *lêp* 'V1 with X', 'V1 while taking X' (comitative)

Even though in these cases V2 is a lexical verb, these SVCs are analysed as asymmetrical because V2 can only be one of two choices. This type of SVC is always contiguous, with the O argument of the construction following V2. The shared argument is S of V1 and A of V2. A few examples are given in (13) to (16).

---

64 In addition, there are SVCs where *lêp* follows a transitive verb. These are rare and only occur in a very specific context with the adverbial phrase *la pwên*, discussed in Section 10.2.3.1.

(13) *iro reng touu*
 i$_A$=to   [teng tou]$_{SVC}$=u$_O$
 3sg=CONT cry   give=3du
 'He was crying after them.' [*lit.* 'cry-give'] (MS250311_0017)

(14) *ipe ngui rou ngoyai*
 i$_A$=pe   [ngui tou]$_{SVC}$ [ngoyai]$_O$
 3sg=PFV snarl give   cuscus
 'He snarled after the cuscus / he went after the cuscus while snarling at him.' [*lit.* 'snarl- give'] (LL010711_0069)

(15) *ipe wop lêp la kau rinan*
 i$_A$=pe   [wop lêp]$_{SVC}$ la   ka-u   tina-n
 3sg=PFV fly take   go.to CLF.FOOD-3du mother-PERT
 'He flew and took [it] for him and his mother to eat.' [*lit.* 'fly-take'] (KW290611_0038)

(16) *numwai re iangou lêp mangat tai...*
 numwai te   i$_A$=[angou lêp]$_{SVC}$ [mangat ta-i]$_O$
 old.man REL 3sg=arrive take   work   CLF.POSS-3sg.PERT
 'The old man, who came to us with his work...' [*lit.* 'arrive-take'] (BK040311_0007)

9.1.4 SVCs with a posture verb

The forms *to* and in particular *tu* (both meaning 'stay') are encountered following lexical verbs. They derive from posture verbs, but are most common as preverbal particles indicating secondary aspect. Their semantics and use as aspectual particles is discussed in detail in Section 6.2.1.2. *tu* seems to be used postverbally or as V2 when a change of state referred to by the main verb is permanent. Two examples are given in (17) and (18).

(17) *kamou rang ime nêm tu areo*
 [kamou ta-ng]$_S$ i=me   [nêm   tu]$_{SVC}$ a=te-yo
 speech CLF.POSS-1sg.PERT 3sg=come be.finished stay   OBL=LIG-DEM.INT
 'My talk has come to an end at this point.' [*lit.* 'come be finished- stay there'] (KM060111_0104)

(18) *tinan no mat tu nii*
[tina-n]$_A$ no   [mat tu]$_{SVC}$ ni=i$_O$
mother-PERT IPFV die   stay   away=3sg
'Her mother died and left her.' [*lit.* 'die-stay away from her.']
(OL201210_0146)

To is also encountered following directionals; it then indicates that something has moved and come to rest at a certain point. Only when used in combination with *lak* 'go' and *sak* 'come up' is it certain that the posture verb follows the directional, since these are the only directionals that have a different form when they are used as main verbs heading a predicate and do not introduce any constituents (see Section 3.3.1.3). In other cases, the SVC could also be analysed as a directional preceding a main posture verb. A few examples appear in (19) and (20) below.

(19) *tapan lak to ai nurunan musau*
tap$_S$=an   [lak to]$_{SVC}$ a=i   nuruna-n   musau
1pl.INCL=PRF go be   OBL=3sg belongings-PERT foreign.place
'We have started to use [*lit.* 'have gone to be at'] Western goods.'
(SP190311_0071)

(20) *net teo in sak to poyengan*
[net]$_S$ te-yo   i=an   [sak   to]$_{SVC}$ poyenga-n
sea   LIG-DEM.INT 3sg=PRF come.up be   throat-PERT
'The sea had come up to his throat.' (WL020711_0117)

The events represented by the main verb and the posture verb are quite easily conceptualised as occurring separately from each other. Even so, they are analysed as SVCs because the two verbs occur within a single intonation contour, and the subject is never indexed on the second verb, indicating that together they form one verb complex.

This type of SVC is usually contiguous, and both V1 and V2 are usually intransitive. There is, however, one example with a transitive V1 and an object bound pronoun occurring between V1 and V2. This example is from a procedural text about planting yam. Yam vines are referred to with the numerals for animates and are accordingly indexed on the verb complex. In (21), the speaker explains how a yam vine should be tied to a beanstalk-like device to prevent it from resting on the soil and getting burned by the sun.

(21) woroui ro wat
wo_A=[tou=i_O to]_SVC wat
2sg=put=3sg be   up.high
'You put it to be up high.' (NK290311_1_0035)

9.1.5 SVCs with a directional

Directionals are used as V2 in SVCs to indicate the direction of a path with main verbs, most typically (but not only) verbs of motion. More information about their semantics can be found in Section 3.3.1.3. Almost any verb describing an action that has motion or transfer inherent in it can be accompanied by a directional; this is the preferred option. This type of serialisation is very common in Oceanic languages (Crowley 2002). Directionals can be serialised with either transitive or intransitive verbs. Each will be discussed in turn below. Directionals are also attested as preverbal particles; these are not discussed here, but in Section 6.3.

9.1.5.1 Directionals with intransitive verbs

Directionals are very commonly serialised as V2 with an intransitive main verb. Typically, the main verb is a motion verb, but directionals are also frequently encountered with other intransitive verbs, e.g. those referring to perception or bodily functions. When the directional does not introduce a constituent, it merely specifies a path; the destination or goal of the movement is left unspecified, but it will be retrievable from the discourse context. A directional can also introduce an Oblique constituent of an intransitive clause, preceding a NP headed either by a local or a common noun, or a spatial adverbial or demonstrative. A NP headed by a common noun needs to be additionally marked by the preposition *a=*.

This type of SVC is of the same-subject type. It does not have to be contiguous: an adverb can be inserted between the main verb and the serialised directional. Table 78 shows a number of intransitive motion verbs that are frequently followed by a directional in a SVC.

Sentences (22) and (23) are examples of directionals modifying the main verb without introducing a Locative constituent; the destination of the motion can usually be inferred from the discourse context. In (22), the inherent endpoint can be understood to be the current location of the speaker (the deictic centre).

Table 78: Some motion verbs frequently followed by directionals in SVCs.

Form	Paraphrase
tet	move along a path, a relatively short distance; grow, spread – not used for movement on water
wau	move along a path, a relatively long distance (longer than what is indicated by *tet*) – used for movement on both land and water
liliu	return
terepelek	run
tarak	climb upward, ascend (on hill, ladder, tree etc.)
worup	climb downward, descend (on hill, ladder, tree etc.)

(22) *ong kope liliu me*
 wong$_S$ ko-pe   [liliu me]$_{SVC}$
 1sg.FREE IRR.1sg-PFV return come
 'I will come back (here).' (KW290311_0007)

(23) *uno ro wau sot*
 u$_S$=no   to   [wau sot]$_{SVC}$
 3du=IPFV CONT move go.up
 'They were going uphill / inland.' (LK100411_0059)

Sentences (24) and (25) exemplify directionals introducing a Locative constituent of an intransitive predicate, expressed by a local noun.

(24) *uro aluk liliu me panu*
 u$_S$=to   [[aluk liliu]$_{SVC}$ me]$_{SVC}$ [panu]$_{LOC}$
 3du=CONT paddle return   come   place
 'They were paddling back home.' (NP210511_1_0021)

(25) *imumut sot net a pian*
 i$_S$=[mumut sot]$_{SVC}$ [net]$_{LOC}$ a pian
 3sg=vomit go.up   sea   and good
 'He vomited into the sea and [it was] fine.' (MK060211_0016)

At first sight, sentence (25) seems to be paradoxical. However, what *sot* means here is not that the vomiting was done upwards, but that it was done at the side of the canoe that faced towards the shore (the incident happened at sea). The speaker and thus the deictic centre were located inside the canoe, and the vomiting was done in a direction away from the speaker, towards the shore, which is an inland direction. The use of *sot* is therefore appropriate.

In sentences (26) to (28) a directional, serialised with an intransitive verb, introduces a spatial adverb. Which directional is used depends once again on the position of the speaker and thus the deictic centre. In (26), the meat of the yam is supposed to grow downwards into the ground, away from the speaker. In (27), the action referred to is falling out of a tree, and thus towards the position of the speaker. Sentence (28) describes the growth of the sprout of a yam plant, upwards from the ground towards the speaker, and then up into the air, away from the speaker. It is possible that the instance of *la* here is bleached of all motion meaning, as is the case with a number of other grammaticalisations of *la*, and the entire phrase *la wat* merely functions as an adverbial constituent.

(26) *kanen kipe ret suwot paye*
 [kane-n]₍S₎ ki-pe [tet suwot]₍SVC₎ [paye]₍LOC₎
 meat-PERT IRR.3sg-PFV grow go.down down.below
 'Its meat [i.e. of the yam] will grow in a downwards direction.'
 (NK290311_1_0022)

(27) *upe lot si paye*
 u₍S₎=pe [lot si]₍SVC₎ [paye]₍LOC₎
 3du=PFV fall come.down down.below
 'They fell down [to the ground].' (KW290611_0053)

(28) *kumun teo kipe sa wau la wat*
 [kumun te-yo]₍S₎ ki-pe sa [wau la]₍SVC₎ [wat]₍LOC₎
 sprout LIG-DEM.INT IRR.3sg-PFV come.up move go.to up.high
 'The sprout will come up high.' (NK290311_1_0028)

In (29), an intransitive V1 is followed by an adverb, which in turn is followed by a directional functioning as V2.

(29) *ila ro mwangmwang pulek la almaru*
 i₍S₎=la to [mwang.mwang pulêek la]₍SVC₎ almaru
 3sg=go.to CONT REDUP.watch too go.to right
 'He is also looking to the right.' (Game2_021012_0177)

Occasionally, a spatial adverbial demonstrative follows the directional. In (30), it seems slightly redundant, because *me* already implies movement to the current deictic centre. However, the demonstrative specifies the current location as 'this place' in addition to 'where I am' specified by the deictic directional.

(30) ngawau me arepwo
nga_S=[wau me]_{SVC} [a=te-pwo]_{LOC}
1sg=move come OBL=LIG-DEM.PROX
'I came here [to this place].' (OL201210_0009)

9.1.5.2 Directionals with transitive verbs

9.1.5.2.1 Indicating path for verbs of perception
Indicating a path is common not only with verbs of motion or transfer, but also with verbs of perception, such as *ning* 'see' or *yong* 'hear'. In these cases, a locative constituent is sometimes introduced by the directional, as in (31) below. In other cases, only the direction of the perception event is indicated, as in (32).

(31) iriy maloan suwor ai laliyon
i_A=[tiy [maloa-n]_O suwot]_{SVC} a=i [laliyon]_{LOC}
3sg=observe reflection-PERT go.down OBL=3sg pool
'He saw his reflection down in the pool.' (LL010711_0052)

(32) ngabe ningo sak
[nga]_A=pe ning=[o]_O sak
1sg=PFV see=2sg come.up
'I saw you up [on the veranda].' (ANK020995_0007)

9.1.5.2.2 Specifying Locative or Animate Goal argument
Directionals are also used to introduce an additional Goal argument in a clause containing a transitive main verb. They thus introduce another participant into the clause, one that the main verb is usually not subcategorised for, in order to express a three-participant event (Margetts and Austin 2007). The SVC is of the switch-subject type, with the O of V1 functioning as the S of V2. This type of SVC is always discontinuous: V2 follows the shared argument if the latter is overtly expressed.[65] The types of Oblique argument specified by this type of SVC are

---

[65] When the O argument is not overtly expressed, as is usually the case when it refers to an inanimate O (cf. Chapter 8), it is still assumed that there is a zero trace of the O argument between V1 and V2.

9.1.5.2.2.1 Locative Goal

A Locative Goal (LG) argument is typically specified with transfer verbs such as *tou* 'put, bring' or *yokat* 'carry', which encompass movement of an object or person from a source to a goal, generally inanimate. The form the construction takes depends on whether the location is expressed by means of a local or a common noun. For a local noun or a spatial adverbial (*wat* 'high' or *paye* 'low') the directional by itself is sufficient. A common noun, however, needs to be introduced by the preposition *a=*, which is added as a proclitic to the third singular free pronoun *i*. Sentences (33) to (35) give an indication of how a LG is expressed. An inanimate O argument is usually elided when it refers to the discourse topic, as is the case in these sentences. The discourse topic is *yapi* 'sago' in (33) and (34), and *kasun* 'coconut cream' in (35).

(33) *orou si ai nisip purukei liliu*
    wo$_A$=[tou=∅$_O$ si]$_{SVC}$ a=i   [nisip   purukei]$_{LG}$ liliu
    2sg=put=3sg.ZERO come.down OBL=3sg other.INANIM bowl   again
    'You put [it] into yet another bowl.' (CA120211_1_0028)

(34) *orou liliu sor ai kapol leo*
    wo$_A$=[[tou liliu]$_{SVC}$=∅$_O$ sot]$_{SVC}$ a=i   [kapol te-yo]$_{LG}$
    2sg=put   return=3sg.ZERO go.up OBL=3sg dish   LIG-DEM.INT
    'You put [it] back into the frying dish.' (CA120211_1_0030)

(35) *ope rou lai yapi*
    wo$_A$=pe [tou=∅$_O$ la]$_{SVC}$ a=i   [yapi]$_{LG}$
    2sg=PFV put=3sg.ZERO go.to OBL=3sg sago
    'You will put [it] into the sago.' (CA120211_1_0036)

The examples above are all from the same recording about the procedure of frying sago. Sentence (33) indicates a movement towards the speaker; she puts the sago into another bowl closer to her than the one it comes from. (34) entails an upward movement (and thus triggers the use of *sot*) because the sago is put into a frying dish which is located above the fire. In (35), the presence of motion, together with an indication of the goal of that motion, is probably considered more important than the exact direction, so here *la* is used based on pragmatic grounds. The examples give an idea how directionals are used with small-scale motion. When "canonical" motion towards or from the shore is indicated, usually *si* is used for movement towards the shore and *sot* for motion away from it, since most recordings were made at people's houses near the shore and thus the deictic centre was located there. Some more examples of directionals introducing a LG are given in (36) and (37).

(36) *ilêp payanpôl leo onga isoyek sen parung*
i$_A$=lêp   [payan.pôl]$_O$ te-yo   onga   i=[soyek=∅$_O$
3sg=take   dry.coconut   LIG-DEM.INT   and.so   3sg=move=3pl.ZERO
sen]$_{SVC}$ [patu-ng]$_{LG}$
move.up   head-1sg.PERT
'He took the dry coconuts and hung [them] around my neck.' (LL030611_0048)

(37) *urêyokari si suk*
wurê$_A$=[yokat=[i]$_O$ si]$_{SVC}$ [suk]$_{LG}$
1pc.EXCL=carry=3sg   come.down   shore
'We carried him down to the shore.' (NP210511_2_0063)

Sentence (38) shows that both another verb and an object bound pronoun can intervene between V1 and V2. When there is another verb or adverb present, this always precedes the O argument.

(38) *irou liliuip la panu*
i$_A$=[[tou   liliu]$_{SVC}$=ip$_O$ la]$_{SVC}$ [panu]$_{LG}$
3sg=bring   return=3pl   go.to   place
'He sent them back to their village.' (LL300511_1_0022)

Examples (39) and (40) come from the same story. They show that a directional is not always obligatory in a Locative Goal construction, at least not with all verbs (the Tok Pisin loan *pris* 'jetty' is regarded as a local noun by the speaker, and thus is not marked with the preposition *a=*). The semantics of transfer verbs, such as *tou* 'put, bring' and *yokat* 'carry', encompasses motion on a horizontal plane, hence the needs for such verbs to be accompanied by a directional. The semantics of *top* 'drop', on the other hand, seems not to include this; the verb can therefore do without a directional. In addition, this verb is punctual. Thus, whether or not a directional is used depends partly on the semantics of the main verb.

(39) *ipe roui sa patan pris*
 i$_A$=pe [tou=i$_O$ sa]$_{SVC}$ [pata-n pris]$_{LG}$
 3sg=PFV put=3sg come.up on.top-PERT jetty
 'She put him on top of the jetty.' (MK060211_0047)

(40) *tarêkala ropirê pris*
 tarê$_A$=ka-la top=irê$_O$ [pris]$_{LG}$
 1pc.INCL=IRR.NS-go.to drop=3pc jetty
 'We will go drop them off at the jetty.' (MK060211_0045)

9.1.5.2.2.2 Animate Goal

An Oblique argument which indicates an Animate Goal, such as Recipient or Beneficiary (traditionally often called "indirect object") is obligatorily preceded by possessive/locative *ta-* (which takes a pronominal suffix specifying number and person of the referent), or, alternatively, when the O argument is intended to be eaten by the receiver (see Section 4.2 for a more detailed discussion), by the alimentary classifier *ka-* (also taking a suffix). This kind of Oblique argument is typically encountered with verbs of transfer that involve motion of a Theme to a human Recipient, such as *tou* 'give' and *pul* 'tell'.[66] In some cases, a beneficial or maleficial overtone is clearly indicated ('do X for the benefit/to the detriment of Y') but this can be regarded as a subtype of Recipient.

Thus, the Animate Goal argument receives two types of marking: a directional indicating the path of the motion, and either *ta-* or *ka-*. Some examples are given in (41) to (43).

(41) *ipe sui yapi reo la kau rinan*
 i$_A$=pe [sui [yapi te-yo]$_O$ la]$_{SVC}$ ka-u [tina-n]$_{AG/BEN}$
 3sg=PFV fry sago LIG-DEM.INT go.to CLF.food-3du.PERT mother-PERT
 'She fried the sago for him and his mother [to eat].' (KW290611_0036)

---

66 The verb *tou* figures prominently in the previous section as well. *Tou* could probably best be regarded as a verb of transfer with very general semantics (encompasssing English 'put', 'give', 'bring, escort' and also 'give birth, bring forth; create'). It is disambiguated by context and by the type of marking on the Goal argument.

(42) *ipe la puk mapian la ran asoan*
i$_A$=pe la [puk [mapia-n]$_O$ la]$_{SVC}$ ta-n
3sg=PFV go.to open knowledge-PERT go.to CLF.POSS-PERT
[asoa-n]$_{AG/BEN}$
husband-PERT
'She went to inform her husband.' [*lit*. 'open the knowledge of it to her husband'] (LK100411_0098)

(43) *ikipe si rou kokon la rararê*
i$_A$=ki-pe si [tou [kokoni]$_O$ la]$_{SVC}$
3sg=IRR.3sg-PFV come.down give money go.to
ta-tarê$_{AG/BEN}$
CLF.POSS-1pc.INCL.PERT
'He should come here and give money to us.' (OBK040311_0193)

The most frequently attested directional is *la*, but occasionally another directional is found expressing an Animate Goal argument.

(44) *ngapwa kopul sot tao la remenin telo*
nga$_A$=pwa ko-[pul=∅ sot]$_{SVC}$ ta-o$_{AG}$ la
1sg=want IRR.1sg-tell=3sg.ZERO go.up CLF.POSS-2sg.PERT go.to
temenin te-lo
like LIG-DEM.DIST
'I am going to speak to you as follows.' (PK290411_3_0076)

(45) *ipto apek ip payanpôl leo suwot tang*
ip$_A$=to [apek [ip payan.pôl te-yo]$_O$ suwot]$_{SVC}$
3pl=CONT hit 3pl dry.coconut LIG-DEM.INT go.down
[ta-ng]$_{AG/MAL}$
CLF.POSS-1sg.PERT
'They were hitting me with the dry coconuts.' [*lit*. 'they were hitting the dry coconuts downwards at me'] (LL030611_0055)

Most cases of directionals other than *la* are with the verb *pul* 'tell'. The speaker of sentence (44), which was uttered during a public ceremony, stands on a lower level than his addressee, and thus it is more appropriate to use the directional *sot*, specified for the absolute Frame of Reference (see Section 3.3.1.3), instead of the unspecified *la*. It appears that when the cardinal direction of the motion indicated by the main verb is known, an appropriate directional has to be used. However, when the cardinal direction of the action relative to the deictic centre is unknown or backgrounded, *me* or *la* can be used. Thus, the use or otherwise of a directional specified for absolute FoR seems to be in part dependent on pragmatic factors.

9.2 Symmetrical SVCs

In symmetrical SVCs, both members come from the open class of lexical verbs and not from a grammatically defined subclass. Symmetrical SVCs are therefore only defined via negative criteria: a SVC that does not have a stative verb or directional as one of its members is necessarily symmetrical. Some examples of symmetrical SVCs are given in Table 79 below.

Table 79: Some symmetrical SVCs.

Verb combination	Meaning	Meaning V1	Meaning V2	Shared arguments
apui ngan	cook and eat (tr)	cook (tr)	eat (tr)	A and O
lang saui	revive (tr)	lift up (tr)	lift up (tr)	A and O
song yik	hunt (tr)	run away (intr)	search for (tr)	A and O
sungêek lêp	collect	heap together (tr)	take (tr)	A and O
tiu sungêek	gather	collect (tr)	heap together (tr)	A and O

In the case of *lang saui* 'revive' and other combinations encountered, two verbs that are (almost) synonymous have been serialised. This is an instance of symmetrical serialisation also encountered in several other languages (Aikhenvald 2006: 30). Sometimes, the combination will have an unpredictable meaning, but in other instances the two verbs combined merely reinforce each other's meaning.

The main properties of symmetrical SVCs are that they are always contiguous, and that the two verbs involved always share all their arguments. When both verbs are transitive, they share the A and O argument. Across the board, it can be concluded that symmetrical verb serialisation is a rather marginal phenomenon in Paluai compared to asymmetrical serialisation, as it is hard to find examples in the corpus that unambiguously consist of two independent verbs where the resultant meaning is transparent. In the majority of cases, what could once have been a productive symmetrical SVC has become (semi-)lexicalised, a process discussed below in Section 9.2.2.

9.2.1 Sequences with *liliu* 'return; again'

Sequences with *liliu* form a special case. *Liliu* is a heterosemous form (Lichtenberk 1991): it is attested as an independent motion verb meaning 'move back; return' (entailing motion away from a point followed by motion back to the same point) and as an adverb meaning 'again'. It is often encountered twice in the same sentence:

(46) *taukaaluk liliu la panu liliu*
 tau$_S$=ka-[aluk liliu]$_{SVC}$ la panua liliu
 1pl.INCL=IRR.NS-paddle return go.to village again
 'Let's paddle back home again.' (NP210511_1_0020)

In (46), the first instance of *liliu* modifies the verb *aluk* 'paddle' in a symmetrical SVC. The two verbs share the same S argument. The second instance of *liliu* modifies the entire event 'paddle back home' and functions as a sentential adverb. As an adverb, *liliu* can also modify verbless clauses:

(47) *i ranisip telo, not sê reo liliu*
 [i ta=nisip te-lo]$_S$ [not sê te-yo]$_{NVPRED}$ liliu
 3sg SPEC.COLL=other.INANIM LIG-DEM.DIST child DIM LIG-DEM.INT again
 'The next one, (it's) the small child again.' (Game4_280812_0154-55)

When *liliu*, in its capacity as a verb, is serialised with another verb, the valency of V1 determines the transitivity of the entire SVC. When V1 is intransitive, the entire construction can be analysed as a same-subject SVC similar to the adverbial type. An example would be (46) above. Both the 'paddling' and the 'returning' action are carried out by the same participant. On the other hand, when *liliu* is serialised with a transitive V1, the entire construction is transitive. Note that *liliu* behaves differently from directionals indicating motion by the O argument (discussed in Section 9.1.5.2.2). It is always contiguous with V1, and the object bound pronoun cliticises onto it when the O has an animate referent, as in (48). Directionals, on the other hand, always follow the O argument when it is overtly expressed.

(48) *ngala rou liliui Lorengau*
 nga$_A$=la [tou liliu]$_{SVC}$=i$_O$ Lorengau
 1sg=go.to bring return=3sg L.
 'I went and brought him back to Lorengau.' (ANK020995_0011)

(49) *ippe me rou liliu la kep*
 ip$_A$=pe me [tou liliu]$_{SVC}$=∅ la ka-ep
 3pl=PFV come give return=3sg.ZERO go.to CLF.food-1pl.EXCL.PERT
 'They came and gave [it] back to us.' (KM060111_0087)

liliu can also directly follow a (transitive or intransitive) verb and act as an adverb. Since, in this case, it does not take any arguments, there is no argument sharing between *liliu* and the verb that precedes it. This is therefore not an instance of a SVC. An example is given in (50).

(50) *urêno ro pe liliu kou wot*
 wurê$_A$=no to pe liliu [kou]$_O$ wot
 1pc.EXCL=IPFV CONT do again fishing go.level
 'We were fishing again for a while.' (MK060211_0017)

Generally speaking, *liliu* may form a symmetrical SVC when V1 is a verb that entails motion: either an intransitive motion verb such as *tet* 'move; walk' or *aluk* 'paddle', or a transitive verb that represents a 'giving' event (and thus motion of a Theme argument) such as *tou* 'give, bring'. Alternatively, *liliu* may be an adverb. Combined with a verb that does not entail motion, *liliu* can only function as an adverb.

9.2.2 Lexicalisation of symmetrical SVCs

Symmetrical SVCs are prone to lexicalisation; as indicated above, unambiguous instances are hard to find. The process of lexicalisation is illustrated in this section by sequences that *as a whole* have verbal status and have a form *tou* as their second member. An overview of *tou* sequences found in the corpus is given in Table 80. Synchronically, *tou* is not attested as a predicate head, so it can probably be considered an adverbial with the general meaning of 'be fastened'.[67] Consequently, the sequences discussed here are not SVCs. It is unclear how productive and widely used *tou* is synchronically. There may be more combinations with *tou* as a second part than currently attested, but it is probably not fully productive. As mentioned, because *tou* is not a verb, the sequences with *tou* are not SVCs. What, then, are they?

---

[67] The homophonous transitive verb *tou* 'put, give', which can occur as an independent verb and as V2 in valency-increasing SVCs, is probably unrelated.

Table 80: A selection of combinations with *tou*.

Combination	Meaning	Valency	Transparent	First part attested as independent verb
sok tou	tie	Tr	no	yes, 'shoot' (tr)
neng tou	stop walking	Intr	no	yes, 'climb' (tr)
yek tou	put (s.t. to stay)	Tr	no	yes, 'hit' (tr)
tuk tou	summon people by drum beat	Tr	no	yes, 'beat' (tr)
san tou	end a story	Intr	no	yes, 'cut' (tr)
tik tou	withhold	Tr	no	yes, 'carry' (tr)
arei tou	hold with teeth	Tr	yes	yes, 'bite' (tr)
kap tou	tie, bundle	Tr	n/a	*kap* only attested as 1st part of other lexicalised SVC, *kap tep*
têk tou	put, build	Tr	maybe	yes, 'build' (tr)
sang tou	cross over	Tr	n/a	*sang* only attested as 1st part of other lexicalised SVCs
sum tou	catch	Tr	no	yes, 'cover' (tr)
pang tou	adopt	Tr	no	yes, 'feed' (tr)
kum tou	store in the mouth	Tr	yes	yes, 'hold in mouth' (tr)
wei tou	bundle, tie together	Tr	yes	yes, 'coil' (tr)
souek tou	stop punting a canoe	Intr	no	yes, 'push' (tr)
samsam tou	ease (rain)	Intr	n/a	no
tat tou	tighten	Tr	no	yes, 'pull up' (tr)
nêk tou	grab	Tr	n/a	no
pulu tou	stick	Intr	n/a	no

Interestingly, combinations with *tou* seem to be at different stages of lexicalisation. For some sequences, such as *nêk tou* or *pulu tou*, the first part is not found anywhere else at all. In other cases, such as *neng tou* or *souek tou*, the first part is attested as an independent verb, but with a different valency than in the combination with *tou*. *Neng*, for instance, is a transitive verb meaning 'climb (a tree)'. However, when *neng* is combined with *tou*, the entire combination is intransitive and means 'halt, stop walking'. *Tou* is not an intransitivising device, however, since in other cases it does not affect the valency of V1 (e.g. *pang tou*, *wei tou*). *Neng* may have two senses, one as an independent verb and one as part of a lexicalised verb sequence. Sometimes, the meaning of the *tou* construction can be straightforwardly derived from the meaning of the first part, but in other cases the meaning of the combination is unpredictable.

In yet other cases, such as *kap tou* 'tie up' and *sang tou* 'cross', the first part of the combination is attested only as the first part of a limited number of verb sequences and not as an independent form. The only other instance of *kap* is in *kap tep* (which has a meaning similar to *kap tou*), and one other instance of *sang* is in *sang mwal* 'step over (something bad)', where *mwal* is not attested as an independent form either. Thus, it seems that the first part of a sequence can also be affected by loss of status as an independent form, although to a lesser degree than the second part. Historically, what may have happened in such cases is sketched in Table 81:

Table 81: Potential historical stages of SVC lexicalisation.

Stage 1	Verb – Verb	Productive SVC
Stage 2	Verb – Adverb	Element 2 loses independent status
Stage 3	Semi-verb – Adverb	Element 1 loses independent status and becomes disconnected from its independent counterpart (i.e. the form becomes heterosemous)
Stage 4	Idiomatic collocation	Element 1 has become obsolete as an independent form and is only found in one or more lexicalised sequences. Element 2 may have ceased to be productive as an adverb

This table shows why the number of combinations where the second part is not attested as a full verb is much higher than the number of combinations where this is the case for the first part. The variation in verb-like sequences in Paluai is due to the fact that most combinations are at a different stage of lexicalisation along the historical cline. A complicating factor is that the same form may be encountered in combinations that are at different stages of lexicalisation on the cline. *Tou*, as shown above, is attested as the second element in combinations that are either at Stage 2 (*souek tou*), Stage 3 (*neng tou*) or Stage 4 (*nêk tou, kap tou*).

For forms attested as the first part in these combinations, this may also be the case. *Neng*, for instance, can occur together with the independent stative verbs *mat* 'die' and *pui(pui)* 'be soft'; the resulting combinations *neng mat* and *neng pui* are transitive and mean 'kill by standing on' and 'crush with foot; trample', respectively. They are at Stage 1. In addition, *neng* occurs together with *tou* as described above; this combination is at Stage 2. Transitivity of the two combinations at Stage 1, on the one hand, and the one at Stage 2, on the other, differs, and this is not caused by the second elements they take (*mat*, *pui* or *tou*) since these are all intransitive or non-valency changing. Thus, as already mentioned above, *neng* may have two senses, one as an independent verb and the other as a dependent form that needs a second element. It is thus heterosemous. The two senses may be in the process of becoming increasingly disconnected from each other.

The examples with *tou* are instructive because this form is so common; however, the process is most likely much more widespread. It shows that lexicalisation of Paluai SVCs is not a straightforward matter, because they may not always lexicalise as units. The question is also whether lexicalisation or grammaticalisation would be a more adequate label for the process described above. Meanings of parts and combinations become more conventionalised and unpredictable, which is a hallmark of lexicalisation. However, while there is no change from a lexical item into a grammatical item, items do get more restricted in their use, which is usually seen as a criterion for grammaticalisation. Although still belonging to an open word class, adverbs can be seen as more grammaticalised than verbs, because they are dependent on other forms for their use.

9.3 SVCs: conclusions

In what follows, the main conclusions with respect to SVCs will be put forward. This discussion will focus on asymmetrical SVCs, since these are much more common and come in a variety of types.

9.3.1 Formal properties of asymmetrical SVCs

Table 82 on pages 295 gives an overview of the types of asymmetrical SVCs encountered and of their formal properties. There are some interesting differences between SVCs with directionals, on the one hand, and other types of SVCs, on the other.

With cause-effect SVCs, the order is iconic, since V1 always represents the cause and V2 the effect. With adverbial SVCs, the modifying verb (V2) always follows the one (V1) that indicates the action. With all asymmetrical SVCs, the more restricted member always follows the less restricted one. Cause-effect, adverbial and valency-increasing SVCs are always contiguous: bound object pronouns always follow V2, and no adverbs or other elements can be situated between them.

Cause-effect SVCs and SVCs specifying an Oblique argument in a three-participant event are of the switch-subject type, whereas all other SVCs are of the same-subject type. The difference between the two switch-subject types is the position of the O argument. With the former type, the O argument always follows V2, whereas with the latter type, it is placed between V1 and V2. SVCs with directionals are different from other types of SVC because they do not

Table 82: Properties of asymmetrical Serial Verb Constructions.

Type of SVC	Shared arguments	Valency of components	Construction transitivity	Constituent order	Contiguity	Semantics	Productivity	Gram. word	Phon. word
1. Cause-effect (9.1.1)	O of V1 = S of V2	V1 tr, V2 stat intr	transitive	iconic	yes	cause-effect, resultative	semi-prod.	yes	no
2. Adverbial (9.1.2)	S of V1 = S of V2	V1 tr or intr, V2 intr	valency of V1	modifying verb follows	yes	(manner) adverbial	prod.	yes	no
3. Valency-increasing (9.1.3)	S of V1 = A of V2 (new O)	V1 intr, V2 tr	transitive	transitivising verb follows	yes	transitivising	prod.	yes	no
4. Posture verb (9.1.4)	S of V1 = S of V2	V1 + V2intr	intransitive	posture verb follows	no	motion/event and rest	restricted	no	no
5. Intransitive directional (9.1.5.1)	S of V1 = S of V2	V1 intr, V2 dir	intransitive	directional follows	no	path, direction	prod.	no	no
6a. Transitive directional (9.1.5.2.1)	O of V1 = S of V2	V1 tr perception, V2 dir	transitive	directional follows	no	locative	prod.	no	no
6b. Ditransitive directional (9.1.5.2.2)	O of V1 = S of V2	V1 tr transfer, V2 dir	ditransitive	directional follows	no	goal (locative or animate)	prod.	no	no

need to be contiguous. As shown below, they may also behave differently with respect to nesting of SVCs and derivational processes.

Members of SVCs cannot be negated or questioned separately. If this is attempted, a biclausal construction, sometimes with a different meaning, is the result (cf. (55)). In that case, the subject needs to be cross-referenced on each verb and the construction is no longer a SVC. This is illustrated by means of a cause-effect SVC in (51) to (57), but the same condition applies to all SVC types.

(51) *nganeng pui pwayap*
 nga$_A$=[neng pui]$_{SVC}$ [pwayap]$_O$
 1sg=step.on be.soft pawpaw
 'I stood on the pawpaw, thereby crushing it.'

(52) *ngamaneng pui pwayap pwên*
 nga$_A$=ma=[neng pui]$_{SVC}$ [pwayap]$_O$ pwên
 1sg=NEG$_1$=step.on be.soft pawpaw NEG$_2$
 'I did not stand on the pawpaw and crush it.' [i.e. the entire action 'stand-crush' did not take place]

(53) **nganeng mapui pwayap pwên*
 *nga$_A$=[neng ma=pui]$_{SVC}$ [pwayap]$_O$ pwên
 1sg=step.on NEG$_1$=be.soft pawpaw NEG$_2$
 Intended: 'I stood on the pawpaw without crushing it.'

(54) **ngamaneng pwên pui pwayap*
 *nga$_A$=ma=[neng pwên pui]$_{SVC}$ [pwayap]$_O$
 1sg=NEG$_1$=step.on NEG$_2$ be.soft pawpaw
 Intended: 'I crushed the pawpaw, [but] not by standing on it.'

(55) *nganeng pwayap a imapui pwên*
 nga$_A$=neng [pwayap]$_O$ a i$_S$=ma=pui pwên
 1sg=step.on pawpaw and 3sg=NEG$_1$=be.soft NEG$_2$
 ? 'I stood on the pawpaw/climbed the pawpaw tree and it was not soft.'

(56) **woneng pui le pal pwayap?*
 *wo$_A$=[neng pui]$_{SVC}$ le pal [pwayap]$_O$
 2sg=step.on be.soft or break pawpaw
 Intended: 'Did you step on and crush, or crack the pawpaw?'

(57) *woneng le yek pui pwayap?
*wo_A=neng le [yek pui]_SVC [pwayap]_O
2sg=step.on or hit be.soft pawpaw
Intended: 'Did you step on, or hit and crush the pawpaw?'

Sentence (58) is a good example of the difference between juxtaposed and coordinated predicates, on the one hand, and SVCs, on the other. The sentence consists of two subordinate clauses within the first main clause, which is coordinated with two other main clauses. Each verb complex has to be preceded by a separate instance of the irrealis prefix. The serialised verbs, which form a single predicate and thus a single verb complex, do not each receive a prefix.

(58) *pian te igat naluai re ipkano ret lak, kano la yil, a kano si a kano apui ngan*
[pian te i=gat naluai [te ip_S=ka-no [tet lak]_SVC]_SC1
good SUB 3sg=have garden.food SUB 3pl=IRR.NS-IPFV move go
[ka_A-no la yil=∅_O]_SC2]_MC1 a [ka_S-no si]_MC2
IRR.NS-IPFV go.to dig=3sg.ZERO and IRR.NS-IPFV come.down
a [ka_A-no [apui=∅_O eat=3sg.ZERO]_SVC]_MC3
and IRR.NS-IPFV cook=3sg.ZERO ngan=∅_O
'It is good that there is garden food, for they could go and dig [it] up, and they could come down, cook [it] and eat [it].' (WL020611_0067)

9.3.2 Combinations involving more than two verbs

Occasionally, sequences of more than two verbs are attested (preverbal particles, including directional particles, are disregarded here since they are not considered full verbs). These sequences can in fact always be analysed as one SVC (of two verbs) nested inside another SVC, i.e. a SVC serialised with a third verb. In many cases, a SVC with *liliu* (see Section 9.2.1 above) is serialised with a directional SVC:

(59) *uro aluk liliu me panu*
u_S=to [[aluk liliu]_SVC me]_SVC [panu]_LOC
3du=CONT paddle return come place
'They were paddling back home.' (NP210511_1_0021)

Within productive SVCs, there is a difference between SVCs that have to be contiguous and SVCs that do not. The former will be nested inside the latter. Thus, a sequence with *liliu*, as shown above in (59), or a valency-increasing SVC as shown below in (60), is nested inside a directional SVC. More lexicalised and more iconic combinations are nested inside more grammaticalised combinations, and symmetrical SVCs are nested inside asymmetrical SVCs, not the other way around.

(60) *ipe wop lêp la kau rinan*
　　 i$_A$=pe　　[[wop　lêp]$_{SVC}$=Ø$_O$　　la]$_{SVC}$　ka-u　　　　　　tina-n
　　 3sg=PFV　fly　　take=3sg.ZERO　　go.to　CLF.FOOD-3du.PERT　mother-PERT
　　 'He flew and took [it] for him and his mother to eat.' [*lit*. 'fly-take']
　　 (KW290611_0038)

In other cases, a semi-lexicalised collocation (see Section 9.2.2) is serialised with another verb in a symmetrical SVC, as in (61).

(61) *imainap yamat kipe yop sangtou pwên*
　　 i=ma=inap　　　[yamat]$_S$　ki-pe　　　　[yop　sang.tou]$_{SVC}$=Ø　pwên
　　 3sg=NEG$_1$=able　person　　IRR.3SG-PFV　jump　cross=3pl.ZERO　　NEG$_2$
　　 'It was not possible for people to jump over [the walls].' (LM190611_0016)

In example (61), *sang tou* is a lexicalised idiom involving the element *tou* as discussed above in Section 9.2.2. This in turn is serialised with the main verb *yop* 'jump'. *yop* most often occurs with another verb following it in a symmetrical SVC, but it is attested by itself as well. The sequence can be regarded as an adverbial SVC with *yop* as V1 and *sang tou* as V2. The internal hierarchy of *yop sang tou* shows that sequences with *tou* are indeed more compound-like, and can function on a par with single verbs in SVCs.

9.3.3 Grammaticalisation of SVC components

Parts of asymmetrical SVCs may be subject to grammaticalisation, in particular if they are from a small restricted class. Over time, the meaning of some verbs used as V2 in cause-effect/resultative or adverbial SVCs may be extended, resulting in a more aspectual meaning. In other words, the second verb of such a SVC may grammaticalise into an aspect marker (cf. *nêm*, discussed in Section 6.2.2.1).

From Section 9.1.5.2.2, it is clear that with regard to the marking of peripheral arguments, Paluai directionals are used in much the same way as adpositions are used in many other languages. The question is whether the directionals are on a grammaticalisation cline towards full-fledged prepositions. Evidence to the contrary includes that they can still be used as main verbs. Durie (1988: 5) mentions that if a verb is used independently, this "inhibits semantic bleaching and subsequent reanalysis". Another indication that directionals are still verbs is the fact that the argument they introduce (if not headed by a local noun) still needs to be marked additionally by *a=*, *ta-* or *ka-*, like any other non-core argument.

When a directional is serialised with an intransitive verb without introducing a constituent, it merely specifies a path, and when it is serialised with a transitive verb, there is a switch-subject SVC, with the O argument of V1 functioning as the S of V2. Moreover, it seems that the directionals are still predominantly used as V2 in serialisations when a genuine sense of motion on a horizontal plane is present. A case in point is the contrast between (39) and (40), where the semantics of the main verb *tou* 'put' entail motion on a horizontal plane, but the semantics of the main verb *top* 'drop' do not.

An argument in favour of grammaticalisation is that directionals are currently not only used for large-scale motion events that are clearly either directed towards, parallel to, or away from the shore and can thus be related to the cardinal directions quite easily, but also for small-scale motion events, where the 'level' component of their meaning mainly refers to movement up or down a vertical cline, as is the case in sentences (33) and (34). It could be argued that this is an example of semantic bleaching and extension/context generalisation, two principal mechanisms involved in grammaticalisation (Heine and Kuteva 2002). It is likely that each directional has become heterosemous and is attested both with the full lexical semantics of 'move uphill away from deictic centre' etc., and as a grammaticalised form, with a more generalised meaning 'upwards away from deictic centre'. However, there is no formal difference apparent between the two senses of the same directional. In addition, sentences like (25), where *sot* 'go up' is used to indicate landward direction for the downwards motion of vomiting, indicate that the absolute dimension can still be relevant even for small-scale movements.

Another argument that may support the claim that directionals are grammaticalising towards prepositions is the fact that they are not obligatorily cross-referenced by subject bound pronouns, and thus generally carry no verbal morphology. This would be in line with a third principle of grammaticalisation: decategorialisation. Verbal categories such as subject indexes, TAM and irrealis marking are expressed only once per VC; they are not repeated on each verb. However, with switch-subject directional SVCs (type 6 in Table 82), the directional V2 is considered to be placed outside of the VC because it follows the O argument in case it is overtly expressed. In these cases, one may expect a subject bound pronoun, since the serialised verb forms a new VC. The fact that there is no bound pronoun indicates that the directional may have grammaticalised into a more preposition-like form. However, if a subject bound pronoun were present, the sequence would have to be analysed as biclausal and would not classify as a SVC.

An exception to the general claim that directionals have not fully grammaticalised may be the use of *la* as a marker of adverbial phrases, discussed in Section 3.6.2.2. This use of *la* could have developed from its use as a change-of-state marking copula 'become' (see Section 7.7.2). It is quite possible that in these cases, *la* is reanalysed as a non-verbal form, and thus is not part of a SVC.

9.3.4 SVCs as grammatical and phonological words

Grammatical wordhood of SVCs is relatively straightforward. When a SVC can be replaced in the same slot by another (single) verb, this indicates that it is a grammatical word. This is the case for valency-increasing SVCs (for instance, *wau lêp* (lit. 'walk-take') could be replaced by *yokat* 'carry') and cause-effect SVCs, which could be replaced by a causative construction. With adverbial SVCs, grammatical wordhood is not certain; in any case, the adverbial V2 could be left out and the sentence would still be grammatical. The remaining types of SVC do not form grammatical words: they cannot be replaced by a single verb.

SVCs do not form phonological words. Speakers can pause or hesitate (producing *uh* and *aw*) within contiguous SVCs. Alternation processes that only occur within phonological word boundaries, such as assimilation of /t/ to a preceding /l/, are not attested within contiguous SVCs.

Furthermore, morphological evidence points in the direction of the verbs within SVCs being separate phonological words. From *nêk tou* 'grab', discussed in Section 9.2.2, the nominalisation *nêktouan* can be derived with the suffix *-an*. Since it is nominalised as an entity, it probably forms one phonological word. As explained, however, *nêk tou* does not classify as a SVC, since neither of the parts is attested independently. There is limited evidence of derivation from semi-lexicalised collocations, such as the adjective *yopayison* 'slippery' derived from *yop ayit* 'slip'. These processes are nonetheless marginal, and not productive. Most importantly, they only occur with sequences that are far advanced on the lexicalisation cline, such as *nêk tou* and *yop ayit*.

Reduplication always affects only one element of a SVC. Reduplication of V1 has either an intransitivising or a nominalising function, just as "regular" reduplication of a single verb (see Section 3.3.2.1.2). Reduplication of V2, on the other hand, appears to have an intensifying function. Some examples of SVCs with one of their elements reduplicated are given in (62).

(62) *sok tou* 'tie' → *soksok tou* 'be tangled'
 tok si 'sit down' → *toktok si* 'sitting' (N)
 wau pit 'gather' → *wauwau pit* 'gathering' (N)
 yek pui 'crush' → *yek puipui* 'beat to a pulp'
 arei mat 'bite dead' → *arei matmat* 'mangle'

Because phonological and derivational processes do not affect SVCs as a whole, it can be concluded that SVCs do not form phonological words. Cause-effect, adverbial and valency-increasing SVCs are therefore a part of the language where the phonological and the grammatical word do not coincide; in these cases, one grammatical word is made up of two different phonological words.

10 Speech act distinctions and polarity

The two categories discussed in this chapter, mood and polarity, relate at the same time to the VC and to a higher level of organisation such as the clause or the sentence. Mood is defined as the grammatical expression of illocutionary force; polarity refers to the positive or negative value of a proposition.

10.1 Mood

Three moods can be distinguished: declarative, imperative and interrogative. Each type of mood is most prototypically connected to one type of speech act or illocutionary function: declarative mood is normally used for statements, imperative mood for commands, and interrogative mood for questions. Paluai does not have morphological marking for any of these, nor is mood distinguished by grammatical means such as constituent order. Mood is mainly indicated by means of prosody, with different intonation patterns for a question or a statement (see Section 2.3.2.1). In addition, grammatical criteria for declarative and imperative clauses are slightly different, and content questions can be recognised by the presence of a question word. In what follows, features of declarative, interrogative and imperative sentences will be discussed, with special attention to the use of the interrogative and the imperative moods. Differences between the latter two and the declarative mood, which can be considered the unmarked or default option, will be highlighted.

10.1.1 Interrogative mood

Interrogative mood applies to the speech act "question", which is most typically an information-seeking speech act. Three types of questions can be distinguished: content questions, polar questions and tag questions. These will be discussed in turn.

10.1.1.1 Content questions
Content questions contain an interrogative word and make inquiries about a particular object or proposition (see Section 4.7 for an overview of interrogative words and what type of element they query). The interrogative word remains in situ: its syntactic slot is exactly that of the queried constituent in the question's declarative counterpart. Content questions are thus recognisable

by the presence of an interrogative word, and also by a specific intonation contour that differs from the one associated with declarative statements. The focused element of a question receives phrasal stress. The question 'What are you doing?' can be realised in two ways (with the stressed element printed in bold):

(1) *oro pe sa?*
    wo_A=to   pe   [sa]_O
    2sg=CONT do/make what
    a) [oro pɛ **sa**]   '**What** are you doing/making?'
    b) [oro **pɛ** sa]   'What are you **doing**?'

Although (1a) and (1b) query the same entity (the object of the V *pe* 'do'), the focus in (1a) lies on the nature or result of the action, while in (1b) it lies on the action itself (see Section 12.3.4 for more information on the pragmatic category "focus"). Since *pe* is ambiguous between 'do' and 'make', (1a) is more likely interpreted as questioning the identity of an object the addressee is producing than as questioning an action of the addressee. In more familiar languages like English, a sentence like (1b) could for instance be uttered when the speaker is dismayed about seeing the addressee doing something she doesn't agree with. This doesn't seem to be the case in Paluai: whether the stress is on the interrogative word (which is usually sentence-final) or the element preceding it seems not to be dependent on speaker attitude, but may have to do with the pragmatic context.

The sentences below give a number of examples of content questions with *sê* 'who'. Like *sa* 'what', *sê* is usually found in a non-verbal predicate (see Chapter 7 for discussion). Both query the identity of the NP that, in the declarative counterpart of the question, fills the non-verbal predicate that provides the answer. The NP referring to the queried entity will be represented as the subject of the non-verbal predicate, and is often topicalised and thus left-dislocated, as in (2); see Section 12.3.3 for more information. In this case, a pronoun copy is often preceding the interrogative word, which forms the head of the predicate.

(2) *i reo, i sê?*
    [i   te-yo]_S   i   [sê]_NVPRED
    3sg LIG-DEM.INT 3sg who
    'She there, who is she?' (052b_0009)

(3) *tarepwo ran sê?*
    [ta=te-pwo]₍S₎   [ta-n   sê]₍NVPRED₎
    SPEC.COLL=LIG-DEM.PROX CLF.POSS-PERT who
    'Whose is this?' (WL020611_0044)

(4) *puyunum teo ran sê?*
    [puyuṉum te-yo]₍S₎   [ta-n   sê]₍NVPRED₎
    toilet   LIG-DEM.INT CLF.POSS-PERT who
    'The toilet, whose is it?' (052b_0268)

samai- 'what of', questioning part-whole and kinship relations, and *samnon* 'how many', questioning quantities, also usually form the predicate of a verbless clause. Examples are given in (5) and (6). The answer to questions containing these will be a (directly possessed) kinship or body part term or a numeral, respectively, and will also constitute the predicate of a verbless clause.

(5) *Pokut i samaim?*
    [Pokut]₍S₎ i   [samai-m]₍NVPRED₎
    P.   3sg   what.of-2sg.PERT
    'How is Pokut related to you?' [*lit.* 'What is Pokut of you?']
    (052b_0015)

(6) *kerin tao samnon?*
    [kerin ta-o]₍S₎   [samnon]₍NVPRED₎
    year   CLF.POSS-2sg.PERT how.many
    'How old are you?' [*lit.* 'How many are your years?']

pari ai sa and *tenepa* question relations between propositions: *pari ai sa* questions purpose (see (7)), and *tenepa* questions manner (see (8)). They are not verbal forms, since they cannot take subject bound pronouns or TAM particles, and thus form the predicate head of a verbless clause. Since they are not questioning entities but propositions, the answer to this type of content question is not represented by a non-verbal predicate, but an entire (verbal) clause. *tenepa* can also question manner of action (introduced by *la*).

(7) *tareo pari ai sa?*
    [ta=te-yo]₍S₎   [pari   a=i   sa]₍NVPRED₎
    SPEC.COLL=LIG-DEM.INT   belonging.to   OBL=3sg what
    'What is all this for?' / 'What's the purpose of this?' (LL030611_0043)

(8) *tarepwo renepa?*
[ta=te-pwo]_S　　　　　[tenepa]_{NVPRED}
SPEC.COLL=LIG-DEM.PROX　how
'How/why is this happening?' (LK100411_0094)

For more information about the meaning difference between *(la) tenepa* and *la sa*, see Section 4.7.5. *la sa* 'how' questions manner of an action described by a verb, and occurs as an adverbial phrase modifying this verb, as in (9). *pa* 'where' and *kapi* 'when' question place and time, respectively, and also manifest as adverbial modifiers to verbal predicates (see (10) and (11)), unless they are used in an elliptical question 'When?' or 'Where?'. *pa* usually occurs with either *la* 'go' or *to* 'be', whereas *kapi* modifies a verb describing an event of which the speaker wants to know when it happened or when it will happen.

(9) *auro pangai la sa?*
au_S=to　　pangai　la sa
2du=CONT　think　　how
'What do you think?' (OBK040311_0106)

(10) *pwai rao ra Ngi tepwo, ila pa?*
[pwai　ta-o　　　　　　　　ta=Ngi　　　　　te-pwo]_S　　i=la
cousin　CLF.POSS-2sg.PERT　SPEC.COLL=Ngi　LIG-DEM.PROX　3sg=go.to
a=pa
OBL=where
'Your cousin Ngi, where did he go?' (052b_0313)

(11) *Ponaun kipe angou kapi?*
[Ponaun]_S　ki-pe　　　　angou　kapi
P.　　　　　IRR.3sg-PFV　arrive　when
'When will Ponaun arrive?'

10.1.1.2 Polar questions

A polar question essentially contains a proposition for which the speaker seeks confirmation or rejection. This type of question is often called 'yes/no question', because it can be answered with 'yes' or 'no', but in reality it is often answered by a positive or negative (or neutral/undecided) response in another form; the term "polar" is therefore more appropriate (cf. Dixon 2012: 377). In

Paluai, constituent order of polar questions is not different from that of statements, nor are they morphologically marked. The only way to distinguish them from statements is by prosodic cues: polar questions have a sharp rise and fall in pitch on the final element, in contrast to statements, which show gradually falling pitch (see Section 2.3.2 for more information). Examples of polar questions are given in (12) and (13).

(12) *ogat kel?*
wo$_A$=gat [kel]$_O$
2sg=have canoe
'Do you have a canoe?' (052b_0302)

(13) *igat nik laro net pulek?*
i$_A$=gat   [nik]$_O$ la   to net pulêek
3sg=have fish   go.to be sea too
'Is there fish in the sea too?' (052b_0308)

A variety of the polar question is the "alternative question", where the two possible answers are already given, either by coordinating two NPs or clauses with *le* 'or', as in (14), with *le pwên* 'or not', as in (15), or with *le papwên* 'or not yet', as in (16). In the former case, *le* is sometimes repeated after the two alternatives.

(14) *i reo, iro patan nan le iro patan yoy, le?*
[i   te-yo]$_S$ i=to   pata-n   nan   le i$_S$=to pata-n
3sg LIG-DEM.INT 3sg=be on.top-PERT ground or 3sg=be on.top-PERT
yoy   le
stone or
'This [thing], is it on the ground or is it on top of the stone, or..?' (Game1_021012_0046)

(15) *aupa kanesek tultul sesap le pwên?*
au$_A$=pwa   [ka-nesek   tultul   [sesap]$_O$]$_{Compl:Pot}$ le pwên
2du=want.to IRR.NS-reveal advice something   or NEG
'Do you want to give some advice, or not?' (OBK040311_0162)

(16) *tinang, ngainap ai songan um le papwên?*
 tina-ng nga_S=inap a=i song-an wumwa le
 mother-1sg.PERT 1sg=able OBL=3sg run.away-NOM house or
 papwên
 not.yet
 'Mother, am I ready for marriage or not yet?' (KW290611_0031)[68]

10.1.1.3 Indirect questions

Indirect or embedded questions (e.g. of the type 'I'm not sure who left') can be formed by a complement clause linked to the negated verb *mapwai* 'know'; see (17) for an indirect content question and (18) for an indirect polar one. The content question word remains in situ in the complement clause. There are very few instances of indirect questions in the corpus, so more targeted elicitation would be needed to find out more about their syntax.

(17) *ngamaro mapwai re ila apa pwên*
 nga_A=ma=to mapwai [te i_S=la a=pa]_{Compl:O} pwên
 1sg=NEG₁=CONT know COMP 3sg=go.to OBL=where NEG₂
 'I'm not sure where she went to.' (WL020611_0016)

(18) *ngamaro mapwai re kapwa i re i kulusun kei re ipan san pwên*
 nga_A=ma=to mapwai [te kapwa [i te-yo]_S i
 1sg=NEG₁=CONT know COMP if 3sg LIG-DEM.INT 3sg
 [kulusu-n kei [te ip_A=an san=∅_O]_{RC}]_{NVPRED}]_{Compl:O} pwên
 rubbish-PERT tree REL 3pl=PRF cut=3sg.ZERO NEG₂
 'I'm not sure whether this is sawdust [*lit.* 'rubbish of a tree that they've sawn].' (Game1_021012_0324)

10.1.1.4 Tag questions

A tag question is in fact not a question, but rather a statement for which the speaker seeks confirmation (and expects that it will be given). Tag questions in Paluai end with the tag particle *e* (realised [ɛ] ~ [ɛh]) and have an intonation

---

[68] The V + N collocation *song wumwa* (lit. 'run away house') is used to refer to getting married. The reason for this expression, as people told me, is that the woman leaves her own house and moves in with her husband's family (patrilocal system), and thus literally "runs away from" her house.

pattern that differs markedly from "regular" questions, both content and polar. On the tag, there is a dip in the pitch contour: it falls sharply and then rises again. Thus, only the actual interrogative part, the tag, has a pitch contour that differs from that of statements. There is also a slight pause between the statement and the tag. The tag can be translated by means of English 'right?' or 'isn't it?' or Tok Pisin *laka*. Sentence (19) gives an example.

(19) *arêan yong, e?*
 arê$_A$=an yong=Ø$_O$ e
 2pc=PRF hear=3sg.ZERO TAG
 'You have heard [it], right?' (PK290411_1_0049)

In addition, a particle *osa* is sometimes found at the end of statements. Since it contains the element *sa*, it probably has an interrogative-like function. It seems to be used as a discourse particle similar to a tag, but with a strong rhetorical function, as in examples (20) and (21).

(20) *iro pwa, "parian osa?"*
 i$_A$=to pwa   [paria-n osa]$_{Compl:O}$
 3sg=HAB think wife-PERT TAG
 'He used to think, "She is his wife, after all."' (WL020711_0055)

(21) *mui iro gat kamou ran ngoyai osa?*
 [mui]$_A$ i=to   gat   [kamou]$_O$ ta-n   ngoyai osa
 dog   3sg=HAB have speech   CLF.POSS-PERT cuscus TAG
 'After all, the dog has a conflict with the cuscus.' (LL010711_0089)

It seems that *osa* is used to mark information that is self-evident and beyond any doubt; as such, it is typical of strong rhetorical questions, i.e. questions that cannot be answered negatively and that are used to reinforce the point of view of the speaker. Examples (20) and (21) indicate that all things considered, it is evident that siblings' spouses have to be respected, and that the dog and cuscus are in an antagonistic relationship with each other.

10.1.2 Imperative mood

Imperative mood pertains to the speech act type 'command'. Second person imperative is marked by omission of the subject bound pronoun for the singular, and by an irrealis prefix for other numbers. Negative imperative is expressed either by the marker *napunan* or the modal particle *sa* and negative polarity; more details are provided below. Second person imperatives are the most common ones. There are no examples of any first person imperatives in the data (although there are a few examples of first person negative imperatives). There may be a number of constructions that can be analysed as third person imperatives, but these are rare.

10.1.2.1 Commands and requests

Context (which can contain non-linguistic cues such as hand gestures) and prosodic cues (such as tone of voice, loudness, possibly pitch) are usually the only means of distinguishing imperative from declarative mood for the second person singular. There are two exceptions to this. Firstly, in imperatives the subject bound pronoun can be omitted, which is not the case for declaratives. Some examples:

(22) *toksi!*
 tok.si
 sit.down
 'Sit down!'

(23) *ipwa, "mwêk." tareo ipul me rang*
 i_A=pwa [mwêk]$_{Compl:Quot}$ [ta=te-yo]$_{TopO}$ i_A=pul=Ø$_O$
 3sg=say be.quiet   SPEC.COLL=LIG-DEM.INT 3sg=tell=3sg.ZERO
 me [ta-ng]$_{AG}$
 come CLF.POSS-1sg.PERT
 'He said, "Be quiet." That, he told me.' (LL030611_0044)

Often, though, the bound pronoun is included in the imperative, as in example (24):

(24) *wola lêp kong payanpôl sip te ila ro arelo me*
 wo$_A$=la lêp [ka-ng   payan.pôl sip   [te
 2sg=go.to take CLF.food-1sg.PERT dry.coconut one.INANIM REL
 i_S=la to a=te-lo]$_{RC}$]$_O$ me
 3sg=go.to be OBL=LIG-DEM.DIST come
 'You go and take my dry coconut [for me to eat] that is over there, and bring it here.' (LK 100411_0063)

Secondly, in imperatives that contain a directional SVC (see Section 9.1.5), V1 is often elided, which would not be acceptable for a declarative sentence. Examples are given in (25) and (26).[69]

(25) *sip pame me!*
 [sip pame]$_O$ me
 one.INANIM betel.nut come
 'Give me one betel nut!' [*lit.* 'one betel nut come']

(26) *samel me!*
 [samel]$_O$ me
 knife come
 'Give me the knife!'

These instances of imperatives can be considered requests, but they are uttered with the same intonation contour as declaratives. Therefore they are analysed as imperatives rather than interrogatives (in any case, a request can often be regarded as a polite command). The elided verb could possibly be *tou* 'give', but this is not certain, since the abovementioned constructions are much more commonly produced than their more elaborate counterparts.

Paluai does not seem to distinguish different politeness levels for imperatives. Commands like the above seem to be quite brusquely stated when they are translated in English, but this is probably not perceived as such by native speakers. It is just that the language doesn't provide a means to mitigate a direct command by using, for example, a modal. The structures above are probably perceived as neutral, rather than impolite.

For second person non-singular, irrealis clauses can have an imperative interpretation, but strictly speaking they are always ambiguous between a declarative and imperative reading. Remember that reality status is formally unmarked for second person singular; this may be due to the fact that imperative and irrealis have merged for this category. In examples (27) and (28), marked for paucal and dual person and thus showing an overt irrealis marker, it is relatively certain that an imperative meaning is intended:

---

[69] Verb elision is assumed here because the objects requested are inanimate and thus not expected to move by their own volition. It seems therefore logical to assume an implicit instigator of the action and thus an implicit transitive main verb.

(27) kay, arê not te arê manak teo, arêkaret la wat liliu; arêkala yikyik
 kay [arê not te arê manak te-yo]$_S$
 okay 2pc child REL 2pc elder LIG-DEM.INT
 arê=ka-tet la wat liliu arê$_S$=ka-la yik.yik
 2pc=IRR.NS-move go.to up.high again 2pc=IRR.NS-go.to REDUP.search.for
 'Okay, you boys that are grown up, you (will) go up again; you (will) go (in order to) search.' (NP210511_2_0019)

(28) aukala liliu um tau
 au$_S$=ka-la liliu wumwa ta-o
 2du=IRR.NS-go.to return house CLF.POSS-2sg.PERT
 'You two (must) return to your [sg] house.' (KM060111_0027)

There seem to be no grammatical restrictions on the type of verb that is allowed to occur in an imperative clause. Example (23), for instance, shows a stative verb. Semantic and pragmatic restrictions do probably exist. For example, the stative verb *lot* 'fall' will not generally occur in an imperative clause, but a discourse context is conceivable in which a speaker orders someone to fall on purpose. The stative verb *tet* 'grow', on the other hand, is unlikely to occur in an imperative clause, since things cannot grow on command. (However, this does not necessarily stop speakers from using an imperative, which would be comparable to someone telling his kettle: "Boil!").

10.1.2.2 Negative imperatives

Negative imperatives or prohibitives are accompanied by the marker *napunan* (*lit.* '(it is) forbidden to'), which is related to the noun *napu-* '(food) taboo'. *napunan* is always clause-initial and the predicate following it is always marked as irrealis. Examples are given for second, first and third person singular, respectively, in (29) to (31).

(29) napunan wosui nik a woyektou to pulen kone
 napunan wo$_A$=sui [nik]$_O$ a wo$_A$=yek.tou=Ø$_O$
 NEG.IMP 2sg=catch fish and 2sg=put=3sg.ZERO
 to pulen.kone
 be sand.beach
 'You are not allowed to catch fish with a line and put [it] on the beach.' (LL030611_0022)

(30) *napunan kowau la kason nik*
   napunan ko$_S$-wau la kaso-n nik
   NEG.IMP IRR.1sg-move go.to near-PERT fish
   'It was forbidden for me to go near the fish.' (LL030611_0026)

(31) *napunan aso-om kiro ret touo*
   napunan [asoa-m]$_A$ ki-to tet tou=o$_O$
   NEG.IMP husband-2sg.PERT IRR.3sg-CONT move give=2sg
   'Your husband cannot follow you.' (KS030611_1_0027)

Another way to express a negative imperative is by negating a *sa* clause. *sa* normally refers to ability, but when negated, the clause can get prohibitive overtones, as discussed in Section 6.5.2.2, and thus the use of *sa* can be seen as a negative imperative strategy. Two examples given in Section 6.5.2.2 are repeated below.

(32) *osa lêpi pwên*
   wo$_A$=sa lêp=i$_O$ pwên
   2sg=MOD take=3sg NEG
   'You cannot take [i.e. adopt] him [I won't allow it].' (LM240611_0046)

(33) *osa yuai naluai pwên*
   wo$_A$=sa yuai [naluai]$_O$ pwên
   2sg=MOD call garden.food NEG
   'You cannot/should not call out for a tuber to plant.'(NP220611_1_0024)

This way of phrasing a negative imperative may be a little milder and thus more polite than using *napunan*. However, in the stories that (32) and (33) are taken from, clauses with *napunan* are also used to refer to the same situation. Thus, the distinction between the two may be stylistic more than anything else.

10.1.3 Dependencies between mood and verbal categories

There are a number of dependencies between mood (a category that relates to the clause) and verbal categories such as aspect, reality status and modality, as discussed in Chapter 6. There appear to be no restrictions as to which grammatical categories can be combined with interrogative mood: there are questions attested with marking for irrealis, imperfective, perfective, perfect, secondary aspect and agentive modality (ability). Interrogative mood is also attested with negative polarity (see Section 10.2.2.4 below).

As mentioned above, content questions are often verbless clauses and thus do not contain any TAM marking. In general, questions seem to be less often marked for perfective or imperfective aspect than statements. This could be explained because a question seeks information about the nature of an event and thus aspect tends to be neutral. An exception is formed by questions about an event that is unfolding in the present, for example 'What are you doing?' shown in (1), or (34) below. These questions generally carry continuative marking, to indicate that the speaker is seeking information about an action that is ongoing at the here-and-now of the speech event.

(34) *woro ning naêmwan le pwên?*
 wo$_A$=to ning [naêmwa-n]$_O$ le pwên
 2sg=CONT see back-PERT or NEG
 'Can you see [*lit*. 'are you seeing'] his back or not?' (Game2_280812_0052)

True commands, such as the one in (22), do not show any marking on the verb and thus are not marked for TAM or any other categories. In addition to this, there are clauses marked as irrealis that are ambiguous between a declarative and an imperative reading, such as (27) and (28). Negative imperatives always have irrealis marking. Situations referred to in the imperative or the prohibitive (negative imperative) are not actualised yet: they still belong to the realm of thought, making irrealis marking appropriate. For second person singular, which will be most frequently used with imperative, irrealis is zero-marked. This means that in these cases the formal difference between a declarative and an imperative is only evident by presence or absence of the bound pronoun (apart from intonation and non-linguistic clues).

10.2 Polarity

Polarity refers to the positive (affirmative) or negative status of a constituent, which can be a phrase, a predicate or an entire clause. A positive realis predicate can be negated by means of the discontinuous elements *ma=* . . . *pwên*; see below for details. An irrealis predicate cannot be negated by *ma=*. Instead it takes a modal particle *sa*, with *pwên* added at the end of the clause. *ma=* and *sa* are mutually exclusive; in realis clauses, too, they do not co-occur. The negation has scope over the material that is located between the two elements *ma=* . . . *pwên* or *sa* . . . *pwên*. Only *pwên* is used for non-predicative negation. It can also be used by itself as answer to a question.

Negation touches upon every aspect of the grammar, since (in theory) any construction has a potential negative counterpart. In what follows, negative and affirmative answers to polar questions will be discussed first, followed by negation of realis and irrealis predicates, respectively. Existential and possessive predicates and their negation will be discussed briefly, followed by the interrelation between negation and verbal categories. Next, non-predicative negation and its peculiarities will be discussed.

10.2.1 Answer to a polar question

pwên can be used as a negative answer to a polar question (see Section 10.1.1.2). It can be translated in English as 'no', but also as 'nothing':

(35) woro pe sa? – pwên
 wo$_A$=to pe [sa]$_O$ pwên
 2sg=CONT do what NEG
 'What are you doing?' – 'Nothing.'

There is no lexical item to answer a polar question affirmatively (like English 'yes'), except perhaps the interjection *uu*. An affirmative answer can be given by means of an affirmative sentence, the interjections *ah* [aː], *uu* [uː] or *mmhm* [mhm] together with prosodic cues, or non-linguistic cues such as nodding the head. Increasingly, the Tok Pisin loan *yes* is used as well.

10.2.2 Predicative negation

10.2.2.1 Negation of realis and verbless predicates

Predicative negation in Paluai is discontinuous. With verbal predicates, the bound form *ma=* directly follows the subject bound pronoun, preceding the verb complex; *pwên*, which is also encountered independently, follows the negated predicate. See Chapter 6 for a discussion of verbal categories and the structure of the affirmative verb complex. Exactly which slot in the VC is filled by *ma=* is not entirely clear. At least two analyses are available: it either occupies the slot that can also be taken by the irrealis prefix, or it takes the first preverbal TAM slot. An argument for the first analysis is that *ma=* never occurs together with the irrealis prefix. An argument for the second analysis is that *ma=* interferes with the TAM possibilities of the clause: it cannot co-occur with any of the particles that can fill this slot in an affirmative clause, and thus a negated clause cannot be marked for core aspect.[70] However, *ma=* can also occur with non-verbal predicates, which may indicate that it actually occupies neither of the slots in the VC, but is a component of a different order.

Elements following the verbal predicate, such as objects, complement clauses, adverbial phrases and Oblique constituents, occur before *pwên*. Determining precisely over which part of the sentence the negation has scope is straightforward: it is always the material occurring between *ma=* and *pwên*. Non-verbal predicates, including possessive predicates, are negated in the same way, with *ma=* placed between the subject and the predicate material and *pwên* following the predicate material (see Chapter 7 for more information on non-verbal predicates). Any modifiers to the head noun or adjective, such as intensifiers, appear between *ma=* and *pwên*. The sentences in (36) and (37) give examples of a negated verbal and verbless clause, respectively.

(36) *tapmaro ning muyan pwên*
 tap$_A$=**ma**=to ning [muya-n]$_O$ **pwên**
 1pl.INCL=NEG$_1$=HAB see skin-PERT NEG$_2$
 'We cannot see [*lit.* 'are not seeing'] his skin.' (LK250111_0087)

(37) *naman i maaloen pun pwên*
 naman i$_S$ **ma**=[aloen pun]$_{NVPRED}$ **pwên**
 perhaps 3sg NEG$_1$=long INTF NEG$_2$
 'Perhaps it is not very long.' (KW290311_0029)

An example of a negated non-verbal predicate with a possessive classifier is given in (38).

(38) *panu repwo marao pwên*
 [panua te-pwo]$_S$ **ma**=[ta-o]$_{NVPRED}$ **pwên**
 place LIG-DEM.INT NEG$_1$=CLF.POSS-2sg.PERT NEG$_2$
 'This place is not yours.' (NP210511_2_0057)

In contemporary Paluai, possession is often expressed verbally by the Tok Pisin loan *gat* 'have', which can also be negated. An example, in the form of an often-heard complaint, can be found in (39).

---

[70] The modal particle *sa*, which can also occupy this slot, can function in a negative predicate with just *pwên*. Since *sa* never co-occurs with *ma* this is further evidence that they may occupy the same slot in the VC.

(39) *ngamagat pame pwên*
nga$_A$=**ma**=gat [pame]$_O$ **pwên**
1sg=NEG$_1$=have betel.nut NEG$_2$
'I don't have any betel nut.'

In (40), *pwên* appears at the end of the complement clause to *inap* 'be able'; the negation thus has scope over the entire predicate including the complement clause. In (41), *pwên* appears at the end of a very long object NP, including several relative clauses. The whole sequence is in the scope of the negation.

(40) *ipmainap kape roui la ai pwên*
ip$_A$=**ma**=inap [ka$_A$-pe tou=i$_O$ la a=i]$_{Compl:O}$ **pwên**
3pl=NEG$_1$=able IRR.NS-PFV put=3sg go.to OBL=3sg NEG$_2$
'They were not able to put him into it.' (Game1_021012_0490)

(41) *imagat sesap te iro kason te koning wen te kipe ngan pwên*
i$_A$=**ma**=gat [sesap [te i$_S$=to kaso-n [te
3sg=NEG$_1$=have anything REL 3sg=be near-PERT REL
ko$_A$-ning=Ø$_O$ wen [te ki$_A$-pe
IRR.1sg-see=3sg.ZERO move.level REL IRR.3sg-PFV
ngan=Ø$_O$]$_{RC}$]$_{RC}$]$_{RC}$]$_O$ **pwên**
eat=3sg.ZERO NEG$_2$
'There isn't anything near to him that I could see that he could eat.'
(Game1_021012_0074)

In addition, an entire subordinate clause can be in the scope of a negation. In (42), the clause *yi mate kapwa i pwapwa mangsilan pwên* forms a complement clause to the verb *ning*. In turn, it contains a subordinate clause. However, this type of construction is rare.

(42) *ining pun te i mare kapwa i pwapwa mangsilan pwên*
i$_A$=ning pun [te i=**ma**=[te kapwa i$_S$
3sg=see INTF COMP 3sg=NEG$_1$=SUB as.if 3sg
[pwapwa mangsilan]$_{NVPRED}$]]$_{Compl:O}$ **pwên**
story meaningless NEG$_2$
'He truly saw that it was not as if this was a meaningless story.'
(LK250111_0082)

10.2.2.2 Negation of irrealis predicates

The element *ma=* is incompatible with irrealis and with the modal particle *sa*. Negated irrealis clauses and clauses with *sa* therefore do not contain *ma=*, but either show the irrealis *kV-* prefixed to *sa*, or *sa* by itself, combined with *pwên* following the negated elements. The meaning expressed here is that something may not happen in the future. This kind of clause is discussed in Section 6.5.2.3. Irrealis predicates do not just get a "neutral" negative meaning when negated, but have modal overtones such as 'not able to' or 'not allowed to'.[71] Two examples from Chapter 6 are repeated in (43) and (44).

(43) *ip lau kasa ro ning muyom pwên*
 [ip lau]_A ka-**sa** to ning [muya-m]_O **pwên**
 3pl people IRR.NS-MOD CONT see skin-2sg.PERT NEG
 'The people will not be able to see your skin.' (LK250111_0073)

(44) *ikisa ningong ai kunawayen pwên*
 i_A=ki-**sa** ning an-sê=ong_O a=i kunawaye-n **pwên**
 3sg=IRR.3sg-MOD see PART-DIM=1sg OBL=3sg life-PERT NEG
 'She should/ought not (be able to) see me again for the rest of her life.' (WL020711_0078)

10.2.2.3 Negation of existential predicates

Existential predicates are discussed in more detail in Section 3.3.1.5. An affirmative existential predicate often contains the existential verb *tok* 'be, stay; exist'. However, these predicates do not seem to be negated by *ma=* ... *pwên*. Rather, to express predicatively that something is *not* there, speakers seem to prefer the Tok Pisin loan *igat* 'there is' (often pronounced [ikat], more in line with Paluai phonology) used with a third singular dummy subject *i=*, which is then negated.[72]

---

71 In the present-day language, the Tok Pisin loan *inap* 'be able to' is frequently used instead of *sa*. However, *inap* can be negated by using *ma=* and takes a complement clause which is always marked for irrealis (for an example, see (40)). Complement clauses with *inap* are discussed in more detail in Section 11.1.2.5.

72 *gat* is also a verb expressing possession and is used as such by speakers of Paluai, both when speaking Tok Pisin and when using it as a borrowed form when speaking Paluai. This subsection focuses on the existential *igat* construction.

(45) *imagat yon pwên*
  i_A=**ma**=gat   [yanu]_O **pwên**
  3sg=NEG_1=have water   NEG_2
  'There was no water.' (NS220511_1_0032)

(46) *imagat som yamat te iro pwalingan ip pwên*
  i_A=**ma**=gat   [som   yamat [te i_S=to pwalinga-n ip]_RC]_O
  3sg=NEG_1=have one.ANIM person REL 3sg=be with-PERT   3pl
  pwên
  NEG_2
  'There was not a single human being who lived among them.'
  (LL300511_1_0005)

Alternatively, negative existential predicates are found showing only *pwên*, sometimes combined with the possessive classifier *ta-*. *pwên* seems to function here as a negative existential main verb; compare examples (47) to (49).

(47) *som not mwen pwên tai*
  [som   not   mwen]_S pwên ta-i
  one.ANIM child man   NEG   CLF.POSS-3sg.PERT
  'He didn't have any sons.' [*lit.* 'a male child did not exist to him']
  (LM240611_0029)

(48) *i reo pwên tararap tole*
  [i   te-yo]_S   pwên ta-tap   tole
  3sg LIG-DEM.INT NEG   CLF.POSS-1pl.INCL.PERT EMP
  'This [snake] does not exist here [*lit.* 'exist to us'] for sure.'
  (Game1_021012_0090)

(49) *a i re nganngan pwên ai...*
  a   i_S [te [nganngan]_S pwên a=i]_RC
  and 3sg REL food   NEG   OBL=3sg
  'And the one in which there is no food...' (Game1_021012_0319)

It is quite likely that *pwên* has verbal origins, an analysis also supported by the lexical verb *pepwên* 'be finished, be empty', which may be related to *pwên*. However, constructions in which *pwên* appears to be a main verb are rather marginal in the present-day language, since expression of both possessive and existential meanings has largely been taken over by the Tok Pisin loan *gat*, which is negated in the "regular" way with *ma=* ... *pwên*.

10.2.2.4 Dependencies between negation and verbal categories

As already mentioned, predicative negation interrelates with several verbal categories. This is mainly due to the form *ma=*. First of all, *ma=* is incompatible with the irrealis prefix *kV-* and is thus only found with realis predicates. *ma=* can be used to negate non-verbal predicates, as sentence (37) shows. Secondly, *ma=* cannot occur together with the modal particle *sa* or with core aspectual particles *no* (imperfective), *pe* (perfective) and *an* (perfect). Thus, negated predicates are neutralised with respect to core aspect. *ma=* does occur together with the secondary aspectual particles *to* and *tu* (it is not attested in the corpus together with *yen*, which is much rarer than *tu* and *to*). Since *ma=* appears to occupy the same slot in the VC and interrelates with modality, aspect and reality status, it can be considered part of the TAM system. *pwên*, on the other hand, shows no relation to the TAM paradigm or other verbal categories. As we will see below, *pwên* is used, by itself, for non-predicative negation.

Interrogative mood and negative polarity can occur together, as shown in example (50). This sentence, with interrogative intonation, was used when discussing a picture showing a toy cow positioned on top of the cabin of a toy truck. It is assumed that the speaker tries to express his conviction that in a natural state of affairs, the cow ought to fall off once the truck starts moving, since it is not tied up. Thus, pragmatically this utterance is somewhat marked, since the speaker is mainly expressing surprise at the unusual situation represented in the picture. Questions with negative polarity do not occur very often; usually an alternative question will be used, such as example (15) above.

(50) *ipkaru ret lêpi a isa lot pwên?*
 ip$_A$=ka-tu [tet lêp]=i$_O$ ya i$_S$=sa lot pwên
 3pl=IRR.NS-STAT.CONT move take=3sg then 3sg=MOD fall NEG
 'When they are taking him he won't fall off?!' (Game1_021012_0485)

Questions with negative polarity are answered in opposite ways in Paluai and related languages, on the one hand, and in English, on the other. Answering a negative question affirmatively indicates that the content of the question, the negative proposition, is indeed true. Answering it negatively indicates that the negative proposition is not true. This is exemplified in (51) and (52).

(51) *tepwo omalai skul pwên? – uu*
 te-pwo wo$_S$=ma=la a=i skul pwên uu
 LIG-DEM.INT 2sg=NEG$_1$=go.to OBL=3sg school NEG$_2$ yes
 'You didn't go to school today?' – 'Yes [indeed, I didn't go to school].'

(52) *tepwo omalai skul pwên? – pwên*
 te-pwo wo_S=ma=la a=i skul pwên pwên
 LIG-DEM.INT 2sg=NEG$_1$=go.to OBL=3sg school NEG$_2$ NEG
 'You didn't go to school today?' – 'No [I did go to school].'

10.2.3 Non-predicative negation

For non-predicative negation, *ma=* is not used. A negative adverbial phrase that can modify a verb will be discussed first. After that, there will be a brief discussion of negated NPs.

10.2.3.1 The negative adverbial phrase *la pwên*

The phrase *la pwên* means 'in vain'. In the data, this phrase is only attested in combination with SVCs that have a transitive V1 and *lêp* as V2. In turn, this SVC is only attested modified by *la pwên*. The directional *la* 'go to' is used to form adverbial phrases from adjectives (see Section 3.6.2.2); the phrase *la pwên* may be comparable to these adverbial phrases. Its English counterpart, after all, also contains an adverbial.

(53) *iyik lêp tinan la pwên*
 i$_A$=[yik lêp]$_{SVC}$ [tina-n]$_O$ la pwên
 3sg=search.for take mother-PERT go.to NEG
 'He searched in vain for his mother.' [*lit.* 'He did not search-get his mother.'] (KW290611_0054)

(54) *ma urêkol lêpi la pwên pun*
 ma wurê$_A$=[kol lêp]$_{SVC}$=i$_O$ la pwên pun
 but 1pc.EXCL=wait.for take=3sg go.to NEG INTF
 'But we waited for him in vain.' [*lit.* 'But we did not wait-get him at all.'] (NP210511_2_0025)

Importantly, the phrase *la pwên* only negates the second part of the SVC, *lêp*: what the sentences mean is that searching or waiting was done, but that it was without result. As discussed in Chapter 9, elements of SVCs cannot be negated separately in the usual way, with *ma=* ... *pwên*. It seems that this construction is a means to circumnavigate this restriction.

10.2.3.2 Negated noun phrases

There is no derivational morphology, such as privative, to negate a noun phrase. Many cases of what look like negated NPs can alternatively be analysed as negative existential predicates, with *pwên* as a main verb; compare Section 10.2.2.3 above.

In a limited number of cases, NPs negated with *pwên* seem to have become lexicalised; examples of these are shown in Table 83 below. These phrases are most commonly used predicatively in a non-verbal predicate, and not as noun modifiers. However, they can be used next to a generic noun to make a further specification, for example in *muyou ken pwên* 'death adder' (*lit.* 'snake leg-PERT NEG').

Table 83: Some lexicalised negative phrases.

Form	Meaning
mara-n pwên	blind; blunt (eye-PERT NEG)
layenga-n pwên	deaf (ear-PERT NEG)
kolo-n pwên	mute (voice-PERT NEG)
ke-n pwên	legless (*lit.* 'leg-PERT NEG')
tama-n pwên	(eldest) son of deceased person during mourning (*lit.* 'father-PERT NEG')

10.2.4 Intensification and moderation of negated clauses

Negation can be intensified by the general intensifier *pun* placed outside the scope of the negation, following the negated elements. This yields the meaning 'not at all':

(55) *ma urêkol lêpi la pwên pun*
    ma   wurê$_A$=[kol   lêp]$_{SVC}$=i$_O$ la   pwên pun
    but 1pc.EXCL=wait.for take=3sg   go.to NEG   INTF
    'But we waited for him in vain.' [*lit.* 'But we did not wait-get him at all.']
    (NP210511_2_0025)

Another way to intensify negation is to modify the negated element with a diminutive marker *(n)ansê*, inside the scope of the negation, which has intensification as its net result. This type of verbal modifier is only attested in clauses with negative polarity. *(n)ansê* occurs in negated clauses with and without *ma=*.

(56) *imaro worupê nansê si paye pwên o*
 i_S=**ma**=to   worup   nan=sê   si   paye   **pwên**   yo
 3sg=NEG₁=HAB descend PART=DIM come.down down NEG₂ DEM.INT
 'He never climbed down anymore at all.' (LL010711_0081)

(57) *ngamalêp nansê masiom pwên*
 nga_A=**ma**=lêp   [nan=sê   masia-m]_O   **pwên**
 1sg=NEG₁=take PART=DIM appreciation-2sg.PERT NEG₂
 'I don't appreciate what you did at all.' [*lit.* 'I do not take a bit of your appreciation.'] (LL010711_0062)

(58) *ningong nansê pian pwên*
 [ninga-ng]_S   [nan=sê   pian]_NVPRED   pwên
 appearance-1sg.PERT PART=DIM good   NEG
 'I don't look good at all.' [*lit.* 'My looks are not a bit good'] (LL010711_0052)

Conversely, when the negated element is modified by the intensifier *pun*, inside the scope of the negation, the negation is moderated, and the net meaning is 'not very'.

(59) *naman i maaloen pun pwên*
 naman   i_S   **ma**=[aloen pun]_NVPRED   **pwên**
 perhaps 3sg NEG₁ =long INTF   NEG₂
 'Perhaps it is not very long.' (KW290311_0029)

10.2.5 Ellipsis of *ma=*

The element *ma=* does not have to be present in elliptical coordinated clauses. Two examples are given in (60) and (61).

(60) *no Ngi reo ila university; wong tepwo pwên*
 no   [Ngi te-yo]_S   i=la   university wong   te-pwo
 only N. LIG-DEM.INT 3sg=go.to   1sg.FREE LIG-DEM.PROX
 pwên
 NEG
 'Only Ngi went to university; I didn't.' (OL201210_0127)

(61) *a kei sei re iro minan teo, nansêk to pei a nansêk pwên?*
 a [kei sei [te i_S=to mina-n]_{RC} te-yo]_{TopS}
 and tree one.long REL 3sg=be hand-PERT LIG-DEM.INT
 [nan=sêk]_S to pei a nan=sêk pwên
 PART=half.long CONT appear and PART=half.long NEG
 'And (as for) the stick that is in his hand, one end is showing and one end isn't?' (Game2_280812_0113)

Because the second part of sentence (60) is an elliptical sentence (the predicate *la university* 'go to university' is deleted because it is identical to the predicate of the preceding clause), *ma=* is not present; *pwên* is sufficient. This is also the case in (61), where *to pei* is deleted because it is identical to the predicate of the preceding clause. It appears that the use of *ma=* is closely connected to the (realis) predicate, whether it is verbal or non-verbal.

11 Clausal relations and clause combining

A clause is a linguistic unit that consists of a predicate plus its core and peripheral arguments. A predicate is most often headed by a verb or SVC, but its head can also be a noun, adjective, numeral, preposition or interrogative word. See Chapter 6 for a discussion of verbal predicates, Chapter 7 for a discussion of non-verbal predicates, and Chapter 8 for a discussion of grammatical relations and alignment of arguments within the clause.

A main clause can stand alone, whereas a dependent clause usually does not make sense by itself and needs to be embedded within a main clause. This is the case with relative clauses and complement clauses, with the exception of quotational complement clauses after the verb *pwa* 'say'. Alternatively, a dependent clause may be linked to a main clause by means of a subordinating conjunction. In addition, two main clauses can be linked to each other by means of a coordinating conjunction. In what follows, relations between main and dependent clauses will be discussed first, followed by relations between two main clauses.

11.1 Dependent clauses

A dependent clause modifies or specifies a main clause or a constituent in a main clause. The following types of dependent clauses can be distinguished:
1. Relative clause, modifying a noun;
2. Complement clause, functioning as a core or peripheral argument to a verb, or as a modifier to an adjective functioning as a predicate head in a non-verbal predicate;
3. Adverbial subordinate clause:
    a. Temporal (simultaneous and successive);
    b. Manner;
    c. (Possible) consequence;
    d. Concessive;
    e. Conditional.

Dependent clauses are introduced by the marker *te*, with the exception of some types of modal complement clauses and conditional subordinate clauses. In very general terms, *te* marks the fact that a particular constituent slot is filled by a clause rather than by a phrase. It may have the same origin as the formative *te-* encountered in morphologically complex demonstratives (see Section 4.3).

The various types of dependent clauses and their grammatical features will be discussed in turn.

11.1.1 Relative clauses

A relative clause (RC) functions as a modifier to a noun and has an argument coreferential with the noun it modifies (the common argument or CA; cf. Dixon 2010b). In what follows, the formal properties of relative clauses are discussed, followed by the syntactic functions the CA can have in the main clause (MC) and relative clause.

A relative clause immediately follows the noun it modifies and is introduced by the marker *te*. It can contain either a verbal or a non-verbal predicate. Relative clauses are very common in Paluai; it seems there are virtually no restrictions on which type of nominal can be a relative clause head. NPs with a pronoun, a local, common or personal noun or a numeral as head are all attested with RCs. Apart from the marker *te*, there is no special marking on relative clauses. However, each type of relative clause looks slightly different depending on the combination of functions of the common argument. The formal properties of the various types of relative clause will therefore be discussed in more detail below. The fullest realisation of the CA is almost always in the MC; in the RC the CA is often only expressed by a pronoun (in the case of a S/A or animate O argument, or an Oblique) or a suffix (in the case of a Possessor). Inanimate O arguments are not cross-referenced at all.

There are quite a number of main clauses in the corpus in which two relative clauses are embedded. They are at times difficult to analyse, as it is not always clear whether the second RC is embedded within the first, or whether it also shares an argument with the MC, rather than with the first RC.

Paluai relative clauses can be both restrictive and non-restrictive (cf. Dixon 2010b: 314). A restrictive relative clause gives the noun it modifies specific reference, whereas a non-restrictive relative clause modifies a noun that already has specific reference. There is no formal difference between both types and there also seems to be not much of a prosodic difference. Most of the relative clauses encountered are restrictive; non-restrictive ones are generally encountered when a name or other additional information about a discourse participant is given. The discourse context should suffice to disambiguate between the two types of relative clauses. Whenever an example below is non-restrictive, this is indicated.

11.1.1.1 Possible syntactic functions in the RC of the common argument

The common argument can have a range of functions both in the MC and RC; see Table 84. In a verbless relative clause, however, the CA can only have the function of subject; it can never head the predicate. Interestingly, with most combinations, the function of Possessor is only found in RCs in direct possessive constructions, not in indirect ones (see Sections 3.2.2 and 5.5 for more information on possessives). The CA is encountered in an indirect possessive construction in the RC only when it has Oblique function in the MC. The CAs most commonly found in the corpus are those that function as S/A or O in the RC. In the sections below, examples of each syntactic function of the CA in the RC are given, and their formal properties are discussed.

Table 84: Functions of CA in MC and RC.

Function in MC	Functions attested in RC
Intransitive / transitive subject (S/A) – also of verbless clause	S/A – also of verbless clause
	O
	Possessor in direct possessive construction
	Oblique
Transitive object (O)	S/A – also of verbless clause
	O
	Oblique
	Possessor in direct possessive construction
Oblique	S/A – also of verbless clause
	O
	Oblique
	Possessor (both direct and indirect)
Nominal predicate	S/A – also of verbless clause
	O
	Oblique
	Possessor in direct possessive construction
Possessor (in indirect possessive construction)	S/A – also of verbless clause
	Possessor in direct possessive construction

11.1.1.1.1 Common argument is S/A in RC

When the CA is the S/A of the RC, a subject bound pronoun, coreferential with its referent in the MC and agreeing with it in person and number, has to be present. When the RC contains an irrealis predicate, use of the irrealis prefix is sufficient, since it also contains person/number information. This situation is identical to the one observed for main clauses. In general, the RC is immediately followed by the intermediate demonstrative *(te)yo*. When a demonstrative form is present, it always occupies the final position in the NP (see also Chapter 5). Thus, the demonstrative following a RC indicates that it should be considered a modifier to the head noun. A number of examples of RCs with the CA as S/A are given in (1) to (3). The CA is printed in bold in both MC and RC.

(1) *yamat te iro wau panuan teo iwop pal sapon ai parei*
 [**yamat** [te **i**$_S$=to wau panua-n]$_{RC}$ te-yo]$_A$ **i**=wop pal
 person REL 3sg=CONT move front-PERT LIG-DEM.INT 3sg=fly break
 [sapo-n]$_O$ a=i parei
 top-PERT OBL=3sg pole
 'The person who was going in front of her hit the top of her head with a pole.' (KW290611_0012)

(2) *a manuai re iru riki reo kipe nêm*
 a [**manuai** [te **i**$_S$=tu tik=i]$_{RC}$ te-yo]$_S$ **ki**-pe
 and eagle REL 3sg=STAT.CONT creep.on=3sg LIG-DEM.INT IRR.3sg-PFV
 nêm
 be.finished
 'And [the rash from] the eagle [food taboo] that was covering him will be finished.'[73] (NP260511_0022)

(3) *pau re kime pei reo kipe piypiy*
 [**pau** [te **ki**$_S$-me pei]$_{RC}$ te-yo]$_S$ **ki**-pe
 coconut.oil REL IRR.3sg-come appear LIG-DEM.INT IRR.3sg-PFV
 piy.piy
 REDUP.splash
 'The coconut oil that will come up will spatter.' (CA120211_2_0027)

(4) and (5) show examples of a CA that functions as O in the MC and S/A in the RC. What is striking in these cases is that the bound pronoun in the RC often does not agree in number with the O of the matrix clause, but is almost invariably the third singular subject bound pronoun *i=*, as in (4).

---

[73] Paluai believe that each individual has a particular animal as food taboo, which is inherited from the mother. If a food taboo is broken, the person will feel unwell and develop skin rashes, sores and itches. The taboo can be eliminated by applying herbs specific for that particular food taboo.

(4) owom antek ip kurun kei namwi re iro lalon
     wo_A=wom antek   [ip kutun kei namwi [te i_S=to   lalo-n]_RC]_O
     2sg=chop put.away 3pl small tree small   REL 3sg=be inside-PERT
     'You chop away the small trees that are inside.' (KM190211_0011)

(5) manuai re iriki reo, ngan pe andek ai sapon kein pari ai manuai
     [manuai   [te   i_A=tik=i]_RC   te-yo]_TopO   nga_A=an pe
     eagle   REL   3sg=creep.on=3sg LIG-DEM.INT   1sg=PRF   do
     antek=Ø   a=i   sapo-n   kei-n   pari   a=i
     put.away=3sg.ZERO OBL=3sg top-PERT herb-PERT belonging.to OBL=3sg
     manuai
     eagle
     'The eagle [rash] that covered him, I've removed [it] with herbs meant for the *manuai* [food taboo].' (NP260511_0024)

Examples in which the CA has Oblique function in the MC and S/A function in the RC are given in (6) and (7).

(6) isuwot ai peilêp te iro poyon pirou reo
     i_S=suwot   a=i   [peilêp   [te   i_S=to   poyo-n   pirou]_RC
     3sg=go.down OBL=3sg canoe   REL   3sg=be under-PERT callophylum
     te-yo]_Obl
     LIG-DEM.INT
     'He went down to the canoe that was underneath the callophylum tree.' (LK100411_0039)

(7) i reo i stori pari ai naluai re iangou Paluai
     [i   te-yo]_S   i   [stori pari   a=i   [naluai   [te
     3sg LIG-DEM.INT 3sg story belonging.to OBL=3sg garden.food REL
     i_S=angou   Paluai]_RC]_Obl]_NVPRED
     3sg=arrive P.
     'That is the story of the garden food that arrived on Baluan.' (KS030611_1_0050)

Examples (8) and (9) show a commonly encountered combination in which the CA is subject of a verbless MC and S/A in the RC. The RC immediately follows the noun heading the subject NP, which is cross-referenced by a bound pronoun in the RC. The non-verbal predicate then follows the relative clause. Example (8) shows a non-restrictive relative clause. Example (9) may be analysed as such as well, but it could also be restrictive. A person can have several classificatory mothers according to the kinship system (see Section 3.2.2.1) and could also be adopted. The relative clause in (9) limits reference to only one person, the biological mother. It is therefore probably restrictive.

(8) *youn ip te iro rok taip teo i molmolean kanen ip o*
[**you-n ip** [te **i**$_S$=to tok ta-ip]$_{RC}$ te-yo]$_S$ i
tail-PERT 3pl REL 3sg=CONT stay CLF.POSS-3pl.PERT LIG-DEM.INT 3sg
[molmolea-n kane-n ip yo]$_{NVPRED}$
decoration-PERT body-PERT 3pl DEM.INT
'Their tail, that is with them, it is a decoration of their bodies.'
(LL300511_1_0079)

(9) *tinan te iroui o, i reo i punpot tai rea*
[**tina-n** [te **i**$_S$=tou=i]$_{RC}$ yo]$_S$ i te-yo i
mother-PERT REL 3sg=give.birth=3sg DEM.INT 3sg LIG-DEM.INT 3sg
[punpot ta-i]$_{NVPRED}$ te-ya
side.of.mother CLF.POSS-3sg.PERT LIG-then
'His mother who gave birth to him, she [i.e. her relatives] is his *punpot* [receiving side in mortuary payments].' (SP190311_0022)

A relative clause with an S/A as CA can also modify a noun that heads a non-verbal predicate. This is exemplified in (10).

(10) *i reo i panu nuanan te irok*
[i te-yo]$_S$ i [**panua nuanan** [te **i**$_S$=tok]$_{RC}$]$_{NVPRED}$
3sg LIG-DEM.INT 3sg place true REL 3sg=be
'This was a real place that existed.' (LK100411_0132)

An example of a CA that has the function of Possessor in the MC and S/A in the RC appears in (11). Only indirect possession is encountered. The relative clause is non-restrictive. The Possessor CA forms part of the fronted O argument.

(11) *no kamou tan numwai re ipul la remenin...*
no [kamou ta-n **numwai**]$_O$ [te **i**$_A$=pul=Ø la
only speech CLF.POSS-PERT old.man REL 3sg=tell=3sg.ZERO go.to
temenin]$_{RC}$
like
'Only the words of the old man, who spoke [them] as follows...'
(OL201210_0101)

11.1.1.1.2 Common argument is O in RC

When the CA functions as object in the RC, it is generally not overtly expressed, unless it has an animate referent. Again, the RC is followed by the intermediate demonstrative *(te)yo*. Examples where the CA is S/A in the MC are given in (12) and (13).

(12) *kapol le worou sot yep teo kikôk*
 [**kapol** [te wo$_A$=tou=Ø$_O$ sot yep]$_{RC}$ te-yo]$_S$ **ki**-kôk
 dish REL 2sg=put=3sg.ZERO go.up fire LIG-DEM.INT IRR.3sg-heat.up
 'The dish that you put on the fire will heat up.' (CA120211_1_0024)

(13) ...*te mangat te taupwa kape reo inêm o*
 te [**mangat** [te tau$_A$=pwa ka-pe=Ø$_O$]$_{RC}$ te-yo]$_S$
 SUB work REL 1du.INCL=want.to IRR.NS-do=3sg.ZERO LIG-DEM.INT
 i=nêm yo
 3sg=be.finished DEM.INT
 'For the work that the two of us were going to do is finished.'
 (LL010711_0068)

An example in which the CA is also O in the MC is given in (14). Again, this type of relative clause is often followed by the intermediate demonstrative *(te)yo*.

(14) *ippe lêp poyan nganngan te epmangan nêm pwên teo*
 ip$_A$=pe lêp [**poyan nganngan** [te ep$_A$=ma ngan
 3pl=PFV take left.over food REL 1pl.EXCL=NEG$_1$ eat
 nêm=Ø$_O$ pwên]$_{RC}$ te-yo]$_O$
 be.finished=3sg.ZERO NEG$_2$ LIG-DEM.INT
 'They received the leftover food that we did not eat up.'
 (KM060111_0086)

A CA that functions as Oblique in the MC and as O in the RC is also not overtly shown in the RC; see (15).

(15) *ngapwa kotok sori ai nayei re ngan pe*
 nga$_S$=pwa ko-tok.sori a=i [**nayei** [te nga$_A$=an
 1sg=want.to IRR.3sg-apologise OBL=3sg lie REL 1sg=PRF
 pe=Ø$_O$]$_{RC}$]$_{Obl}$
 make=3sg.ZERO
 'I want to apologise for the lies that I have told.' (PK290411_1_0039)

Occasionally, a combination is found where the CA is subject of a verbless MC and O in the RC:

(16) *mangat te kope pe repwo i mangat pari ai matmat tan tamong*
 [**mangat** [te ko$_A$-pe pe=Ø$_O$]$_{RC}$ te-pwo]$_S$ i [mangat
 work REL IRR.1sg-PFV do=3sg.ZERO LIG-DEM.PROX 3sg work
 pari a=i matmat ta-n tama-ng]$_{NVPRED}$
 belonging.to OBL=3sg grave CLF.POSS-PERT father-1sg.PERT
 'This work that I will do, it is the work belonging to the grave of my father.' (SY100411_0003)

In example (17), the O argument in the RC is coreferential with a Possessor in the MC. The O argument is indexed on the verb complex because it is animate. The relative clause is non-restrictive.

(17) *koropunan tareo i ran Memelan, te ip silal ipyokari reo*
 [koropunan ta=te-yo]$_S$ i [ta-n **Memelam** [te
 origin.of SPEC.COLL=LIG-DEM.INT 3sg CLF.POSS-PERT M. REL
 ip silal ip=yokat=**i**$_O$]$_{RC}$ te-yo]$_{NVPRED}$
 3pl spirit 3pl=carry=3sg LIG-DEM.INT
 'The origin of those [i.e. of creepy-crawlies] lies with Memelam, who was carried away by spirits.' (LM190611_0057)

11.1.1.1.3 Common argument is S of verbless RC

Common arguments functioning as subjects in a verbless RC are frequently encountered in Paluai, although there are not many straightforward examples where the CA is also subject in the MC. The sentence in (18) may be a case in point; the relative clause is non-restrictive, since the reference to *wong* is already specific.

(18) *wong te wong poyon tepwo, ngamainap kope sot liliu kôlôlôi pwên*
 [**wong** [te **wong**$_S$ [poyo-n]$_{NVPRED}$]$_{RC}$ te-pwo]$_S$
 1sg.FREE REL 1sg.FREE under-PERT LIG-DEM.PROX
 nga$_S$=ma=inap ko-pe sot liliu kôlôlôi pwên
 1sg=NEG$_1$=able IRR.1SG-PFV go.up again gathering.place NEG$_2$
 'Me, who is "underneath" [i.e. a person of lower status], I will not be able to go and stand in front of a crowd [to speak publicly].' (OBK040311_0037)

This type of relative clause may be used to put special emphasis on additional information about a discourse participant. For more information on this, see Chapter 12 on discourse practices.

A constituent functioning as O in the MC can also be the subject of a verbless RC:

(19) *ipkalêp tut ip sapon kei re nangin o*
 ip$_A$=ka-lêp tut [ip sapo-n kei [te Ø$_S$
 3pl=IRR.NS-take completely 3pl top-PERT herbs REL 3sg.ZERO
 [nangin]$_{NVPRED}$]$_{RC}$ yo]$_O$
 scent DEM.INT
 'They will take all the herbs that have a (nice) scent.' (NP220611_1_0011)

It seems that the subject of a verbless RC is often not cross-referenced by a pronoun, especially when it is third person. In this regard, verbless clause subjects differ from subjects of verbal predicates, which are obligatorily expressed. The reason for this difference is that only verbs can take bound pronouns, and thus verbless clauses contain only free pronouns. These are not obligatory, in contrast to their bound counterparts.

There is also a type of RC where the CA is Oblique in the MC and subject of a verbless RC. An example is given in (20), where the Oblique argument forms the Animate Goal of a ditransitive clause.

(20) *ipno rou pangpangai raip la ran mui re i paran menengan*
 ip$_A$=no tou [pangpangai ta-ip]$_O$ la
 3pl=IPFV give thought CLF.POSS-3pl.PERT go.to
 ta-n [mui [te i$_S$ [paran menengan]$_{NVPRED}$]$_{RC}$]$_{AG}$
 CLF.POSS-PERT dog REL 3sg very big
 'They shared their thoughts with the dog that was really big.'
 (LL300511_1_0014)

In example (21), the RC modifies a Possessor in the MC, the proper noun *Pwekei*. The relative clause is non-restrictive.

(21) *ngapwa korêk pwapwa ran Pwekei re i kôn panu pari Nengpul*
 nga$_A$=pwa ko-têk [pwapwae ta-n Pwekei]$_O$ [te i$_S$
 1sg=want.to IRR.1sg-tell story CLF.POSS-PERT P. REL 3sg

[kôn   panua   pari   Nengpul]_NVPRED]_RC
spirit   place   belonging.to   N.
'I am going to tell the story of Pwekei, who is a bush spirit from Nengpul.'
(PN100411_0003)

11.1.1.1.4 Common argument is Possessor in RC

This type of RC is particularly common with *ngaya-* 'name', but is also occasionally encountered with another directly possessed noun. In examples (22) and (23), the CA is subject in the MC; (22) is non-restrictive. In the RC, the CA is co-referenced by a pertensive suffix.

(22) *pein lapanum te ngayan Alup Kasinaman teo...*
 pein   **lapanum**   [te   ngaya-n   Alup Kasinaman]_RC   te-yo
 woman   first.marriage REL   name-PERT   A.K.   LIG-DEM.INT
 'His first wife, whose name was Alup Kasinaman...' (LK100411_0103)

(23) *...sômsômu re kanen pwên?*
 sômsômu   [te   kane-n   pwên]_RC
 plate   REL   meat-PERT NEG
 '[Is it] the plate that has no food [*lit.* 'its food (is) not']?'
 (Game1_021012_0281)

In example (24), the CA functions as O in the MC and as Possessor in a direct possessive construction in the RC.

(24) *ipto pe pun kaye sê re ngaro rêk pwapwaen tepwo*
 ip_A=to   pe   pun   [**kaye** **sê** [te   nga=to   têk pwapwae-n]_RC
 3pl=HAB   do   INTF   habit   DIM   REL   1sg=CONT tell story-PERT
 te-pwo]_O
 LIG-DEM.PROX
 'They really show this behaviour the story of which I was telling today.'
 (LL300511_1_0091)

CAs that have a function as Oblique in the MC and as Possessor in the RC are encountered in constructions of both the direct (25) and the indirect type (26). In the former example, the relative clause is restrictive, while in the latter, it is non-restrictive.

(25) *ila ro nansê masoan ai pil le laêngan yuêp teo?*
i_S=la   to   nan=sê   masoa-n   a=i   [pil   [te laênga-n
3sg=go.to be   PART=DIM separate-PERT OBL=3sg ladle REL ear-PERT
yuêp]_RC   te-yo]_Obl
two.INANIM LIG-DEM.INT
'Is it a bit removed from the bowl that has two handles [*lit.* 'ears']?'
(Game1_280812_0045)

(26) *ope la ran Toung Kulupwe te kanen kop tai ila ro rengteng Lêp*
wo_S=pe   la   ta-n   [**Toung Kulupwe** [te [kane-n   kop
2sg=PFV go.to CLF.POSS-PERT   T.K.   REL meat-PERT lime
ta-i]_S   i=la   to   teng.teng Lêp]_RC]_Obl
CLF.POSS-3sg.PERT 3sg=go.to CONT REDUP.cry Lêp
'You will go to Toung Kulupwe, whose lime gourd is making noise in Lêp.' (LalauKanau100411_0014)

11.1.1.1.5 Common argument is Oblique in RC

Oblique arguments in RCs are always cross-referenced by the *i* marker following the preposition *a=*, as they are in main clauses. This is illustrated in (27).

(27) *ipto yik sal le ipkayop ai rea*
ip_A=to   yik   [**sale** [te   ip=ka-yop   a=i]_RC]_O   te-ya
3sg=CONT search.for road   REL 3pl=IRR.NS-break.out OBL=3sg   LIG-then
'Then they were searching for a way to escape.' [*lit.* 'a road on which they could break out'] (LM190611_0051)

A CA that is Oblique in both the MC and RC mostly refers to a Locative in both:

(28) *upe la lopwan te ukala yektou lau rau ai*
u_S=pe   la   [**lopwa-n**   [te   u_A=ka-la   yek.tou [lau
3du=PFV go.to place-PERT   REL 3du=IRR.NS-go.to put   fishing.net
ta-u]_O   a=i]_RC]_Obl
CLF.POSS-3du.PERT OBL=3sg
'They went to the place where they were going to throw out their fishing net.' (NP210511_1_0011)

(29) *i pari ai sipen ponat sê re Koleyep sen to ai repwo*
  i_S   [pari   a=i   [**sipen ponat sê** [te [Koleyep]_S
  3sg belonging.to OBL=3sg part.of soil   DIM REL K.
  sen   to a=i]_{RC} te-pwo]_{Obl}]_{NVPRED}
  move.up be OBL=3sg LIG-DEM.PROX
  'He was from the small bit of land where Koleyep is located now.'
  (MS250311_0012)

11.1.1.2 Some further types of relative clauses

Two types of relative clauses are worth discussing separately. They include:
- Relative clauses with an interrogative word or indefinite quantifier in the main clause;
- Relative clauses containing the phrase *te ipto pwa*.

11.1.1.2.1 MC with an interrogative word or indefinite quantifier

In a number of cases, a MC contains an interrogative word or indefinite quantifier as the CA. The relative clause marker *te* is always present: an interrogative word or indefinite quantifier by itself cannot function as a relative pronoun. Examples (30) and (31) illustrate interrogative words as CAs. *Sap* 'which' introduces a noun modified by a relative clause, while *sê* 'who' is modified directly. In fact, in this function, *sap* is more frequently attested in the corpus than as an interrogative. Apart from *sê* and *sap*, there are no other cases of interrogatives modified by a relative clause.

(30) *kipat sap kain te kino pat ai*
  ki_A-pat   [**sap kain** [te ki-no   pat=∅_O   a=i]_{RC}]_O
  IRR.3sg-plant which kind REL IRR.3sg-IPFV plant =3sg.ZERO OBL=3sg
  'He will plant whichever kind [of plants] he can plant in there.'
  (KM190211_0038)

(31) *i sê re ila ru pe kotkot ai um tang?*
  [i   **sê** [te i_A=la   tu   pe   [kotkot]_O]_{RC}]_S [a=i
  3sg who REL 3sg=go.to STAT.CONT make smoke   OBL=3sg
  wumwa ta-ng]_{Obl}
  house CLF.POSS-1sg.PERT
  'Who is it that is making smoke at my house?' (KS030611_1_0021)

Example (32) illustrates the use of an indefinite quantifier as CA:

(32) *imagat sesom te kipe ro awênong pwên*
 i$_A$=ma=gat [**sesom** [te **ki**$_A$-pe to awên=ong$_O$]$_{RC}$]$_O$ pwên
 3sg=NEG$_1$=have anyone REL IRR.3SG-PFV HAB help=1sg NEG$_2$
 'There wasn't anyone who would be helping me.' (KM060111_0016)

11.1.1.2.2 RC containing the phrase te ipto pwa

Relative clauses containing the phrase *te ip=to pwa* (*lit.* REL they=HAB say) are very common in Paluai. The phrase can be translated in English as '(which is) called …'. This type of RC is an exception to the tendency for the CA to be more fully expressed in the MC, since the CA is expressed mostly as a full NP in both the MC and RC, with the one in the RC mostly containing a proper noun. The CA functions as an O argument to the verb *pwa*. Examples appear in (33) and (34). Relative clauses with *te ipto pwa* are usually of the non-restrictive kind, but in example (33), for instance, a restrictive interpretation is more likely.

(33) *tepwo pun igat yuêp kain mun te ipto pwa panuatu*
 te-pwo pun i$_A$=gat [**yuêp kain mun** [te ip$_A$=to
 LIG-DEM.PROX INTF 3sg=have two.INANIM kind banana REL 3pl=HAB
 pwa [**panuatu**]$_O$]$_{RC}$]$_O$
 say Vanuatu
 'Nowadays, there are two kinds of banana that are called "Vanuatu".' (KS030611_1_0044)

(34) *lapanan ip Maput, te ipto pwa Pobei…*
 lapana-n ip Maput [te ip$_A$=to pwa [**Pobei**]$_O$]$_{RC}$
 chief-PERT 3pl M. REL 3pl=HAB say P.
 'The chief of the Titan people, who was called Pobei…' (MS250311_0056)

11.1.1.3 Relative clauses: conclusions

With regard to relativisation possibilities, there is a cross-linguistically well established accessibility hierarchy, first discussed by Keenan and Comrie (1977):

(35) Subject > Direct Object > Indirect Object > Oblique > Genitive > Object of comparative

According to this hierarchy, if in a given language, Obliques for instance can be relativised (i.e. if the CA can function as Oblique in the RC), any type of argument higher up in the hierarchy can also be relativised. This is mostly borne out for Paluai. All types of arguments in the hierarchy, except Object of comparative, may be found inside a relative clause. The fact that Object of comparitive is not found may be due to the infrequent use of comparative constructions in the language. Relativisation of nominal predicates is not found either.

11.1.2 Complement clauses

A complement clause (CC) functions as an argument to a verb. Similarly to a relative clause, it is introduced by the general subordinate clause marker *te*. Only complement clauses functioning as O or Oblique argument to a verb were found in the corpus; it is therefore likely that a complement clause cannot function as S or A argument.

All complement clauses, whether O or Oblique, always immediately follow the verb. As such, they can be distinguished from relative clauses, which always follow a noun. O complement clauses only take the dependent clause marker *te*, whereas with Oblique complement clauses the general preposition *a=* is cliticised to *te*. The distinction between O and Oblique complement clauses appears roughly to reflect a semantic distinction between Fact and Activity complement clauses (cf. Dixon 2010b: 389–394), although there is no one-to-one correspondence. A further semantic type of complement clause is called "Potential", as it refers to an unrealised state of affairs. The verb *pwa* 'say, think' (which also grammaticalised into a desiderative marker 'want to') is unusual because the complement clauses it takes are never marked by *te*.

Verbs that can take complement clauses fall into several semantic subclasses; this has implications for the form of the complement clause. An overview of semantic subclasses is given in Table 85. Each of the semantic subtypes of verbs and the forms of the complement clauses they take will be discussed in turn. Occasionally, a complement clause is found modifying an adjective functioning as the predicate head of a non-verbal predicate (see Section 7.2). This type of dependent clause is never found modifying an adjective that functions as a noun modifier. In addition, a dependent clause modifying a noun functioning as head of a non-verbal predicate is always analysed as a relative clause, and not as a complement clause. After discussion of the various complement-taking verbs, some examples of complement clauses modifying adjectives will be discussed.

Table 85: Semantic subtypes of complement-taking verbs.

Subclass	Examples	Argument status of CC	Marking
Attention	*ning* 'see', *yong* 'hear', *yêk* 'feel'	Object or Oblique	*te* or *a=te*
Thinking/ speaking general	*pangai* 'think', *mapwai* 'know'; *pul* 'tell'; *pwapwasek* 'speak about'	Object	*te*
Quotation	*pwa* 'say, think'	Object	–
Liking	*yangyang* 'like', *kaêrêt* 'fear'	Oblique	*a=te*
Modal	*pwa* 'want to, be about to'	Object	Irrealis
	inap 'be able; be possible'	Object or Oblique	*te* or *a=te* plus irrealis

11.1.2.1 Verbs of Attention

Examples of complement clauses that function as O argument to a transitive verb of Attention are given in (36) and (37).

(36) *woning te pau reo in mwat*
 wo_A=ning [te [pau te-yo]_S i=an mwat]_{Compl:O}
 2sg=see COMP coconut.oil LIG-DEM.INT 3sg=PRF be.cooked
 'You see that the coconut oil is ready.' (CA120211_2_0039)

(37) *ila yêk te pat teo ila konun*
 i_A=la yêk [te [pat te-yo]_S i=la konun]_{Compl:O}
 3sg=go.to feel COMP bed LIG-DEM.INT 3sg=go.to heavy
 'He felt that the bed had become heavy.' (LM190611_0026)

These instances of complement clauses to Attention verbs can be considered Fact complement clauses. They refer to the factual result of a change of state that occurred prior to the time of the event expressed by the complement-taking verb. Perfect aspect is often used in this type of complement clause, although the directional *la* 'go to' as a copula marking change of state (see Section 7.7.2) is also encountered.

Attention verbs can also take Oblique complement clauses. These can often be analysed as Activity complement clauses, referring to an activity that is still ongoing at the time the action encoded by the complement-taking verb takes place, as in (38) and (39).

(38) *urêpe yong are yamat teo ipe kolon wat*
wurê_A=pe yong [a=te [yamat te-yo]_A i=pe [kolo-n]_O
1pc.EXCL=PFV hear OBL=COMP person DEM.INT 3sg=make voice-PERT
wat]_{Compl:Obl}
up.high
'We heard somebody shouting up high.' (NP210511_2_0024)

(39) *ipto ning are panu ro masai*
ip_A=to ning [a=te [panua]_S to masai]_{Compl:Obl}
3pl=CONT see OBL=COMP place CONT become.clear
'They saw the place getting clear [i.e. the dawn break].' (LM190611_0050)

The complement clause of Attention verbs can be marked as irrealis, but only if the main clause is also irrealis, as in (40) and (41). This is different from what was observed with Modal complement-taking verbs (see Section 11.1.2.5).

(40) *no re kining are mui nisom kime ret*
no te ki_A-ning [a=te [mui nisom]_S ki-me tet]_{Compl:Obl}
when IRR.3sg-see OBL=COMP dog other.ANIM IRR.3sg-come move
'When it will see another dog coming...' (LL300511_1_0083)

(41) *ope ning are pau reo kime pei*
wo_A=pe ning [a=te [pau te-yo]_S ki-me
2sg=PFV see OBL=COMP coconut.oil LIG-DEM.INT IRR.3sg-come
pei]_{Compl:Obl}
appear
'You will see coconut oil appearing.' (CA120211_2_0026)

However, perfect aspect is also encountered within Oblique complement clauses, as shown in (42). This indicates that the semantic distinction does not map one-to-one onto the formal distinction, since these clauses are better analysed as Fact complement clauses.

(42) *ining are kumun teo in sa pei*
i_A=ning [a=te [kumun te-yo]_S i=an sa pei]_{Compl:Obl}
3sg=see OBL=COMP sprout LIG-DEM.INT 3sg=PRF come.up appear
'He saw that the sprout had appeared.' (NP210511_1_0061)

11.1.2.2 Verbs of Thinking/Speaking

Verbs of Thinking and Speaking only take Object complement clauses, corresponding to Fact complement clauses in the semantic classification. Fact complement clauses are usually realis, even if the main clause is marked as irrealis, as in (43). However, they can also be marked as irrealis when the main clause is realis, indicating a future intention. In (44), the complement clause to *pangpangai* is realis and in its turn contains another complement clause to *pwa*, which is irrealis.

(43) *kapwa ipto mapwai re ngagat, a ipkape me lêp*
kapwa ip$_A$=to   mapwai [te   nga$_A$=gat=Ø$_O$]$_{Compl:O}$ ya
if   3pl=CONT know   COMP   1sg=have=3pl.ZER   then
ip$_A$=ka-pe   me   lêp=Ø$_O$
3pl=IRR.NS-PFV come   take=3pl.ZERO
'If they know that I have [them], they will come to get [them].'
(AK160411_2_0029)

(44) *ngaro pangpangai re ngapwa korou mosapen*
nga$_A$=to pang.pangai [te   nga$_A$=pwa   [ko$_A$-tou
1sg=CONT REDUP.think COMP 1sg=want.to   IRR.1sg-give
mosape-n]$_{Compl:Pot}$]$_{Compl:O}$
bride.price-PERT
'I was thinking that I wanted to distribute her bride price.'
(KM060111_0043)

Complement clauses to verbs of Thinking or Speaking can contain an interrogative, as in (45). The interrogative stays in situ and the complement marker *te* has to be used.

(45) *ngape mapwai re tinang iwau apa*
nga$_A$=pe mapwai [te   [tina-ng]$_S$ i=wau   a=pa]$_{Compl:O}$
1sg=PFV   know   COMP mother-1sg.PERT   3sg=move OBL=where
'I knew where my mother was from [i.e. which family].'
(OBK040311_0083)

11.1.2.3 Verbs of Quotation

The only complement-taking verb in this semantic subclass is *pwa* 'say'. It was already discussed in Section 8.2, where examples can also be found. A complement clause to *pwa* is never marked with *te*; in addition, this clause could occur by itself, as a main clause. Only realis complement clauses to *pwa* can have a direct quotation function. Indirect speech reports are rare, but see Section 6.5.1 for a number of examples of utterances that can be interpreted as such.

When the complement clause to *pwa* is irrealis and the subject of the complement clause is coreferent with that of the main clause, it has a Potential interpretation and *pwa* functions as a modal verb; see Section 6.5.1.

11.1.2.4 Verbs of Liking

Verbs of Liking take an Experiencer subject and an optional Stimulus argument (see Section 8.1.2.1). The Stimulus argument can be represented by a complement clause, always marked as Oblique. Examples with the verbs *yangyang* 'like, love', *mwamwasêk* 'be ashamed' and *sosol* 'be sad' are given in (46) to (48).

(46) *epmaro yangyang are woru pe songe pwalingan ep pwên*
ep$_S$=ma=to   yangyang   [a=te   wo$_S$=tu   pe   songe
1pl.EXCL=NEG$_1$=CONT   like   OBL=COMP   2sg=STAT.CONT   make   play
pwalinga-n   ep]$_{Compl:Obl}$ pwên
with-PERT   1pl.EXCL   NEG$_2$
'We don't like you playing with us.' (WL020711_0037)

(47) *iro mwamwasêk are i mayoun pwên*
i$_S$=to   mwamwasêk   [a=te   i$_S$ ma=[you-n]$_{NVPRED}$ pwên]$_{Compl:Obl}$
3sg=CONT   be.ashamed   OBL=COMP   3sg   NEG$_1$=tail-PERT   NEG$_2$
'He was ashamed of it not being his tail.' (LL300511_1_0086)

(48) *kolun teo ipe ro sosol are iro yongip*
[kolu-n   te-yo]$_S$ i=pe   to   sosol   [a=te   i$_A$=to
inside-PERT   LIG-DEM.INT   3sg=PFV   CONT   be.sad   OBL=COMP   3sg=CONT
yong=ip$_O$]$_{Compl:Obl}$
hear=3sg
'He [*lit.* 'his inside'] was feeling sad at hearing them.' (WL020711_0040)

The Oblique complement clause following some Liking verbs can be irrealis, to express a future intention or wish:

(49) *pian tau o, ngayangyang are taukaret tou*
 [pian ta-u yo]_{TopO} nga_S=yangyang [a=te
 good CLF.POSS-3du.PERT DEM.INT 1sg=like OBL=COMP
 tau_A=ka-tet tou=Ø_O]_{Compl:Obl}
 1du.INCL=IRR.NS-move give=3sg.ZERO
 'The goodness of the two of them, I wish for the two of us to follow [it].'
 (PK290411_3_0026)

11.1.2.5 Modal verbs

There are two modal verbs that can take a complement clause: *pwa*, discussed in Section 6.5.1, and the Tok Pisin loan *inap* 'be able to', which is used to express ability and possibility. The lexical material for the verb *inap* comes from English 'enough', but its semantic and grammatical properties are thoroughly substrate. It can have the meaning '(be) sufficient', but it also has a modal meaning, in which case it has an interpretation of either ability or possibility. In both cases it takes a complement clause, which does however look slightly different for each interpretation.

When *inap* refers to ability, it usually has a complement clause marked not with *te*, but only as irrealis, as in examples (50) and (51) below. In some cases, it is attested with a complement showing a nominalisation, as in (52).

(50) *ngamainap kobe be pwên*
 nga_A=ma=inap [ko_A-pe pe=Ø_O pwên]_{Compl:Pot}
 1sg=NEG_1=able IRR.1sg-PFV do=3sg.ZERO NEG_2
 'I am not able to do [it].' (LL030611_0033)

(51) *ipmainap kabe roui lai pwên*
 ip_A=ma=inap [ka_A-pe tou=i_O la a=i pwên]_{Compl:Pot}
 3pl=NEG_1=able IRR.NS-PFV put=3sg go.to OBL=3sg NEG_2
 'They were not able to put him into it.' (Game1_021012_0490)

(52) *ngainap ai songanum le papwên?*
 nga_S=inap a=i [song-an wumwa]_{Compl:Pot} le papwên
 1sg=able OBL=3sg run.away-NOM house or not.yet
 'Am I able to get married or not yet?' (KW290611_0031)

When *inap* refers to possibility, it has a third person singular dummy subject in the main clause, and the complement clause is marked with *(a=)te* in addition to being marked as irrealis. *inap* can have a CC with either O or Oblique syntactic function, i.e. it can be either marked by *te* or by *a=te*. The semantic distinction between Fact and Activity does not apply in these cases, as their foremost characteristic is that they can be regarded as Potential CC's, and accordingly are marked irrealis. Two examples of *inap* with an O CC, followed by two examples of *inap* with an Oblique CC, are given in (53) to (56).

(53) *kaba iinap te no wo ono abut, ya woabut*
 kapwa i$_A$=inap [te no wo$_A$ wo=no apwut=Ø$_O$]$_{Compl:O}$ ya
 if 3sg=able COMP only you 2sg=IPFV clear.bush=3sg.ZERO then
 wo$_A$=apwut=Ø$_O$
 2sg=clear.bush=3sg.ZERO
 'If it is possible that you clear the bush [for a garden] by yourself, alright then you will clear [it].' (KM190211_0007)

(54) *imainap te muyan kibe palak liliu pwên*
 i$_A$=ma=inap [te [muya-n]$_S$ ki-pe palak liliu]$_{Compl:O}$ pwên
 3sg=NEG$_1$=able COMP skin-PERT IRR.3sg-PFV bad again NEG$_2$
 'It is not possible that his skin will go bad again.' (NP260511_0024)

(55) *woning de iinap are yep kino rer ai*
 wo$_A$=ning [te i$_S$=inap [a=te [[yep]$_S$ ki-no tet
 2sg=see COMP 3sg=able OBL=3sg fire IRR.3sg-IPFV spread
 a=i]$_{Compl:Obl}$]$_{Compl:O}$
 OBL=3sg
 'You see that it is possible for fire to spread through it.' (KM190211_0017)

(56) *imainap are ip silal kabe wau tarak toui a kabe lêbi pwên*
 i$_S$=ma=inap [a=te [ip silal]$_A$ ka-pe wau tarak tou=i$_O$
 3sg=NEG$_1$=able OBL=COMP 3pl spirit IRR.NS-PFV move climb give=3sg
 a ka-pe lêp=i$_O$]$_{Compl:Obl}$ pwên
 and IRR.NS-PFV take=3sg NEG$_2$
 'It was not possible for the spirits to follow him into the tree and get him.' (LM190611_0041)

Thus, there seems to be a clear formal difference between two types of complement clauses introduced by *inap*: those referring to ability and those referring to possibility. The first are marked just by irrealis, and show coreference between the S/A argument of the main and the dependent clause, while the second are marked by both *(a=)te* and irrealis. It is unclear why the latter can take both *a=te* and *te* as complementizers. It is possible that *inap* CC's have variable marking because they are already additionally marked by irrealis. Also, for the only other verb taking a Potential CC, *pwa* 'want to, be going to', the CC is never marked by a complementizer. The fact that there is variation between *te* and *a=te* indicates that *inap* may not yet be fully integrated into Paluai syntax and therefore exhibits unpredictable behaviour.

It can be concluded that *inap* functions as a deontic modal operator (Timberlake 2007: 329) indicating ability and/or possibility. The authority can be the participant designated to carry out the action, who does not have the (physical) ability to do it. Alternatively, the participant may not be able to do it due to external circumstances. *inap* as a modal operator does not fill a hiatus in Paluai grammar; the 'indigenous' form *sa*, discussed in Section 6.5.2, can fulfil the same function. An example of this use of *sa* is given in (57). However, non-negated *sa* has apprehensive overtones, which are absent from *inap*. The first, negated, use of *sa* indicates an impossibility, whereas the second, non-negated, use refers to a possible outcome that would be undesirable. This may be a reason for *inap* gaining ground as a modal operator: the apprehensive overtones that *sa* carries seem to be absent from *inap*.

(57) *ngasa bul la rai pwên, te isa yeki*
nga$_A$=sa pul=Ø la ta-i pwên [te i$_A$=sa
1sg=MOD tell=3sg.ZERO go.to CLF.POSS-3sg.PERT NEG SUB 3sg=MOD
yek=i$_O$]$_{AdvCl}$
hit=3sg
'I cannot tell him, lest he beat her/for he may beat her.' (WL020711_0056)

11.1.2.6 Complement clause modifying adjective

Sentences (58) to (60) examplify complement clauses modifying adjectives that function as head of a non-verbal predicate.

(58) *ma kapwa i menmenengan de wope lêp ip yamat ai...*
ma kapwa i$_S$ [men.menengan [te wo$_A$=pe lêp [ip yamat]$_O$
but if 3sg REDUP.big COMP 2sg=PFV 3pl person OBL=3sg
a=i]$_{Compl:O}$]$_{NVPRED}$
take
'But if it [garden] is of such a size that you should take people to it [to help you]...' (KM190211_0007)

(59) *pian te ngape mapwai pun yamat tang*
 pian [te nga=pe mapwai pun.yamat ta-ng]_Compl:O
 good COMP 1sg=PFV know genealogy CLF.POSS-1sg.PERT
 '[It's] good that I know my genealogy.' (OBK040311_0088)

(60) *ngapwa, "pian are korou ansê mosapen pariong"*
 nga_A=pwa [pian [a=te ko_A-tou an-sê [mosape-n
 1sg=think good OBL=COMP IRR.1sg-give PART-DIM bride.price-PERT
 paria-ng]_O]_Compl:Obl]_Compl:Quot
 wife-1sg.PERT
 'I thought, "It's good if I were to pay the bride price for my wife."'
 (KM060111_0041)

Complement clauses to adjectives are usually of the Fact type, but they can also be of the Potential type, as in (60). In the former case, the complement clause is marked by *te*, whereas in the latter it is marked by *a=te* (plus irrealis). The modified adjective functions as head of a non-verbal predicate. In particular with *pian*, the verbless clause subject is often not overtly expressed.

11.1.2.7 Complement clauses: conclusions

As we have seen, complement clauses come in a number of varieties. As this embedded clause type is relatively uncommon in Paluai, it is not certain whether complement clauses in spontaneous language use can show all the possibilities relating to argument structure and TAM distinctions that main clauses do. All the TAM particles are attested in complement clauses, and they are also encountered containing O arguments, although not with Obliques.

11.1.3 Adverbial subordinate clause

An adverbial subordinate clause is linked to a main clause and provides more information in the same way as an adverbial phrase. There are several subtypes, all marked by different forms. An overview is given in Table 86. When a main and subordinate clause are linked, one clause can be regarded as the "Focal clause" (FC), referring to the central activity or state of the biclausal linking, and the other as the "Supporting clause" (SC) (cf. Dixon 2009), giving information about that state or activity.

Since dependent clauses are only marked by *te* and not, for instance, by different constituent order or TAM possibilities, it is sometimes difficult to

Table 86: Types of adverbial subordinate clauses.

Type of clause	Semantic varieties	SC marker	FC marker
Temporal (relative time)	Before/After	panuan te 'before'	ya 'then/FOC'°
		monokin te 'after'	ya 'then/FOC'°
	Simultaneity, contiguity	no te 'just as, as soon as'	ya 'then/FOC'°
		pwotnan te 'when'	ya 'then/FOC'°
		pêng te 'when'	ya 'then/FOC'°
		taim te 'when'*	ya 'then/FOC'°
	Stretch of time	inap te 'until'*	–
Manner	–	kanpwên 'as if'	–
(Possible) consequence		(a=)te 'for, as'	–
	Reason	pari ai te 'because'	–
		ai sa? ma 'because'	–
		aronan te 'consequently'	–
	Consequence	tangoan te 'consequently'	–
		longoan te 'consequently'	–
	Result	(te) onga 'so that'	–
	Possible consequence	te ... sa 'lest, in case'	–
Concessive	–	ma i te 'but, although'†	ma i te 'but'†
Conditional	"Plain" conditional	Irrealis	ya 'then/FOC'
	((a=)te) kapwa conditional	kapwa 'if'	ya 'then/FOC'
	Counterfactual conditional	kanopwên 'if only'	ya 'then/FOC'

* Tok Pisin loan.
° Marking of the FC with *ya* is optional.
† Either SC or FC are marked, but not both.

decide whether a construction consists of a main and subordinate clause or two main clauses. In any case, the distinction between main and subordinate is most likely a continuum.

What was said above about *te* is also true for its role in adverbial subordinate clauses: whenever a constituent slot is filled by a sequence marked with *te*, this is an indication that the slot is filled by a clause rather than a phrase. Consequently, the presence of *te* indicates that a clause should be considered as an adjunct (i.e. a subordinate clause) to another clause, rather than an independent main clause. In what follows, the various types of adverbial subordinate clauses will be discussed in turn.

11.1.3.1 Temporal subordinate clause
A temporal subordinate clause indicates whether one event happened 1) in sequence with, 2) simultaneous to, or 3) during a stretch of time lasting up to another event. Temporal subordinate clauses probably developed out of relative clauses modifying a temporal noun. When the subordinate clause precedes the main clause, the two clauses are separated by the marker *ya* 'then'.

11.1.3.1.1 Temporal sequence
Sequential relations in time between two events are expressed by *panuan* 'before', as in (61), and *monokin* 'after(wards)', as in (62), both followed by the dependent clause marker *te*. The forms are nominal in origin; they are related to the spatial nouns *panu* 'front' and *monok* 'back, behind'. In some cases, there may be overtones of consequence or result, as in (62): the fact the people quit eating wild bush tubers was a direct effect of yams being introduced. The subordinate clause and main clause can appear in either order for *panuan*. With *monokin*, the subordinate clause always precedes the MC. Often, but not always, the main clause is marked by *ya* 'then'.

(61) *ipkaliliu la lopwan ip panuan te pulêng teo kipe masai*
ip$_S$=ka-liliu   la   lopwa-n   ip   [panuan te   [pulêng
3pl=IRR.NS-return go.to place-PERT 3pl before   SUB dawn
te-yo]$_S$ ki-pe   masai]$_{AdvCl}$
LIG-DEM.INT   IRR.3sg-PFV   be.clear
'They would go back to their houses before the dawn would break.'
(LK250111_0027)

(62) *monokin te suei a mwayen teo ipei, a eppe lusim nganan*
[monokin te   [suei   a   mwayen te-yo]$_S$ i=pei]$_{AdvCl}$ ya
after   SUB short.yam and yam   LIG-DEM.INT 3sg=appear   then
ep$_A$=pe   lusim   [ngan-an]$_O$
1pl.EXCL=PFV leave   eat-NOM
'After long and short yams were introduced, we quit eating [them] [i.e. wild bush tubers].' (KS030611_1_0004)

11.1.3.1.2 Temporal simultaneity/contiguity
There are several forms that indicate co-occurrence or contiguity of events without a strong cause-effect relation, although overtones of such a relation can be present. One of them, *taim* ('when', *lit.* 'at the time that ...'), is a loan from Tok

Pisin. The temporal subordinate clause precedes the main clause, and again the main clause is often marked by *ya* 'then'.

(63) *taim te imwat teo, a wope antek*
[taim te i$_S$=mwat te-yo]$_{AdvCl}$ ya wo$_A$=pe
when SUB 3sg=be.cooked LIG-DEM.INT then 2sg=make
antek=Ø$_O$
put.away=3sg.ZERO
'When it is cooked ready, you will put [it] away.' (CA120211_1_0028)

(64) *taim te wong pa namwi a tamong imat tu niong*
[taim te wong$_S$ pa [namwi]$_{NVPRED}$]$_{AdvCl}$ ya [tama-ng]$_S$
when SUB 1sg.FREE yet small then father-1sg.PERT
i=mat tu ni=ong$_O$
3sg=die stay away=1sg
'While I was little, my father died and left me.' (SY100411_0005)

(65) *potnan te ngoyai iriyriy suwot a ino lêp masian*
[pwotnan te [ngoyai]$_S$ i=tiy.tiy suwot]$_{AdvCl}$ ya
when SUB cuscus 3sg=REDUP.observe go.down then
i$_A$=no lêp [masia-n]$_O$
3sg=IPFV take appreciation-PERT
'When the cuscus looked down, he got happy.' (LL010711_0032)

(66) *no re ining maloan suwot teo a nian teo ino paran palak*
[no te i$_A$=ning [maloa-n]$_O$ suwot te-yo]$_{AdvCl}$ ya
just.as 3sg=see reflection-PERT go.down LIG-DEM.INT then
[nia-n te-yo]$_S$ i=no paran palak
stomach-PERT LIG-DEM.INT 3sg=IPFV really bad
'As soon as he saw his reflection, he got very angry.' (LL010711_0053)

11.1.3.1.3 Stretch of time

The Tok Pisin loan *inap* is used to indicate that one event occurred during a stretch of time up to the moment another event started; since it is a verb, it takes a complement clause. This use of *inap* is not very common, however, and *inap* is encountered more often in its modal sense 'be able; be possible' (see Section 11.1.2.5).

(67) *wono yen yet teo inap pun te kila mwat*
wo$_A$=no yen yet=Ø$_O$ te-yo [inap pun te ki$_S$-la
2sg=IPFV CONT stir=3sg.ZERO LIG-DEM.INT until INTF SUB IRR.3sg-go.to
mwat]$_{AdvCl}$
be.cooked
'You keep on stirring [it] until it will be cooked.' (CA120211_2_0035)

11.1.3.2 Manner subordinate clause

The form *kanpwên* 'as if' is encountered as a marker of subordinate clauses indicating manner. As *kanpwên* is relatively rare in the corpus, it is not certain how exactly it is used. It is unclear, for instance, whether the presence or absence of the subordinate clause marker *te* makes a difference in meaning. In (68) there is no marking with *te*, but in (69) there is. In addition, (68) contains *kapwa* 'if' in addition to *kanpwên*, possibly because the verbless clause cannot be marked as irrealis. *kanpwên* describes unreal or improbable situations and perhaps also the conscious act of pretending, as in (68). The subordinate clause, forming the SC, always follows the main clause.

(68) *iro pei kanpwên kapwa i mumusau, ma i re pwên*
i$_{Ai}$=to pe=i$_{Oi}$ [kanpwên kapwa i$_S$ [mumusau]$_{NVPRED}$]$_{AdvCl}$
3sg=HAB make=3sg as.if if 3sg white.person
ma i te pwên
but NEG
'She acts as if she is a white woman, but [she's] not.' (field notes 31/08/2012)

(69) *kino pungpung youn te onga kanpwên kisuwot pung nonot te i reo i youn le pwên*
ki$_A$-no pung.pung [you-n]$_O$ [te onga kanpwên
IRR.3sg-IPFV REDUP.smell tail-PERT SUB and.so as.if
ki$_A$-suwot pung nonot [te [i te-yo]$_S$ [i
IRR.3sg-go.down smell recognise COMP 3sg LIG-DEM.INT 3sg
you-n]$_{NVPRED}$]$_{Compl:O}$ le pwên]$_{AdvCl}$
tail-PERT or NEG
'He will sniffle at the tail as though he can distinguish whether it is his tail or not.' (LL300511_1_0085)

11.1.3.3 (Possible) consequence clause
This section discusses the various ways of expressing consequence and possible consequence by means of a subordinate clause.

11.1.3.3.1 Reason clause
An "implicit" reason subordinate clause is marked by either *te* or *a=te*. Examples are given in (70) and (71). It could be said that the insertion of *te* or *a=te* functions like a semicolon: the clauses are loosely linked to each other, with one clause sometimes forming an afterthought to the other. The marked subordinate clause is the SC and always follows the main clause. For a stronger, more explicit reason connection between the two clauses, *pari ai te* 'because' (discussed below) is used. It is not clear whether there is a semantic difference between *te* and *a=te*.

(70) *muyou reo imaro patan yoy pwên, te no parun teo iro patan yoy*
 [muyou te-yo]$_S$ i=ma=to   pata-n   yoy   pwên [te [no
 snake LIG-DEM.INT 3sg=NEG$_1$=be top-PERT stone NEG$_2$ SUB only
 patu-n   te-yo]$_S$ i=to   pata-n   yoy]$_{AdvCl}$
 head-PERT LIG-DEM.INT 3sg=be top-PERT stone
 'The snake is not on top of the stone; (as) only its head is on top of the stone.' (Game1_021012_0033)

(71) *imagat not pwên, are imala songan um pwên*
 i$_A$=ma=gat   not   pwên [a=te   i$_S$=ma=la   songan.um
 3sg=NEG$_1$=have child NEG2 OBL=SUB 3sg=NEG$_1$=go.to marriage
 pwên]$_{AdvCl}$
 NEG$_2$
 'She didn't have children; (as) she never got married.' (KW290611_0009)

To suggest a stronger reason relation, the form *pari ai te* (lit. 'belonging to OBL=3sg SUB') is used, as in (72). Again, the marked subordinate clause is SC and follows the main clause, which is FC.

(72) *i repwo soan wat tole, pari ai re ipmaro pe ai min pwên*
 [i   te-pwo]$_S$ [soan wat   tole]$_{NVPRED}$ [pari   a=i
 3sg LIG-DEM.PROX price up.high EMP   belonging.to OBL=3sg
 te   ip$_A$=ma=to   pe=Ø$_O$ a=i   min   pwên]$_{AdvCl}$
 SUB 3pl=NEG$_1$=HAB make=3sg.ZERO OBL=3sg hand   NEG$_2$
 'This here is very expensive, because they don't make [it] by hand.' (Game1_021012_0164–65)

The phrase *ai sa? ma* may be a calque from Tok Pisin; it is predominantly used by younger speakers, below the age of about 50. *ai sa* literally means 'for what' and is used as a rhetorical question to indicate the meaning 'because'; an example is given in (73). The phrase is always followed by *ma*, probably functioning as an emphatic marker. The second clause, marked by *ai sa? ma*, functions as SC, but there is no marker *te* to indicate a dependency relation.

(73) *lau rau reo ila lot maut. ai sa ma nik in pwak ai lau*
[[lau   ta-u   te-yo]₅   i=la   lot   maut]_MC1
fishing.net CLF.POSS-3du.PERT LIG-DEM.INT 3sg=go.to fall sink
[a=i   sa   ma   [nik]₅   i=an   pwak   a=i   lau]_MC2
OBL=3sg what EMP   fish   3sg=PRF   be.stuck   OBL=3sg fishing.net
'Their net went under, because fish had become stuck in the net.'
(NP210511_1_0015/16)

11.1.3.3.2 Consequence clause

There are three forms that can indicate consequence. All of them are nominal in origin, from the directly possessed subclass, and can do double duty as sentential modal adverbs. The forms are *aronan* (from *arona-* 'way'), *tangoan* (from *tangoa-* 'effort; habit') and *longoan* (from *longoa-* 'habit'). When accompanied by the marker *te*, they introduce a consequence dependent clause, meaning 'accordingly, consequently, as a result'. Again, the marked subordinate clause is SC and follows the main clause, which is FC. *tangoan* is illustrated in (74) and (75).

(74) *ikipe lai nganngan; tangoan te ippe rou i relo lalon sômsômu relo*
i₅=ki-pe   [la   a=i   nganngan]_CC   [tangoan   te   ip_A=pe
3sg=IRR.3SG-PFV go.to OBL=3sg food   therefore SUB   3pl=PFV
tou   [i   te-lo]_O   lalo-n   sômsômu   LIG-DEM.INT
put   3sg   LIG-DEM.DIST inside-PERT   dish   te-lo]_AdvCl
'It is probably food; accordingly, they have put it in a dish.'
(Game1_021012_0332–33)

(75) *uro pe pian. tangoan te yaui ran Yêp Ponaun tu nêm hausik Lorengau, Maiau Pongap to pwalingan*
u_A=to   pe   [pian]_O   [tangoan   te   yaui   ta-n   Yêp Ponaun
2du=HAB do   good   therefore SUB   wind CLF.POSS-PERT Y.P.
tu   nêm   hausik   Lorengau   Maiau Pongap   to
STAT.CONT be.finished   hospital   L.   M.P.   be

pwalinga-n]~AdvCl~
with-PERT
'They used to do good [to each other]. Accordingly, (when) Yêp Ponaun died at the Lorengau hospital, Maiau Pongap was with him.'
(PK290411_3_0010–12)

Result clauses are closely related to consequence clauses. The marker *onga* 'thus, so' is used to indicate this type of dependent clause. Its use has been extended, however, and in the present-day language it functions more like a general clause linker and filler. Some of its uses can still clearly be recognised as indicating result. The clause marked by *onga* is SC and follows the FC.

(76) *sam pe kak, te onga Alup Sauka a wong Maiau pwalingan Alup tan Kabon, urêyoppiy la net*
[sam]~S~ pe kak [te onga [Alup Sauka a wong Maiau
outrigger PFV overturn SUB and.so A.S. and 1sg.FREE M.
pwalinga-n Alup ta-n Kapon]~S~ wurê=yop.piy la net]~AdvCl~
with.PERT A. CLF.POSS-PERT K. 1pc.EXCL=fall go.to sea
'The outrigger overturned, so that Alup Sauka, me, and Kabon's Alup, we fell into the sea.' (MK060211_0034)

More often, however, *onga* is found as a linker between main clauses, indicating sequence of events (see Section 11.2.1.1 below).

11.1.3.3.3 Possible consequence clause

Another type of dependent clause indicates possible consequence, usually with apprehensive overtones. This type of clause is discussed in Section 6.5.2, because it includes the modal particle *sa*. The examples given there are repeated in (77) and (78). The marked subordinate clause is SC and follows the main clause, which is FC.

(77) *ngasa pul la rai pwên, te isa yeki*
nga~A~=sa pul=∅~O~ la ta-i pwên [te i~A~=sa
1sg=MOD tell go.to CLF.POSS-3sg.PERT NEG SUB 3sg=MOD
yek=i~ó~]~AdvCl~
hit=3sg
'I cannot tell him, lest he beat her.' (WL020711_0056)

(78) *kapwa karapot nik, napunan kowau la kason nik, te isa poyak*
    kapwa ka$_A$-tapot   [nik]$_O$ napunan ko$_S$-wau   la   kaso-n
    if   IRR.NS-smoke fish   NEG.IMP IRR.1sg-move go.to near-PERT
    nik [te i$_S$=sa   poyak]$_{AdvCl}$
    fish SUB 3sg=MOD be.rotten
    'If they were smoking fish, it would be forbidden for me to go near the fish, lest it spoil.' (LL030611_0026)

Sometimes, a possible consequence clause is very elliptical. Example (79) was used as a warning to a child. Clearly, what is meant is 'cover your head, lest the sun burn you!'. The dependent clause, however, is almost entirely elided, with only the marker *te* and the word for 'sun' overtly expressed.

(79) *wopolpol, te sin!*
    wo$_S$=polpol   te   sin
    2sg=cover.head SUB sun
    'Cover your head, because of the sun!' (field notes 25/09/2012)

11.1.3.4 Concessive clause

Concessive or contrastive dependent clauses are marked by the form *ma i te* 'but, although'. A contrastive relation (between two main clauses) can also be expressed by the linker *ma* 'and; but'; it is not clear whether there is a significant meaning difference between *ma* and *ma i te*. The clause marked with *ma i te* always follows the other clause. It seems to be able to function either as SC, as in (80), or as FC, as in (81).

(80) *ngagat kel pulek, ma i re kel sê rang namwi*
    nga$_A$=gat [kel]$_O$ pulêek [ma i te [kel sê   ta-ng]$_S$
    1sg=have   canoe   too   although canoe DIM CLF.POSS-1sg.PERT
    [namwi]$_{NVPRED}$]$_{AdvCl}$
    small
    'I have a canoe too, although it is a small one.' (052b_0304)

(81) *nian teo ipe palak tai, ma i re ima pul la rai pwên*
    [nia-n   te-yo]$_S$ i=pe   palak ta-i   [ma i te
    stomach-PERT LIG-DEM.INT 3sg=PFV bad   CLF.POSS-3sg.PERT but
    i$_A$=ma   pul=Ø$_O$ la   ta-i   pwên]$_{AdvCl}$
    3sg=NEG$_1$ tell=3sg.ZERO go.to CLF.POSS-3sg.PERT NEG$_2$
    'She got angry with him, but she didn't tell him.' (WL020611_0009)

11.1.3.5 Conditional subordinate clause

A conditional subordinate clause refers to a future or hypothetical event that forms the condition for another event to happen. There are three types in Paluai; the condition (protasis) is always SC whereas the consequence (apodosis) is always FC.

11.1.3.5.1 "Plain irrealis" conditional clauses

With these, the protasis is always irrealis, while the apodosis can be either realis or irrealis. The protasis is never marked by any aspectual particles and only shows irrealis marking; it always precedes the apodosis. The apodosis is often, but not always, marked with *ya* 'then'. Conditional subordinate clauses marked by irrealis and *ya* do not form very strong conditionals, and can often also be translated starting with 'when', rather than 'if'.

When the apodosis is irrealis, we get a potential conditional, as in (82) to (84). Potential conditionals "have a strong affinity with the future" (Timberlake 2007: 323).

(82) *som kime wop a wopwa, "worou som numum si rang"*
 [[som]$_S$ ki-me wop]$_{Pro}$ [ya wo$_A$=pwa [wo$_A$=tou
 one.ANIM IRR.3sg-come fly then 2sg=say 2sg=give
 [numu-m]$_O$ si ta-ng]$_{Compl:Quot}$]$_{Apo}$
 feather-2sg.PERT come.down CLF.POSS-1sg.PERT
 'If/when one comes flying, you say, "Give me one of your feathers."'
 (WL020711_0151)

(83) *urêkala renten, pang kipe rut*
 [wurê$_S$=ka-la tenten]$_{Pro}$ [[pang]$_S$ ki-pe tut]$_{Apo}$
 1pc.EXCL=IRR.NS-go.to speak.to.ancestor rain IRR.3sg-PFV fall
 'If/when we go and speak [to this stone], rain will fall.' (NS220511_1_0070)

(84) *koliliu me poyep a taukape ret*
 [ko$_S$-liliu me poyep]$_{Pro}$ [ya tau$_S$=ka-pe tet]$_{Apo}$
 IRR.1sg-return come afternoon then 1du.INCL=IRR.NS-PFV move
 'When I return in the afternoon, we will go.' (KM050995_0007)

When the apodosis is realis, we get a general or iterative conditional (cf. Timberlake 2007): each time the condition in the protasis is met, the event represented in the apodosis will come to pass. In example (85), the apodosis is also marked for habitual aspect.

(85) *mare kila pei raip a ipto liliu si, a ip saên ip teo, ipto renten taip*
[[mare]$_S$ ki-la   pei   ta-ip]$_{Pro}$ [ya   ip$_S$=to   liliu
sickness IRR.3sg-go.to appear CLF.POSS-3pl.PERT then 3pl=HAB   return
si   a   [ip saê-n   ip   te-yo]$_{TopObl}$ ip$_S$=to
come.down and 3pl father's.sister-PERT 3pl LIG-DEM.INT   3pl=HAB
tenten   ta-ip]$_{Apo}$
speak.to.ancestor CLF.POSS-3pl.PERT
'Whenever they fall sick, they return home and pray to their father's sisters.' (PN100411_0038)[74]

11.1.3.5.2 Conditional clause marked by kapwa

Conditionals can also be formed by marking the protasis with *kapwa* 'if; supposedly'.[75] This may be due to influence from Tok Pisin, which also has periphrastic means to mark conditional clauses. The protasis usually precedes the apodosis. Again, it seems that with an apodosis marked as irrealis, the conditional clause refers to the future (potential), as in (86), and with an apodosis that is realis, it refers to a general or iterative state of affairs, as in (87).

(86) *kapwa ipto mapwai re ngagat, a ipkape me lêp*
[kapwa ip$_A$=to   mapwai [te   nga$_A$=gat=Ø$_O$]$_{Compl:O}$]$_{Pro}$ [ya
if   3pl=CONT know   COMP 1sg=have=3pl.zero   then
ip$_A$=ka-pe   me   lêp=Ø$_O$]$_{Apo}$
3pl=IRR.NS-PFV come take=3pl.ZERO
'If they know that I have [them], they will come to get [them].'
(AK160411_2_0029)

(87) *ma kapwa i pein te imat, igat sanei reloan*
[ma kapwa [i$_S$ [pein   [te   i$_S$=mat]$_{RC}$]$_{NVPRED}$]$_{Pro}$ [i$_A$=gat   [sanei
but   if   3sg woman REL 3sg=die   3sg=have brother

---

[74] Father's sisters and their descendants are believed to have power over people's health and fertility.
[75] *kapwa* may be a complex form consisting of the third person non-singular irrealis marker *ka-* prefixed to the verb *pwa* 'say, think'.

teloan]$_O$]$_{Apo}$
right.now
'But if/when it is a woman who died, there is a widower.'[76]
(SP190311_0039)

Occasionally, a relative or complement clause is marked by *kapwa*. In (88), the relative clause is marked with *kapwa* in addition to being marked with irrealis.

(88) *wosuput luek palak te kapwa kilêpi*
wo$_A$=suput   luek   [palak [te kapwa ki$_A$-lêp=i$_O$]$_{RC}$]$_O$
2sg=send.away come.out evil   REL if   IRR.3sg-take=3sg
'You will send away the evil that may find him.' (YK290411_2_0007)

In (89), a complement clause is marked with *kapwa*.

(89) *ngamaro mapwai re kapwa i re i kulusun kei re ipan san pwên*
nga$_A$=ma=to   mapwai [te   kapwa [i   te-yo]$_S$ i
1sg=NEG$_1$=CONT know   COMP if   3sg LIG-DEM.INT   3sg
[kulusu-n   kei   [te ip$_A$=an   san=Ø$_O$]$_{RC}$]$_{NVPRED}$]$_{Compl:O}$ pwên
rubbish-PERT tree REL 3pl=PRF   cut=3sg.ZERO   NEG2
'I'm not sure whether this is sawdust [*lit.* 'rubbish of a tree that they've sawn].' (Game1_021012_0324)

Sometimes, the protasis follows the apodosis in a subordinate clause, marked by either *te* or *a=te* and *kapwa*. This seems to have the meaning 'in case' and refers to a 'possible reason/condition' clause. It is not clear whether there is any difference in meaning between the use of *te* and *a=te*. The entire construction has strong hypothetical/potential semantics. The clause marked by *(a=)te kapwa* functions as SC, and indicates that the state of affairs represented in it might come to pass, which will then lead to the events represented in the apodosis.

(90) *kiro apek tai, are kapwa kinu*
[ki$_S$-to   apek ta-i]$_{Apo}$ [a=te   kapwa ki$_S$-nu]$_{Pro}$
IRR.3sg-HAB hit   CLF.POSS-3sg.PERT   OBL=SUB if   IRR.3sg-bathe
'She will rub it on herself, in case she will bathe.' (NK290311_2_0023)

---

[76] *sanei* is the official term to refer to the brothers of a deceased man during mortuary ceremonies. However, what is probably indicated here is that the widower of a deceased woman will be grouped together with her brothers during the mortuary exchange ceremonies.

(91) *wolak, te kapwa wopwa wola ning panu reo...*
 [wo$_S$=lak]$_{Apo}$ [te kapwa wo$_A$=pwa [wo$_S$=la ning panua
 2sg=go   SUB if 2sg=want.to 2sg=go.to see place
 te-yo]$_{Compl:Pot}$]$_{Pro}$
 LIG-DEM.INT
 '(When) you will go, in case you want to go and see this place...'
 (LL030611_0012)

kapwa is used with particular frequency for second person singular in order to indicate hypothetical events, probably because irrealis is not formally marked for this person/number combination. Whichever way it is used, it marks information that the speaker is not entirely sure of or cannot vouch for. Thus, in (88) it is inserted in the relative clause because the speaker is not certain that the evil will occur, and the action referred to by the main clause has to be carried out *only* if the events described in the dependent clause will come to pass. Thus, there are conditional overtones in this sentence. In (89), there are no such overtones and the use of *kapwa* indicates that the speaker cannot vouch for the information given.

11.1.3.5.3 Counterfactual conditionals

Counterfactual conditionals are strongly linked to the past. They indicate that if event X had come to pass, this would have led to event Y, but since X did not happen, Y did not happen either. Thus, the speaker asserts that the protasis of this type of conditional is not true (and, as a result, the apodosis isn't either). Counterfactuals are very rare in the corpus, but if they are used, they seem to take the marker *kanopwên*, as in (92) and (93).

(92) *kanopwên ipkano rok, a ip taman ep a ip tupun ep ipkano sa ningip*
 [kanopwên ip$_S$=ka-no   tok]$_{Pro}$ [ya [ip tama-n   ep   a
 CNTF.COND 3pl=IRR.NS-IPFV stay   then 3pl father-PERT 1pl.EXCL and
 ip tupu-n   ep]$_A$ ip=ka-no   sa ning=ip$_O$]$_{Apo}$
 3pl grandparent-PERT 1pl.EXCL 3pl=IRR.NS-IPFV MOD see=3pl
 'If only they had stayed here, then our fathers and grandfathers would have been able to see them.' (NP190511_2_0029)

(93) *kanopwên ip Paluai ipkaliliu lak, a ippe yong kuyun ip lai sul o*
 [kanopwên [ip Paluai]$_S$ ip=ka-liliu   lak]$_{Pro}$ [ya ip$_A$=pe yong
 CNTF.COND 3pl Paluai 3pl=IRR.NS-return go then 3pl=PFV hear

```
    [kuyu-n     ip]ₒ  la      a=i       sul                  yo]_Apo
     noise-PERT  3pl  go.to  OBL=3sg   dry.coconut.leaf      DEM.INT
```
'If only the Paluai had come back, they would have heard their noise on the dry coconut leaves.' (LK100411_0110)

11.2 Combining main clauses

When two main clauses are linked, no marker *te* is present. Moreover, there is no distinction between FC and SC for linked main clauses. The semantic types distinguished are shown in Table 87 below.

Table 87: Semantic types of main clause linking.

Type of clause	Semantic varieties	Form of linker
Addition	Temporal succession	*onga* 'and then'
	Same-event addition	*a* 'and'
	New-event addition; Unexpectedness	*ma* 'and'
	Elaboration	Juxtaposition
	Contrast	*ma* 'but'
Alternatives	Disjunction	*le* 'or'

11.2.1 Addition

There are several subtypes of addition, which will be discussed below.

11.2.1.1 Temporal succession
In (94), which consists of four main clauses, the first two clauses are linked by *onga*. There is no linker between the second and third clause; the third and fourth are linked with *a*. Frequent repetition or near-repetition is a pervasive feature of Paluai narrative; see also Schokkin (2014). The repeated clauses are usually linked with *onga* to indicate temporal succession. The order of *onga* temporal succession is iconic: clauses describing earlier events precede clauses describing later events.

(94) *epsi ret onga epsi panu; epsi panu a sin ilol...*
```
     [eps=si                  tet]_MC1   [onga    eps=si                 panu]_MC2
      1pl.EXCL=come.down      move        and.so  1pl.EXCL=come.down     village
```

[eps=si panu]MC3 [a [sin]s i=lol]MC4
1pl.EXCL=come.down village and sun 3sg=be.dark
'We went and then we came home; we came home and it was dark...'
(LL030611_0078)

11.2.1.2 Same-event and new-event addition

When there is no focus on temporal succession, either *a* or *ma* is used to connect two main clauses. Usually, *a* is used when the second clause somehow "logically follows" from the first one, and *ma* when this is not the case and the second clause is "unexpected". However, this does not hold across the board. This phenomenon may have led to an extension in meaning for *ma* as an emphatic marker (see Section 12.3.4.3). *ma* is mostly used to indicate a relation of contrast; this is discussed below. Use of a different marker for unexpected addition when there is strictly speaking no contrast relation is attested for at least one other Oceanic language; see Lichtenberk (2009b) on To'aba'ita. Clauses linked by *a* may have overtones of temporal succession, or even of a reason-consequence relation. Example (95) shows two main clauses linked by *a*; a temporal succession and even a consequential relation is implied as the event in MC2 ('Stephen Pokut telling S that Ponaun arrived') could not have happened without the event in MC1 ('meeting Stephen Pokut').

(95) *ngapwak tan Stephen Pokut a ipwa, "Ponaun in si"*
[ngas=pwak ta-n Stephen Pokut]MC1 [a iA=pwa [Ponaun
1sg=meet CLF.POSS-PERT S.P. and 3sg=say P.
i=an si]Compl:Quot]MC2
3sg=PRF come
'I ran into Stephen Pokut and he said, "Ponaun has arrived."'
(ANK020995_0013)

Examples (96) and (97) each show two main clauses linked by *ma* without contrastive function. In neither case, there is a relation between the two MCs linked by *ma*, whether temporal or consequential, although there is such a relation between MC1 and MC3 in (97): the cuscus was cooked first and then eaten. It is most likely that *a* is used for same-event addition (and elaboration on an event; see the next section) and that *ma* indicates that the two main clauses must be seen as representing two separate, unrelated events. In some cases, the new event that is represented in the second main clause may be unexpected.

(96) *ila poyep, ma uliliu si*
 [i_S=la poyep]_MC1 [ma u_S=liliu si]_MC2
 3sg=go.to afternoon and 3du=return come.down
 'It had become afternoon and they returned home.' (LM190611_0006)

(97) *uapui ngoyai a ula roksi, ma ula ngan*
 [u_A=apui [ngoyai]_O]_MC1 [a u_S=la tok.si]_MC2 [ma u_A=la
 3du=cook cuscus   and 3du=go.to sit.down and eat=3sg.ZERO
 ngan=Ø]_MC3
 3du=go.to
 'They cooked cuscus and they went and sat down, and they went and ate [it].' (WL020611_0051)

11.2.1.3 Elaboration

Repetition is very common in Paluai narratives because of a discourse strategy named tail-head linkage, on which see also Schokkin (2014). Tail-head linkage means that the last part of a clause is repeated as the first part of the next clause. "By repeating information, tail-head linkage gives speakers the time to process [i.e. "prepare" – DS] the new chain, and addressees the time to process the information contained in the previous chain" (de Vries 2006: 817). Thus, the second MC of example (98) below is repeated, and MC2 and MC3 are juxtaposed without a linker (for a similar example, see MC2 and MC3 in example (94)). However, this may not count as elaboration proper, because in fact the repeated main clause does not give any additional information; it just repeats the information already given in the preceding clause. When additional information is given, the linker *a* is usually inserted between the two clauses, for instance between MC1 and MC2 in (98): *lêp China* is new information. This is further evidence that *a* can probably best be analysed as a same-event addition clause linker, with *ma* reserved for linking separate events.

(98) *wuipe lak a wuipe la lêp China, wuila lêpi re onga...*
 [wui_S=pe lak]_MC1 [a wui_A=pe la lêp [China]_O]_MC2
 1du.EXCL=PFV go   and 1du.EXCL=PFV go.to take C.
 [wui_A=la lêp=i_O]_MC3 te onga
 1du.EXCL=go.to take=3sg SUB and.so
 'We went and we took China [name of a dog], we went and took him and then...' (MK060211_0043)

11.2.1.4 Contrast

When there is a clear relation of contrast between two main clauses, the linker *ma*, also used for new-event addition (as discussed in Section 11.2.1.2), is present. *ma* is homophonous with both an emphatic marker (see Section 12.3.4.3) and with the first element of clausal negation *ma=* (which occurs in combination with *pwên*). They may be diachronically related. Negation is discussed in Section 10.2.

(99) *ngamaakêp nganngan pwên, ma ngano akêp muyou*
[nga$_A$=ma=akêp   [nganngan]$_O$ pwên]$_{MC1}$ [ma nga$_A$=no akêp
1sg=NEG$_1$=pick.up food   NEG$_2$ but 1sg=IPFV pick.up
[muyou]$_O$]$_{MC2}$
snake
'I didn't pick up the food, but I picked up the snake.' (Game1_021012_0510)

11.2.2 Alternatives

The disjunctive coordinator *le* connects two main clauses and can be translated with 'or'. It indicates a choice between two options. Often, *le* is repeated after the second clause.

(100) *i reo, iro patan nan le iro patan yoy le?*
[[i te-yo]$_S$ i=to   pata-n   nan]$_{MC1}$ [le i$_S$=to   pata-n
3sg LIG-DEM.INT 3sg=be on.top-PERT ground   or   3sg=be on.top-PERT
yoy   le]$_{MC2}$
stone or
'This [thing], is it on the ground or is it on top of the stone, or...?'
(Game1_021012_0046)

12 Pragmatics and discourse practices

This chapter discusses pragmatics and discourse phenomena. The subject will not be treated exhaustively, since this would fall well beyond the scope of this work. Rather, a preliminary overview will be given of striking discourse and pragmatic phenomena.

12.1 Preliminaries

Pragmatics is the subfield of linguistics that studies meaning in context, and thus language in use. Because it is engaged with the study of meaning, pragmatics is often seen as being affiliated to, or even indistinguishable from, semantics. Very succinctly, pragmatics differs from semantics because it takes into account setting and context in the derivation of meaning, whereas semantics does not. Fillmore (1976: 83) sees a hierarchical relation between syntax, semantics and pragmatics, which, respectively, encompass form, form and function, and form, function and setting. Lambrecht (1994) argues strongly for a position in which syntax cannot be seen as autonomous from either semantics or pragmatics. Rather, sentence-level discourse organisation, or information structure, determines formal (syntactic and prosodic) outcomes in various ways: this may range from sentence accent, to constituent order, to the organisation of main and dependent clauses. Lambrecht (1994: 6) coins the term *allosentence*, which, on a par with *allophone* or *allomorph*, refers to a member of a set of semantically equivalent, but formally and pragmatically divergent sentences, such as English *My NECK hurts* vs. *My neck HURTS* (small capitals indicate sentence accent).

Although there appear to be a number of universal tendencies, such as a preference to place discourse-salient information in sentence-initial position (cf. e.g. Sgall et al. 1986), languages can differ greatly in how pragmatic constraints play out formally (i.e. in morphosyntax). In what follows, after a terminological preamble, specific phenomena for Paluai will be discussed with regard to information structure management: marking of definiteness, anaphoric and cataphoric reference, topicalisation and focus constructions. The remaining part of the chapter will briefly touch on discourse organisation beyond the sentence level, essentially to discuss the case of narratives (how they are structured and how the most important events are singled out).

12.2 Terminology

The terminology used in this chapter is predominantly based on Lambrecht (1994), Dik (1997) and Krifka (2008). It includes terms such as presupposition and assertion, identifiability, activation, topic and focus.

12.2.1 Presupposition and assertion

One of the primary functions of language (arguably the one that outstrips all others) is the conveyance of information from a speaker to a hearer.[77] Various authors discuss in detail what this transfer of information entails. What is important for the current discussion is that information has a dual nature: it "is itself normally a combination of old and new elements" (Lambrecht 1994: 51). Information "arises by relating something new to something that can already be taken for granted" (Lambrecht 1994: 51). Krifka (2008) uses the term *common ground* (CG) to refer to this dual-natured information shared by interlocutors. The CG "changes continuously, and information has to be packaged in correspondence with the CG" (Krifka 2008: 245). For most authors, an utterance (i.e. sentence) is seen as consisting of two parts: a *pragmatic presupposition* and a *pragmatic assertion* (or *proffered content* (Krifka 2008)). These can be defined as follows (Lambrecht 1994: 52; Krifka 2008: 245-246):

- Pragmatic presupposition: The set of propositions, lexicogrammatically evoked in a sentence, that the speaker assumes the hearer already knows or is ready to take for granted at the time the sentence is uttered, i.e. requirements for the input CG;
- Pragmatic assertion: The proposition, expressed by a sentence, that the hearer is expected to know or take for granted *as a result of* hearing the sentence uttered, i.e. proposed change in the output CG.[78]

Presupposition and assertion can roughly be equated to "given" and "new" information, respectively. What is important to keep in mind is that the presupposition does not entail all the information the hearer possibly knows, or the speaker assumes the hearer possibly knows, but only what is relevant in the current communicative situation (cf. Chafe 1976).

---

[77] Only spoken language will be taken into account in this chapter. Since Paluai does not have a writing tradition, this will basically cover all genres.
[78] Lambrecht (1994) uses the term "proposition" to refer to the kinds of things (situations, events, states of affairs) that are *denoted* by propositions.

12.2.2 Identifiability

Another concept of importance in relation to information structure is *identifiability*. An identifiable referent is "one for which a shared representation already exists in the speaker's and the hearer's mind at the time of utterance, while an unidentifiable referent is one for which a representation exists only in the speaker's mind" (Lambrecht 1994: 77-78). The notion of identifiability thus ties in with that of presupposed and asserted knowledge: the former is identifiable, whereas the latter is not. In many languages, identifiable referents are often marked by a grammatical category of *definiteness*, although there is no one-to-one correspondence. How this plays out in Paluai will be discussed below.

The concept of identifiability is also used to define the semantic category of specificity (Lambrecht 1994: 80-81):

> One way of describing the specific/non-specific distinction in pragmatic terms is to say that a "specific indefinite NP" is one whose referent is identifiable to the speaker but not to the addressee, while a "non-specific indefinite NP" is one whose referent neither the speaker nor the addressee can identify at the time of utterance.

12.2.3 Activation

The conveyance of information in natural language involves not only knowledge, but also consciousness. There are three possible states an identifiable referent could have in the consciousness of the addressee: *active, semi-active/accessible* or *inactive/unused*.[79] When a referent is active, it is in the forefront of a person's focus or consciousness at a particular moment. This is the case for referents currently salient in the discourse, i.e. (given) topics (see below for more discussion of the notion of "topic"). A semi-active referent is in a person's peripheral consciousness; it may be a referent that has been mentioned before in the discourse, but that was backgrounded ("deactivated") due to intervening discourse material. Alternatively, it may be visibly present in the discourse setting and thus be deictically accessible, or be inferred from a "schema". The notion of schema (or "script" or "frame") has a long standing in pragmatics (see Tannen and Wallat 1993 for an overview): it refers to a prototypical situation in which an event sequence, participants, props, preconditions and results are already preconceived.

---

[79] Unidentifiable referents are, by definition, not represented in the consciousness of the addressee, since the addressee will only become aware of them as a result of the sentence in which they are introduced.

Scripts are culturally determined (cf. Nishida 1999); Western scripts involve, for instance, a visit to the doctor, or going to a restaurant. Inactive referents are in a person's long-term memory, and need accentuation (by sentence stress) of the referential expression and full lexical coding in order to be introduced in the discourse and thus become activated.

12.2.4 Topic

The pragmatic category "sentence topic" is thought of by most authors as being "what the sentence is about" (cf. e.g. Erteschik-Shir 2007). The term is defined by Lambrecht (1994: 131) as follows:

> A referent is interpreted as the topic of a proposition if in a given situation the proposition is construed as being *about* this referent, i.e. as expressing information which is relevant to and which increases the addressee's knowledge of this referent. [italics mine]

Dik (1997: 313-315) explicitly makes further distinctions within the category of topic, recognising at least New Topic (first introduced into the discourse), Given Topic, Sub-Topic (inferred by e.g. a schema) and Resumed Topic (comparable to Lambrecht's notion of "semi-active referent"). (Given) sentence topics are preferably expressed by unaccented anaphoric pronouns or even by zero reference. Since topics are necessarily in the presupposition, their information status is active or semi-active/accessible. Many languages exhibit an operation, often called "topicalisation", in which a topic constituent is moved to a canonical position, often to the left of the clause (Lambrecht 1994: 181-182):

> From a certain degree of accessibility on, it is possible in many languages to code a not-yet-active topic referent in the form of a lexical noun phrase which is placed in a syntactically autonomous or "detached" position to the left or, less commonly, to the right of the clause which contains the propositional information about the topic referent.

12.2.5 Focus

The notion of "focus" has received much attention in the literature, but seems to have been notoriously difficult to define. Lambrecht (1994: 207) states that:

> [The focus] is seen as the element of information whereby the presupposition and the assertion differ from each other. The focus is that portion of a proposition which cannot be taken for granted at the time of speech. It is the unpredictable or pragmatically non-recoverable element in an utterance.

This, according to Lambrecht, makes it part of the assertion, but without coinciding with it.

Dik (1997: 326) states: "The focal information in a linguistic expression is that information which is relatively the most important or salient in the given communicative setting". The use of the term *salient* in this definition is unfortunate, however, since this term is also used to refer to constituents with a high degree of givenness (i.e., topics) and it is unclear to what extent, if at all, the two uses of the term refer to something qualitatively different. In Dik's analysis, focus entails not necessarily information that is new to the addressee, but can also involve notions of surprise (when the information is contrary to expectation for the addressee) or contrast. Dik (1997: 330) distinguishes a large range of subtypes of focus, mainly along the two parameters of scope and communicative point, i.e. what pragmatic reasons underlie the assignment of focus.

Focus does not have to refer to a discourse entity (often, it does not); it can also refer to a *relation* between two identifiable entities that is newly established. In many languages, (part of) the sentence material in focus has sentence accent. The reverse, however, does not hold: sentence accent on a particular element does not automatically mean that it is in focus (cf. Lambrecht 1994). Lambrecht (1994: 222) distinguishes three types of focus:

- Predicate-focus structure: unmarked topic-comment sentence type in which the predicate is in focus and topical elements (usually the subject) are in the presupposition.
- Argument-focus structure: the focus identifies the missing argument in a presupposed open proposition.
- Sentence-focus structure: event-reporting sentence type; the focus extends over both the subject and the predicate (minus any topical non-subject elements).[80]

Argument-focus structure can, among others, refer to what is traditionally called "contrastive" focus. In many languages, however, a single sentence can be ambiguous between two or more types of focus.

---

[80] "Predicate", in this sense, should be distinguished from the usual sense of "predicate" in this work. Here, the term refers to the part of the utterance that is not the S/A argument. Other arguments therefore can be part of it. In its usual sense within this work, the term "predicate" refers to the obligatory part of any sentence, which is usually filled by a verb. In this sense, it does not include any arguments.

12.3 Information structure on the sentence level

12.3.1 Identifiability: definiteness and specificity

12.3.1.1 Definiteness

Because Paluai does not have a grammatical category of articles, definiteness is not morphologically indicated. However, definiteness can be seen as a pragmatic rather than a grammatical category (see Section 12.2.2 above, and also Aikhenvald and Dixon 1998, Krifka 2008) and, thus, is evident even in languages that lack a systematic formal expression of it. A NP can be marked as having definite reference in the following ways:
- By a demonstrative modifier following it (in particular the intermediate demonstrative);
- By means of a (direct or indirect) possessive construction.

12.3.1.1.1 Demonstrative as definiteness marker

Without a discourse context, unmarked, bare NPs are ambiguous between a definite and an indefinite interpretation. When a bare NP is mentioned in the discourse for the first time, it can be regarded as indefinite, because the speaker does not regard its referent as identifiable to the hearer. Subsequent references to the same entity will, however, be identifiable to the hearer and thus be definite. These subsequent references are usually represented by an anaphoric pronoun (see below, Section 12.3.2) or by a noun modified by a demonstrative. The latter is exemplified in examples (1) to (4).

(1) *wuikape me lêp kun a sapul [...]*
   wui$_A$=ka-pe   me   lêp   [kun   a   sapul]$_O$
   1du.EXCL=IRR.NS-PFV come take basket and shell.money
   'We will bring a basket and shell money.' (KW290311_0012)

(2) *[...]kape lêp lak [...]*
   ka$_A$=pe   lêp=Ø$_O$ lak
   IRR.NS-PFV take=3sg.ZERO   go
   'We will bring [it]...' ((KW290311_0014)

(3) *[...]kiro wau lêp kun teo*
   ki$_A$=to   wau   lêp   [kun   te-yo]$_O$
   IRR.3sg=CONT move take basket LIG-DEM.INT
   'She will be wearing the basket.' (KW290311_0015)

(4) *a ip yamar ip kape ning a ip kape pwa, "pein teo, rapo ila ran narun Kireng"*
 a [ip yamat]$_A$ ip=ka-pe   ning=Ø$_O$ a   ip$_A$=ka-pe
 and 3pl person 3pl=IRR.NS-PFV see=3sg.ZERO and 3pl=IRR.NS-PFV
 pwa [[pein te-yo]$_{TopS}$ tapo   i=la   [ta-n   naru-n
 say woman LIG-DEM.INT indeed 3sg=go.to CLF.POSS-PERT son-PERT
 Kireng]$_{CC}$]$_{Compl:Quot}$
 K.

'And the people will see [it] and they will say, "This woman, indeed she now belongs to Kireng's son."' (KW290311_0016)

Utterances (1) to (4) form a sequence (with deletion of some repeated material). They are from a narrative about the bride price procedure. This involves the gift of a basket by the groom's family to the bride's family, which the bride will wear as a mark or "proof" of the agreement. The basket is first introduced into the discourse in the utterance given in (1), referred to by a bare NP. To show its status as the new bit of information that this sentence provides, it receives sentence accent. In (2) it is referred to by an anaphoric pronoun, and since it functions as an inanimate direct object, the anaphoric pronoun has a zero form. In (3), it also functions as a direct object, but it is referred to by a full NP. This is probably due to the fact that there is some intervening material between the two references to the same entity. To mark its definite status, as (resumed) sentence topic and being in the presupposition, the noun *kun* is now modified by *teyo*.

Development into definiteness markers is a very common grammaticalisation path for demonstratives (Heine and Kuteva 2002: 4). In addition, *kun* is an important discourse entity, because it is the sign by which, in (4), the people recognise the woman as the bride-to-be. This may be an additional reason for it to be represented again as a full NP (recall from the previous section that the preferred way of referring to topics is by unmarked anaphoric pronouns). The direct object in (4) is elided again, but the act of 'seeing' refers not only to the entity *kun*, but to the entire preceding clause representing the event 'the woman wearing the basket'. In (4), the subject *pein teyo* is topicalised, which is evident from the fact that a clausal adverb *tapo* is inserted between the topicalised NP and the subject bound pronoun cross-referencing it.

12.3.1.1.2 "Anchoring" by possessive construction

Possessive constructions can be used to "anchor" newly introduced referents, since they link the latter to a participant already established in the discourse (who will be the Possessor in the possessive construction). This does not mean that all possessive constructions have this anchoring function: they can also introduce entirely new information, whereby both Possessor and Possessee are newly introduced referents. An example of anchoring is given in (5).

(5) *monokin a wope puk kop tao; wope lêp kom puan tiok...*
monoki-n   ya   wo_A=pe puk   [kop ta-o]_O   wo_A=pe
behind-PERT then 2sg=PFV open lime CLF.POSS-2sg.PERT 2sg=PFV
lêp   [ka-m   puan tiok]_O
take CLF.food-2sg.PERT fruit piper.betle
'After that, you open your lime gourd; you take your *tiok* fruit...'
(LL300511_2_0015)

This sentence comes from an instruction about how to chew betel nut. The second singular pronoun *wo* refers to an established participant, the addressee. The items *kop* 'lime gourd' and *puan tiok* 'piper betle fruit' are mentioned here for the first time. However, because they are mentioned as Possessees in possessive constructions, with the established participant *wo* as Possessor, they refer not to any lime gourd or *tiok* fruit, but to particular items that are (supposed to be) in the possession of the addressee. Since the speaker considers the referents of the NPs to be identifiable to the addressee, they have definite reference.

Another consideration that is relevant for this particular sentence is that *turuop* 'chewing betel nut with *tiok* and lime' is an important activity on Baluan Island and therefore it has a schema or script. The props needed for *turuop*, which include lime and *tiok* fruit, can easily be accessed by anyone familiar with the script. They may therefore be presupposed even before their first mention in the discourse.

12.3.1.2 Specificity

The notion of specificity can be said to occupy a more central role in Paluai discourse than the notion of definiteness; that is to say, there are more elaborate ways to indicate specificity or non-specificity than there are to indicate definiteness. Below is a list of the ways in which specificity of a referent can be indicated in the NP or on the noun:

– Non-specific generic reference: third person non-singular pronoun preceding the head noun (Section 5.2.1)
– Non-specific categorial reference: numeral 'one' preceding the head noun (Section 5.2.2)
– Specific reference, particularly to (an) individuated member(s) from a larger set: the formative *ta=* (Section 4.11)

Since these phenomena were discussed at some length already, the discussion is not repeated here and the reader is referred to the respective sections.

12.3.2 Anaphors and cataphors

An anaphor is "an expression that must be interpreted via another expression (the 'antecedent'), which typically occurs earlier in the discourse" (Cruse 2006: 12). Anaphoric function is often fulfilled by pronouns, demonstratives or other types of deictics. A cataphor has the same function as an anaphor, with the difference that the antecedent occurs later in the discourse.

12.3.2.1 Pronouns as anaphors and cataphors

Once a referent is introduced into the discourse, it is predominantly referred to by third singular anaphoric pronouns, unless it is emphasised by means of topicalisation or a focus construction. Because there is no gender marking in the pronominal system, it can be difficult to track referents of anaphoric pronouns. Consider the following sequence, which consists of lines 64-67 from Text 1 about fighting stones (the entire text can be found in Appendix II). There are three discourse participants in this sequence that are referred to by (zero or overt) anaphors: *Komou* (marked with the index i), *Parulabei* (marked with j) and *asilon sopol te isok sau te ila yen pwapwan* 'the half that he cut off which lies at his side' (marked with k). *Parulabei* is first coreferenced by the suffix -*n* on *asilon*, to indicate that it is his part that got cut off. The next pronoun, *i* (marking the A argument of the first relative clause) refers to Komou, who did the cutting off. The antecedent of the subsequent third singular pronoun *i* (marking the S of the second relative clause) changes to the cut-off piece, which is lying to the side of Parulabei (again referred to only by the suffix -*n*). This entire complex clause, which forms the O argument of the clause in (8) and is topicalised and thus fronted, is the antecedent of the zero third singular object pronoun on the verb *apek*, indicating that it was the cut-off part that Parulabei whacked into Komou. (9) is a bit more straightforward: the third singular subject pronoun refers to Parulabei and the object pronoun to Komou; after that, reference changes and the next subject pronoun refers to Komou, indicating that it was he who fell down the slope of Malsu.

(6) *Komou reo isok sau Parulabei sopol*
 [Komou$_i$ te-yo]$_A$ i_i=sok   sau [Parulabei$_j$ sopol]$_O$
 K.   LIG-DEM.INT 3sg=strike cut P.   one.half.round
 'Komou struck off one half of Parulabei.'

12.3 Information structure on the sentence level — 371

(7) *minak tepwo, kapwa kalak...*
    minak   te-pwo   kapwa ka$_s$-lak
    present.time LIG-DEM.PROX if   IRR.NS-go.to
    'Nowadays, if one goes there...'

(8) *asilon sopol le isok sau re ila yen pwapwan teo, Parulabei reo ipe apek tan Komou reo*
    [asilo-n$_i$ sopol   [te i$_j$=sok sau=Ø   [te i$_k$=la
    side-3sg.PERT one.half.round REL 3sg=hit cut=3sg.ZERO REL 3sg=go.to
    yen pwapwa-n$_j$]]$_k$ te-yo]$_{\text{TopO}}$ [Parulabei$_j$ te-yo]$_A$ i$_j$=pe
    lie side-PERT   LIG-DEM.INT   P.   LIG-DEM.INT   3sg=PFV
    apek=Ø$_k$ ta-n   [Komou$_i$ te-yo]$_E$
    hit=3sg.ZERO CLF.POSS-PERT   K.   LIG-DEM.INT
    'This half part of him [Parulabei] that he [Komou] struck off, which came to lie at his [Parulabei's] side, Parulabei whacked into Komou.'

(9) *iruk antek takaui re ila lot te ila yen lalon Malsu relo*
    i$_j$=tuk   antek   takau=i$_i$ te i$_i$=la   lot   te
    3sg=beat put.away   altogether=3sg SUB 3sg=go.to fall SUB
    i$_i$=la   yen lalo-n   Malsu te-lo
    3sg=go.to lie   inside-PERT   M.   LIG-DEM.DIST
    'He knocked him away altogether, so that he [Komou] fell and came to lie down inside Malsu.'

Occasionally, personal pronouns seem to be used as cataphors as well. This may mainly have a rhetorical purpose. Below, an example is shown, again from the same legend (lines 89-90):

(10) *polpolot teo, u re ukape pe polpolot teo, won yoy re iptaneki la polpolot teo [...]*
    [polpolot   te-yo]$_{\text{Top}}$ u$_A$ te   u=ka-pe   pe [polpolot
    chant.type LIG-DEM.INT   3du   REL 3du=IRR.NS-PFV   do   chant.type
    te-yo]$_O$
    LIG-DEM.INT
    '(As for) this *polpolot*, the two that will chant [*lit.* 'do'] the *polpolot*...'

(11) *Kisawin, Alup Kisawin Pwaril, te i pên tan Ngat Kanawi re taman ining yoy reo, a Alup Pilan Pokut, te ukape pe polpolorin yoy reo.*
[Kisawin Alup Kisawin Pwaril [te i₅ [pên ta-n
K. A.K.P. REL 3sg daughter CLF.POSS-PERT
Ngat Kanawi [te [tama-n]₅ i=ning [yoy]₀]ᴿᶜ]ᴿᶜ te-yo]ₙᵥᴘᴿᴇᴅ]ₐ
Ngat K. REL father-PERT 3sg=see stone LIG-DEM.INT
a [Alup Pilan Pokut]ₐ [te u=ka–pe pe [polpoloti-n yoy
and A.P.P. REL 3du=IRR.NS-PFV do chant.type-PERT stone
te-yo]₀]ᴿᶜ
LIG-DEM.INT
'[It is] Kisawin, Alup Kisawin Pwaril, who is the daughter of Ngat Kanawi whose father has seen these stones, and Alup Pilan Pokut, who are the ones who will chant the *polpolot* of the two stones.'

The personal pronoun *u* in (10) does not have an antecedent; its "antecedent" is only mentioned in the clause following it. This requires greater processing effort on the part of the audience, who will at first not be able to identify the individuals that *u* refers to. This may facilitate a rhetorical effect that draws the attention of the audience to the announcement that a song dedicated to the history of the two stones will be performed.

12.3.2.2 Demonstratives as anaphors and cataphors

Demonstratives were discussed in Section 4.3. For convenience, the table printed in that section, which gave an overview of the various demonstrative forms, is repeated below as Table 88.

Demonstratives are deictic expressions that, in situational deixis, "indicate the location of referents along certain dimensions, using the speaker (and time and place of speaking) as a reference point or 'deictic centre'" (Cruse 2006: 44). With discourse deixis, demonstratives function as anaphors to refer to previously mentioned discourse participants or stretches of discourse, or they act as cataphors and refer forward. Demonstratives are used both for situational and discourse deixis. Basic demonstrative forms and forms prefixed by *te-* cannot occur as independent forms, but always modify a noun or pronoun. The forms starting with *ta=* can occur independently and are thus often used as anaphors to entities previously mentioned in the discourse.[81]

---

[81] The forms starting with *a=* are only used for spatial deixis. They will not be discussed here; examples of their use can be found in Section 4.3.3.

Table 88: Demonstrative paradigm.

Form	Modifies noun	Modifies pronoun	Modifies verb	Independent form
pwo 'this'	sometimes	yes	no	no
yo 'that'	sometimes	yes	no	no
lo 'that (far)'	sometimes	yes	no	no
tepwo 'this'	yes	yes	no	no
teyo 'that'	yes	yes	no	no
telo 'that (far)'	yes	yes	no	no
arepwo 'here'	no	no	yes	no
areyo 'there'	no	no	yes	no
arelo 'over there'	no	no	yes	no
tarepwo 'this'	no	no	no	yes
tareyo 'that'	no	no	no	yes
tarelo 'that (far)'	no	no	no	yes

In discourse deictic function, the proximate demonstrative is often used as a cataphor, referring forward to the immediately following stretch of discourse. This is the case in (12), where *pwapwae tepwo* 'this story' refers to the story immediately following this utterance.

(12) *ngapwa kope pwapwa repwo, pwapwa pari ai pang*
 nga$_A$=pwa ko-pe [pwapwae te-pwo]$_O$ pwapwae
 1sg=want.to IRR.1SG-make story LIG-DEM.PROX story
 pari a=i pang
 belonging.to OBL=3sg rain
 'I am going to tell this story, the story of the rain.' (LM260511_1_0004)

The distal demonstrative, when used as a discourse deictic, is generally also used as a cataphor, in particular in the formula *temenin telo* 'like that', which refers to the contents of an immediately following utterance or stretch of discourse. An example of this use can be found in sentence (13).

(13) *upe pul me rang la remenin telo...*
 u$_A$=pe pul=Ø$_O$ me ta-ng la temenin
 3du=PFV tell=3sg.ZERO come CLF.POSS-1sg.PERT go.to like
 te-lo
 LIG-DEM.DIST
 'They told me as follows...' (LK100411_0056)

The intermediate demonstrative, in its discourse deictic function, seems to have acquired evaluative and recapulative overtones: it occurs frequently at the end of a narrative, following the third singular pronoun *i*. The phrases *i yo* and *i teyo* have a sense of 'Okay' or 'That's it' – see examples (14) and (15). A recapulative analysis is supported by the fact that there is most often a short pause between the phrase with the demonstrative and the main part of the utterance.

(14) *i o, ningning tang o*
   i   yo   ningning ta-ng   yo
   3sg DEM.INT view   CLF.POSS-1sg.PERT DEM.INT
   'That's it. That's my point of view [on the matter just discussed].'
   (OL040311_0047)

(15) *i reo, pwapwaen songanum tui rea*
   i   te-yo   pwapwae-n songan.um ta-ui
   3sg LIG-DEM.INT story-PERT marriage   CLF.POSS-1du.EXCL.PERT
   te-ya
   LIG-then
   'That's it, the story of our marriage.' (KM060111_0098)

Another example of the use of *i yo* is discussed in Section 12.4.2.1.2 below, where it is used to mark the climax of a story. It indicates something like 'there you have it' and is in a way also recapulative: based on what was said in the preceding discourse, things were bound to get pear-shaped, and they did.

Thus, all demonstratives (proximate, intermediate and distal) have dual roles as situational and discourse deictic markers. In discourse, they still do point, but on a more abstract level. Intermediate demonstratives draw the attention of the addressee by emphasising the element they modify and refer back as anaphors, not to an individual entity but to a stretch of discourse. This has probably led to their use as markers of definiteness (see above) and discourse topic (see below). Proximate and distal demonstratives have acquired a slightly different role as cataphors, referring forward.

There may be some instances of the intermediate demonstrative that cannot be classified as anaphors, but that modify newly introduced participants. An example, from line 14 in Text 1 in Appendix II, can be found below.

(16) *te yoy reo, yamat te i pari ai pusungop turê tepwo mwanen teo ipwak ai*
 te [yoy$_i$ te-yo]$_{TopObl}$ [yamat [te i$_S$ [pari a=i
 SUB stone LIG-DEM.INT person REL 3sg belonging.to OBL=3sg
 pusungop ta-urê te-pwo mwanenen]$_{NVPRED}$]$_{RC}$
 clan CLF.POSS-1pc.EXCL.PERT LIG-DEM.PROX straight
 te-yo]$_S$ i=pwak a=i$_i$
 LIG-DEM.INT 3sg=meet OBL=3sg
 'As for these stones, this person who is straight from our clan encountered them.'

What is unusual about example (16) is the second instance of *teyo*, modifying *yamat* 'person'. This discourse participant, who is an ancestor of the storyteller and sees the two stones fighting, is mentioned here for the first time. Normally, new information would not receive any kind of marking; *teyo* is used predominantly to indicate definite reference for discourse entities introduced earlier in the story. However, it seems that when extra emphasis is needed or wanted, speakers can choose to use *teyo* also for new information. In this case, *teyo* may be used to indicate familiarity with a newly introduced participant. It seems that in this function it can only refer to human referents. In colloquial English, a demonstrative (in this case, the proximate one) can be used in a similar fashion: *'I was sitting on the bus, and this girl suddenly started screaming ...'*. In Paluai, its use may be a sign that some knowledge (on the part of the addressee) of the person referred to is presupposed by the speaker. Compare also Lambrecht (1994), who argues that this use of the demonstrative is typical for introduction of referents that are meant to become topics in the subsequent discourse.

The independent demonstratives starting with *ta=*, in particular the intermediate form *tateyo*, are most often attested as discourse demonstratives, referring back to an entity already mentioned before. An example is given below, again from the legend in Text 1 (lines 60-61). The demonstrative refers to the entire clause that precedes it. After that, the event is mentioned again, this time by means of a nominalisation. All this repetition most likely functions to emphasise the important event described by the clause in (17) (see Section 12.4.2.1 below for a discussion of the entire narrative).

(17) *ma ipe nengtou*
 ma i$_S$=pe neng.tou
 but 3sg=PFV stop.walking
 'But he stopped walking.'

(18) *tareo, nengnengtou reo, ino pe*
 [ta=te-yo   neng.neng.tou   te-yo]$_{TopO}$ i=no
 SPEC.COLL=LIG-DEM.INT REDUP.stop.walking LIG-DEM.INT   3sg=IPFV
 pe=Ø
 do=3sg.ZERO
 'He stopped walking there.' [*lit.* 'That, stopping, he did.']

Another example of the use of the intermediate *ta=* demonstrative is given in (19).

(19) *monokin tareo, kape pe yep*
 monoki-n   ta=te-yo   ka$_A$-pe   pe   [yep]$_O$
 behind-PERT SPEC.COLL=LIG-DEM.INT IRR.NS-PFV   make   fire
 'After (all) that, we will make a fire.' (CA120211_1_0022)

Tateyo here refers to the entire discourse up to this point, which describes grating coconuts and rubbing loose sago. It can be concluded that *tateyo*, in contrast to the other demonstrative forms, which modify a noun or a pronoun referring to a participant or entity, often refers to a longer stretch of discourse: to one or several events or actions.

12.3.3 Topicalisation

Topicalisation in Paluai involves fronting of a constituent to sentence-initial position. We have already seen examples of topicalisation in the previous sections. There is a fair range of constituents that are allowed to undergo a topicalisation operation: the core arguments S, A and O, Oblique constituents and Possessors in a direct possession construction. Adverbial adjunct phrases (modifying verbs) are not encountered in sentence-initial fronting position. Topicalised constituents are marked with the intermediate demonstrative *teyo*. The different types of topicalisation in Paluai are discussed below.

12.3.3.1 Topicalisation of S/A

Because S and A arguments already occur in preverbal position, it may not be immediately evident that they have undergone topicalisation, since there is no fronting operation. One indication that they can be topicalised, however, is that a clausal adverb can be inserted between a topicalised constituent and the remainder of the clause. Under normal circumstances, such clausal adverbs occur in sentence-initial position. Generally, an utterance with a topicalised constituent also has a marked intonation pattern, with the topicalised constituent forming a separate minor prosodic unit, and a short pause occurring between the fronted constituent and the remainder of the utterance. An example of S/A topicalisation is given in (20), which is adapted from (4).

(20) *pein teo, rabo ila ran narun Kireng*
 [pein te-yo]$_{TopS}$ tapo i=la [ta-n naru-n Kireng]$_{CC}$
 woman LIG-DEM.INT indeed 3sg=go.to CLF.POSS-PERT son-PERT K.
 'This woman, indeed she now belongs to Kireng's son.' (KW290311_0016)

12.3.3.2 Topicalisation of O

The O argument is very frequently topicalised. When this happens, a full NP is moved from the postverbal object position to sentence-initial position. It thus precedes the A argument and the clause has OAV constituent order. Only when the O argument refers to an animate being will there be a trace element, in the form of a bound object pronoun on the verb complex. In (21), the direct object of *lêp* 'take', which refers to an animate being, is fronted, whereas in (22), the inanimate direct object of *tipêl* 'braid' is.

(21) *a not teo, upe lêpi si suk*
 a [not$_i$ te-yo]$_{TopO}$ u$_A$=pe lêp=i$_i$ si suk
 and child LIG-DEM.INT 3sg=PFV take=3sg come.down shore
 'And the child, they took him down to the shore.' (OL200111_0040)

(22) *samin teo, koripêl la kalomwen*
 [samin$_i$ te-yo]$_{TopO}$ ko$_A$-tipêl=Ø$_i$ la kalomwen
 end.of.rope LIG-DEM.INT IRR.1SG-braid=3sg.ZERO go.to handle
 'The ends of the twines I will braid into the handles [of the basket].'
 (AK160411_2_0024)

12.3.3.3 Topicalisation of an Oblique

Occasionally, an Oblique argument is fronted. It leaves a trace in the form of a third singular personal pronoun, to which the preposition *a=* (which obligatorily marks Obliques) is added as a proclitic. (23) is an example where the Oblique argument of the verb *pwak* 'meet with, encounter' is topicalised (from Text 1 in Appendix II, line 14).

(23) *te yoy reo, yamat te i pari ai pusungop turê tepwo mwanen teo ipwak ai*
   te   [yoy$_i$ te-yo]$_{TopObl}$ [yamat [te   i   pari   a=i
   SUB stone LIG-DEM.INT   person REL 3sg belonging.to OBL=3sg
   pusungop ta-urê   te-pwo   mwanenen]$_{RC}$
   clan   CLF.POSS-1pc.EXCL.PERT LIG-DEM.PROX straight
   te-yo]$_S$ i=pwak   a=i$_i$
   LIG-DEM.INT 3sg=meet OBL=3sg
   'As for these stones, this person who is straight from our clan encountered them.'

12.3.3.4 Topicalisation of a Possessor

In (24), the Possessor NP *mui* 'dog', which would normally directly follow the Possessee NP, is fronted and marked by *teyo*. The Possessor is still marked by a *-n* pertensive suffix on the Possessee NP *nia-* 'stomach'. The Possessor NP *mui* could be regarded as the "psychological subject" (Gabelentz 1869) of the clause, the speech act participant that experiences the emotion 'anger' (brought on by the other protagonist of the story, a cuscus, who in this utterance is only referred to by anaphoric pronouns). The concept of 'be/become angry', however, can only be expressed by the formula *nia-* (TAM) *palak*, with the body part *nia-* 'stomach' (which has to be directly possessed) as its grammatical subject. Thus, the topic of (24) is the psychological subject *mui*, while its grammatical subject is *nian*. Moreover, the anaphoric third singular subject pronoun *i* in the following main clause refers back to the psychological subject of the preceding clause, *mui*, and not to its grammatical subject, *nian*.

(24) *mui reo, nian in paran palak tai a iro ngui roui*
   [mui$_i$ te-yo]$_{TopPoss}$ [nia-n$_i$]$_S$ i=an   paran palak
   dog   LIG-DEM.INT   stomach-PERT 3sg=PRF really bad
   ta-i$_j$ a   i$_i$=to   ngui   tou=i$_j$
   CLF.POSS-3sg.PERT and 3sg=CONT snarl give=3sg
   'The dog, he got very angry with him and he "snarled after" him.'
   [*lit.* 'The dog, his stomach got very bad at him....']
   (LL010711_0067)

Another example of a topicalised possessor, also with the body part *nia-*, is given in (25). Again, the psychological subject *mui* 'dog' is coreferential with the subject of the verb in the second clause, *mumut* 'vomit'.

(25) *mui reo, nian ipe rang kelkeleyek a ipe mumut*
[mui$_i$ te-yo]$_{TopPoss}$ [nia-n$_i$]$_S$ i=pe   tang   kel.keleyek
dog   LIG-DEM.INT   stomach-PERT   3sg=PFV   pick.up   REDUP.turn
a   i$_i$=pe   mumut
and   3sg=PFV   vomit
'[As for] the dog, his stomach got upset and he vomited.' (MK060211_0014)

12.3.3.5 Discourse function(s) of topicalisation
The topicalisation operation is used frequently when speakers want to emphasise a particular participant as important in the discourse. This will almost always be a discourse participant that has been introduced into the discourse prior to the topicalisation operation; it is therefore identifiable to both speaker and addressee and is thus either active or semi-active/accessible. If a topicalisation operation occurred with reference to a "brand new" discourse entity, this would certainly be a very marked phenomenon and would probably be done for reasons other than just emphasis (e.g. rhetorical reasons).

Some discourse contexts may provide an environment that makes topicalisation more likely, when there is a risk of ambiguity or confusion otherwise. In the following sequence, the planting procedures for long and short yams are compared. The sequence is taken from Text 2 in Appendix II (lines 21-26).

(26) *aronan lolomêek teo, suei reo, sôkôm a worou maran teo, kimwangmwang sa wat*
arona-n   lo.lomêek   te-yo   [suei$_i$ te-yo]$_{TopPoss}$
procedure-3sg.PERT   REDUP.plant   LIG-DEM.INT   short.yam   LIG-DEM.INT
sôkôm   ya   wo$_A$=tou   [mata-n$_i$ te-yo]$_O$ ki$_S$-mwang.mwang
some   then   2sg=put   eye-PERT   LIG-DEM.INT   IRR.3sg-REDUP.watch
sa   [wat]$_{LOC}$
come.up   up.high
'As for the way of planting: the short yam, some you will put [into the ground] with their eyes looking up.' [*lit.* 'you put its eye, it will look come up']

(27) *sôkôm a worou maran teo, kila luek paye*
sôkôm   ya   wo$_A$=tou   [mata-n$_i$ te-yo]$_O$ ki$_S$-la   luek
some   then   2sg=put   eye-PERT   LIG-DEM.INT   IRR.3sg-go.to   come.out
[paye]$_{LOC}$
down.below
'Some, you will put [into the ground] with their eyes facing down.'
[*lit.* 'you put its eye, it will go down']

(28) *ma ilai i ramwayen, mwayen teo, woro lomêeki lai re koropun nganngan teo iro au sa wat*
ma i$_S$=la a=i [i ta=mwayen]$_{CC}$ [mwayen te-yo]$_{TopO}$
but 3sg=go.to OBL=3sg 3sg SPEC.COLL=yam yam LIG-DEM.INT
wo$_A$=to lomêek=i$_O$ la a=i [te [koropu-n nganngan
2sg=HAB plant=3sg go.to OBL=3sg SUB base-PERT food
te-yo]$_S$ i=to wau sa [wat]$_{LOC}$]$_{AdvCl}$
LIG-DEM.INT 3sg=HAB walk come.up up.high
'But when it comes to [long] yam, the long yam you will plant in such a way that the bottom of the tuber is facing upwards.'

(29) *a i reo; mwayen teo, iro ret la paye*
a i te-yo [mwayen te-yo]$_{TopS}$ i=to tet la
and 3sg LIG-DEM.INT yam LIG-DEM.INT 3sg=HAB grow go.to
[paye]$_{LOC}$
down.below
'That's it. Long yam usually grows downwards.'

(30) *a mwayen teo, kumun kisa pei, a kanen kipe ret suwot paye*
a [mwayen$_i$ te-yo]$_{TopPoss}$ [kumu-n$_i$]$_S$ ki-sa pei
and yam LIG-DEM.INT sprout-PERT IRR.3sg-come.up come
ya [kane-n$_i$]$_S$ ki-pe tet suwot [paye]$_{LOC}$
then meat-3sg.PERT IRR.3sg-PFV grow go.down down.below
'So as for the [long] yams, when their sprout will appear, their meat will grow downwards.'

(31) *ma suei reo, worou maran la paye, are kanen teo iro pei sa wat*
ma [suei$_i$ te-yo]$_{TopPoss}$ wo$_A$=tou [mata-n$_i$]$_O$ la [paye]$_{LOC}$
but short.yam LIG-DEM.INT 2sg=put eye-3sg.PERT go.to down.below
[a=te [kane-n$_i$ te-yo]$_S$ i=to pei sa [wat]$_{LOC}$]$_{AdvCl}$
OBL=SUB meat-PERT LIG-DEM.INT 3sg=HAB come come.up up.high
'But as for short yams, you put their eyes downwards, as their meat usually grows upwards.'

The sequence above has two distinctly separate but equally salient discourse topics: *suei* 'short yam' and *mwayen* '(long) yam', which are compared with each other. This environment, where two discourse entities are compared and there is much switching back and forth and thus much potential for antecedent confusion, may be a trigger for extensive use of topicalisation. They may be an example of what Lambrecht (1994: 183) calls "contrastive topics", where topicalisation is used to "mark a shift in attention from one to another of two or more already active topic referents". Note that this sense of "contrastive" is different from the one used in relation to focus assignment.

Like most Oceanic languages, Paluai has basic SV/AVO constituent order. By means of topicalisation, constituent order can be changed; it becomes OAV in the case of fronted Objects and XSV/XAVO in case another constituent is fronted. Since Paluai has no passive voice construction, topicalisation is the only way to "promote" an O argument to the privileged clause-initial position.

12.3.4 Focus in Paluai

Firstly, the relation between focus and sentence accent will be discussed. Secondly, two particles that function to foreground or highlight (part of) an utterance, and could thus be analysed as focus markers, will be discussed: *ya* and *ma*.

12.3.4.1 Focus and sentence accent

We have already seen in the previous sections that sentence accent can be used to highlight part of an utterance. A pragmatically unmarked utterance has a topic-comment structure and will carry sentence accent on part of the predicate.[82] As explained in Section 12.2.5, it has predicate focus. Sentences that start with an unaccented pronoun referring to an already established referent have this structure. An example is sentence (1) above, repeated below as (32). Here, it is the O argument that receives sentence accent, but in fact the entire predicate *kape me lêp kun a sapul* refers to new information. Elements receiving sentence accent are indicated in bold.

(32) *wuikape me lêp kun a sapul* [...]
 wui$_A$=ka-pe me lêp [**kun a sapul**]$_O$
 1du.EXCL=IRR.NS-PFV come take basket and shell.money
 'We will bring a basket and shell money.' (KW290311_0012)

---

[82] Note that "pragmatically unmarked" does not mean "pragmatically neutral". In fact, it is not possible for utterances to be pragmatically neutral because they will always be interpreted against a contextual background (cf. Lambrecht 1994).

Predicate focus can also encompass the verb:

(33) *woro pe sa?*
 wo_A=to **pe** [sa]_O
 2sg=CONT do what
 'What are you *doing*?' (052b_0179)

(34) *ngaro nganngan*
 nga_S=to **ngan.ngan**
 1sg=CONT REDUP.eat
 'I am eating.' (052b_0180)

In (33), the personal pronoun *wo* is in the presupposition, and the actions of the referent that *wo* is referring to are focussed. Because the speaker is questioning the action, it is the verb *pe* that is accented. Alternatively, the question can be asked with the action *pe* also in the presupposition. This yields (35), which is an allosentence of (33):

(35) *woro pe sa?*
 wo_A=to pe [**sa**]_O
 2sg=CONT do/make what
 '*What* are you doing/making?'

In this case, we have argument focus. The relation between *wo* and the action *pe* is already presupposed; what the speaker is querying, and thus what is focussed, is the O argument of the verb *pe*. Since *pe* is ambiguous between 'do' and 'make', (35) is more likely to be interpreted as questioning the identity of an object that the addressee is producing, rather than questioning an action of the addressee.

Most often, an utterance has argument-focus structure when it is used to express a relation of contrast, based on what the speaker considers as presupposed by the addressee(s). Consider the examples in (36) and (37), repeated from Section 2.3.2.3:

(36) *ngamaning tareo pwên, ma i re pwapwaen iro rang*
 nga_A=ma=ning [ta=te-yo]_O   pwên ma i te [**pwapwae-n**]_S
 1sg=NEG_1=see SPEC.COLL=LIG-DEM.INT NEG_2 but story-PERT
 i=to ta-ng
 3sg=be CLF.POSS-1sg.PERT
 'I didn't see all that, but the *story* is mine.' (YK290411_1_0030)

(37) ngamaakêp nganngan pwên, ma ngano akêp muyou
nga_A=ma=akêp   [**ngan.ngan**]_O pwên ma nga_A=no akêp
1sg=NEG$_1$=pick.up food   NEG$_2$ but 1sg=IPFV pick.up
[**muyou**]_O
snake
'I didn't pick up the *food*, but I picked up the *snake* (instead).'
(Game1_021012_0510)

In these cases, the contrast is made explicit, but this is not always the case. In (36), it is presupposed that the speaker has a relation to the person that he is talking about (the utterance is from a public speech at a mortuary ceremony). The speaker specifies the relationship by explaining that he did not see events referred to in the preceding discourse, but that the story nevertheless belongs to him. In (37), the contrast relation is more straightforward. The speaker has picked up a picture, since this utterance is part of a *Man and Tree* game (Levinson et al. 1992). The proposition [speaker picked up *x*] is therefore presupposed, and the identity of *x*, which is the O argument of 'pick up', is questioned by the other participant in the discourse (thus, *x* is focussed). He guessed that the speaker had picked up a picture with food in it. The speaker corrects this and contrasts the false proposition [*x* = food] with the correct one [*x* = snake]. This is a clear example of argument focus.

Thus, predicate-focus and argument-focus structures are attested in Paluai. The remaining focus structure, sentence-focus, may also be attested. With sentence focus, the entire utterance (except topicalised non-subject elements) refers to new information and is thus focussed. An example of a sentence-focus structure is the following utterance, repeated from (25) above:

(38) mui reo, nian ipe rang kelkeleyek a ipe mumut
[**mui**$_i$ **te-yo**]_TopPoss [nia-n$_i$]_S i=pe tang **kel.keleyek**
dog LIG-DEM.INT stomach-PERT 3sg=PFV pick.up REDUP.turn
a i$_i$=pe **mumut**
and 3sg=PFV vomit
'[As for] the dog, his stomach got upset and he vomited.'
(MK060211_0014)

This utterance comes from an anecdote about a fishing trip and describes the event of how a dog that was taken aboard the canoe became seasick and vomited; it consists of two coordinated main clauses. Apart from the topicalised NP *mui teyo*, everything in the first main clause refers to brand-new information (and, moreover, to an unexpected event). As was explained above in Section 12.3.3.4, in this sentence the topicalised element is not the grammatical subject (although it is the psychological subject). Thus, this sentence fits Lambrecht's (1994) definition of sentence-focus structures; it is event-reporting and would be a felicitous answer to the question 'What happened?'. Both the grammatical subject *nian* and the predicate *tang kelkeleyek* are new and thus in focus, but *nian* is anchored by means of the pertensive *-n* suffix. Furthermore, the placement of sentence accent is the same as for an utterance with predicate focus: it rests on the predicate.[83] The second main clause does not have sentence-focus structure, since it contains the third singular bound pronoun *i* referring to the given element *mui*. This sentence thus has predicate focus.

12.3.4.2 The particle *ya*

The particle *ya* occurs by itself as well as preceded by the formative *te-*.[84] By itself, it can be used to mark the focal clause of a main-dependent clause sequence, such as the apodosis of a conditional or a main clause modified by a temporal subordinate clause. See Section 11.1.3 for examples.

In other cases, the particle *ya*, by itself, modifies the third singular independent pronoun *i*. The phrase *i ya* could be translated with 'alright, well'. It always occurs sentence-initially and is used to highlight the clause that follows it. An example from Text 1 in Appendix II (line 59) is given in (39). The clause following *i ya* represents an important event in the story and is therefore marked. The marker also links this clause with the material that preceded it. Just before its occurrence, it is mentioned that the subject of the clause heard the noise of the two stones fighting. (39) represents what is "concluded" based on this information.

(39) a i a, imaret lai lopwan mangat tai reo pwên
 a   i   ya   i$_S$=ma=tet   la   a=i   lopwa-n   mangat
 and   3sg   then 3sg=NEG$_1$=move go.to OBL=3sg place-PERT   work
 ta-i   te-yo   pwên
 CLF.POSS-3sg.PERT   LIG-DEM.INT   NEG$_2$
 'And well, he did not go to the place of his work.'

---

[83] This holds when the topicalised element *mui teyo* is regarded not as part of the sentence, but as a left-dislocated element. If *mui teyo* is considered part of the sentence, the structure has sentence accent placement diverging from that of predicate-focus structure, since topicalised elements are also accented.

[84] Paluai *ya* is homophonous to a Tok Pisin emphatic marker, but it is believed that it has its origins in Paluai and is not a borrowed form. This is mainly because it is generally not used as an emphatic marker on nouns, in contrast to its Tok Pisin counterpart.

The complex form *teya* can be used to mark sentence material (usually a discourse participant) in similar fashion. An example can be found in (40) (Text 1 – line 19). Again, the clause modified by *teya* forms a conclusion to preceding material. In the "conceptual arc" building up to line 19, the storyteller explains his family history. This leads to the conclusion that he is the person designated to tell this story, and this is highlighted by the use of *teya*.

(40) *te wong tepwo rea, ngaro rêk pwapwa repwo*
 te [wong te-pwo te-ya]$_{Foc:A}$ nga$_A$=to têk
 SUB 1sg.FREE LIG-DEM.PROX LIG-then 1sg=HAB tell
 [pwapwae te-pwo]$_O$
 story LIG-DEM.PROX
 'So it's *me* alright, I usually tell this story.'

Teya is also encountered following the 3sg free pronoun *i*. Whatever its syntactic function, *teya* is often encountered at the end of narratives and seems to have a recapitulative and concluding function similar to that of the intermediate demonstrative combined with the 3sg pronouns *i yo* or *i teyo* discussed in Section 12.3.2.2. See (41) for an example.

(41) *kalomêek nêm onga inêm tea*
 ka$_S$-lo.lomêek nêm onga i=nêm te-ya
 IRR.NS-REDUP.plant be.finished and.so 3sg=be.finished LIG-then
 'We will finish planting, and so it will be finished then / it will be finished alright.' (KM190211_0037)

Although *ya* and *teya* can be translated with 'then' or with Tok Pisin *nau*, they do not seem to function as temporal deictics in the strict sense. They do not literally refer to a point in time. Rather, they function to highlight certain aspects of the discourse. Therefore, *ya* is analysed as a focus particle. Another example, in which both *i ya* and *teya* are attested, is given in (42).

(42) *woning te inap are yep kino rer ai; i a, ope sul yep ai rea*
 wo$_A$=ning te [inap a=te [yep]$_S$ ki-no tet
 2sg=see COMP able OBL=COMP fire IRR.3sg-IPFV spread
 a=i]$_{Compl:O}$ i ya wo$_A$=pe sulu [yep]$_O$ a=i te-ya
 OBL=3sg 3sg then 2sg=PFV ignite fire OBL=3sg LIG-then
 'You will see that it is possible for fire to spread through it; alright, you will set fire to it then.' (KM190211_0017-18)

12.3.4.3 The particle *ma*

Homophones of *ma* were mentioned several times previously. One functions as the first part of discontinuous predicative negation, as a proclitic to the negated element (see Section 10.2.2) and another one functions as a clause linker indicating contrast and new-event addition (see Section 11.2.1). The element *ma* under discussion here functions as an emphatic particle or perhaps a focus marker. It can be most clearly distinguished from the clause linker counterpart when it is used sentence-finally. This is usually the case after the interrogative word *ai sa* 'for what' (see Section 11.1.3.3.1) or for instance in (43).

(43) hm, yamat ma...
    hm yamat ma
    INTJ person EMP
    'My goodness, this man...' [TP: *man ya tu ya*] (Game1_021012_0420)

In some of its sentence-initial occurrences it may also function as an emphatic marker, but this use is harder to distinguish from its use as a clause linker. Example (44) is again from Text 1 in Appendix II (line 55).

(44) ma pwempwem som to reng a isirak
    ma [pwempwem som]$_S$ to   teng a   i$_S$=sirak
    EMP bird.sp   one.ANIM CONT cry and 3sg=get.up
    'A *pwempwem* bird was crying, and he woke up.'

This sentence is the start of a new episode in the story (for more on the narrative structure of the story, see Section 12.4.2.1 below). There is no connection between the clause represented in (44) and the preceding clause, and thus there is no need for a clause linker. *ma* predominantly seems to fulfil the function of an emphatic/focal particle, which is to signal the start of a new episode.

12.4 Discourse organisation at the paragraph level and beyond

This section focuses on the structure of narratives, i.e. monologues. Of course, conversation, too, is organised beyond the sentence level, but since the present study is based predominantly on data in the form of narratives, relatively little is known about discourse organisation and negotiation of meaning in spontaneous conversation. This remains an area for further study.

12.4.1 The structure of narratives

Narratives come in various genres, which are organised and structured differently. The data set of narratives consists of traditional legends, anecdotes, procedural texts, family histories, children's stories and expositions about a particular theme. Most narratives consist of three parts: an introduction, a body and a coda. In the introduction, the storyteller usually introduces him- or herself and explains the reason or motivation for telling the story. In the case of traditional legends, in particular, because they are clan property, the speaker will elaborate on his line of descent (all legends were told by men) to make it clear that he is entitled to tell the story. Anecdotes usually contain much less introduction, especially when narrated in the company of people who were present at the events, and who are therefore presupposed to be familiar with them. The body of the narration will contain the story itself, usually divided into several episodes. The coda will give a summary and a conclusion, and in the case of a legend, it will normally be used to ensure that the narrated events, which occurred in a distant past, are linked to the present. Often, this will clarify the *raison d'être* of the story. This can for instance be a moral prescription, or an explanation why a certain state of affairs is as it is.

12.4.2 Two examples

When narrating a sequence of events, a narrator has several devices at his or her disposal to structure the narration, in order to make it more accessible for the audience. In particular with legends and other highly conventionalised genres, a number of formulas are frequently used. Below, two cases will be discussed in more detail: one is a legend, the other a procedural text.

12.4.2.1 Example 1: The story of Parulabei and Komou

This story can be found in its entirety in Appendix II. It is a traditional legend about two chiefs, Parulabei and Komou, the latter of whom broke a taboo. As a result, they end up fighting, turn into stones, and Parulabei hits Komou so that he rolls downhill and is forgotten.[85] The narrative consists of an introduction (lines 1-20), the story proper (lines 21-80) and a coda (lines 81-91). The story

---

[85] Legends are not always easy to analyse, since they often contain supernatural events and unpredictable plot twists. It is not clear, for instance, whether the two chiefs turn into stones as a direct result of Komou breaking a taboo, or whether this happened for some other reason.

proper consists of seven episodes. Events are narrated more or less in chronological order, with some foreshadowing and background information thrown in.

12.4.2.1.1 The introduction
In the introduction, the storyteller introduces himself and goes on to say that before one can start to tell a traditional legend, one first has to explain its origin (*lit*. 'hit its road' – line 4). He then continues to introduce himself and his lineage, and explains how he is connected to the story of the two stones: it was his direct ancestor who saw them fighting (line 15). In line 19, he emphasises his entitlement by using a focus clause with the marker *teya*: 'So it is ME who usually tells this story.' The introduction ends in line 20 with the prescribed formula *pwapwa teyo ila temenin telo* 'the story goes like this'. The distal demonstrative *telo* functions as a cataphor for the discourse to follow.

12.4.2.1.2 The narrative proper
Legends occurring in a distant past are always introduced with the formula *palosi teyo ilak*, which is comparable to the English "once upon a time" (line 21). From this line onwards, the temporal setting changes: from the here-and-now of the speech event, the present day, to the temporal setting of the narrative, 'long ago'. The narrative is divided into seven episodes, which can be summarised as follows:
1. Lines 21-25 sketch the background, explain why people had to do their fishing at night, foreshadow the events to come: 'one day, things went awry' (line 25);
2. Lines 26-37 introduce the two protagonists and describe how things went sour; this episode ends with the sun coming up just as the group is about to go down into the crater of the extinct volcano (climax, line 37);
3. Lines 38-48 describe the fight that erupts between Parulabei and Komou;
4. Lines 49-54 are a short digression giving some background on Lalau Alipul, who witnessed the stones fighting;
5. Lines 55-60 describe how Lalau Alipul gets up (change of discourse topic marked by *ma*), and while walking in the bush, hears the stones fighting. A crucial moment is when he stops walking; this is marked by *i ya* and repetition;
6. Lines 61-73 describe the climax of the fight (marked with *ma* in line 68) and the speech that Parulabei gives to Komou afterwards;
7. Lines 74-80 are a wrap-up, describing how Lalau Alipul understands that this is an extraordinary event, and how he acts accordingly.

12.4 Discourse organisation at the paragraph level and beyond — 389

The transition from one episode to the next is usually specified by a discourse marker that functions on the paragraph level: there is either a device indicating the climax of the current subsection or a marker signalling the start of a new episode, or both. Often, change between episodes is also marked by a change in main participants, or a change in spatial or temporal setting (Farr 1999).

Episode 1, which is setting the stage, ends with a clear forewarning of things to come. Immediately following, the two protagonists are introduced. It is clear that Parulabei is the sensible one: he tries to convince Komou that it is time to stop fishing. Komou, however, disagrees, and Parulabei gives in to him.[86] This is an important event, marked by the focalising particle *i ya*. When the group continues fishing, Komou realises that Parulabei is right, and he orders the group to pack up their things and go. They hurry back to their village, but it is too late: when they arrive at the edge of the crater and are about to go down, the sun comes up. This is the climax of the first part of the story and the catalyst for events to come, and thus the event is marked by *i yo*, indicating something like 'well, there you have it'.

This is followed by a description of the quarrel and fight that erupted between the two chiefs, and the mention that they are (turned into) two big stones. This scene is backgrounded for a while from line 49 onwards, when another important character is introduced: Lalau Alipul, the ancestor of the storyteller. There is a short digression on how Lalau Alipul sharpens his knife before he goes to sleep, the function of which is not entirely clear. The discourse topic changes again (marked by *ma*) from line 55 onwards, when Lalau Alipul gets up and hears the tremors and sounds of trees breaking, caused by the fighting stones. Lines 59-61 indicate crucial events: Lalau Alipul stops walking in order to listen to the stones. This is marked by the focus particle *i ya* in line 59, and by repetition of his actions in line 61. Then, he witnesses what happens next: Komou knocks off part of Parulabei and Parulabei reacts by whacking the knocked-off part into him, making him roll down the slope of the crater. In line 68, the particle *ma* is once again used, indicating a change of discourse topic: Parulabei addresses a speech to Komou, explaining to him that, in the future, Komou will lie in the crater, forgotten and overgrown with moss and weeds, but Parulabei will stay on top and people will come and visit him and clear him of weeds. This speech is rendered by direct quotation. The remaining episode, which starts after Parulabei's speech ends in line 74, tells us how Lalau Alipul

---

[86] Parulabei's initial mistake in giving in to Komou may explain why they are both turned into stones.

realises that this is a very important event, and so, instead of going to his garden, he returns home and tells the story to his family.

12.4.2.1.3 The coda
Lines 81-91 wrap up the story and provide a conclusion. The temporal setting changes back from the time of the narrative to the present, with the words *minak tepwo* 'nowadays'. The storyteller repeats the speech that Parulabei gave to Komou and gives the reason for the story's existence in line 83: the story explains why people do not go and see Komou (and thus only hear of him in a story), but they do go and see Parulabei. The reason clause is marked by the particle *ma*. The storyteller concludes by indicating that the story of the two stones has been turned into a *polpolot*, a traditional chant type, and he gives the names of the two women who will perform the *polpolot*. The story concludes with the formula for ending a stretch of discourse: *inêm* 'it's finished'.

12.4.2.1.4 Participants and props
Participants are distinguished from props because they play an active role in the story, whereas props mainly fulfil a supportive role (Farr 1999). The story involves three participants: Parulabei, Komou and Lalau Alipul. There are also the people that went fishing together with Parulabei and Komou, but they do not play an active role in the story and are only referred to with the third person plural pronoun *ip*. They are therefore not considered participants. There are several props in the story, such as 'the dawn' (introduced in line 22), and 'their fish' (line 34). The dawn does play an important role, since it is the catalyst for the events unfolding, but since it is not a sentient being it is regarded a prop rather than a participant. Most props, such as 'their fish' are only mentioned once and thus play a very minor role in the story.

12.4.2.2 Example 2: Planting yams
The second text that can be found in Appendix II is a procedural text about how to plant yams.

12.4.2.2.1 Introduction, narrative and coda
The introduction to this narrative is much shorter, only two lines, and there is virtually no coda except for the last line showing the formula *inêm*. The narrative can be divided into three paragraphs:

1. Lines 3-20: preparatory work;
2. Lines 21-30: comparison of different planting procedures for long and short yam;
3. Lines 31-40: instruction of how to wind the yam vines onto stalks.

The sequence of events is largely chronological, except for the second part, where long yam and various types of short yam are compared with each other, which is largely atemporal.

Because it describes a hypothetical (or habitual, cf. Section 6.4.2.3) event, virtually the entire narrative is cast in irrealis. Most of the time, second person singular is used, which has no formal irrealis marking, but other persons and numbers show an irrealis prefix. There are some conditional clauses, as in line 28, and a few apprehensive clauses, as in lines 36-37. These refer to procedures that have to be followed in order to avoid an undesirable result (e.g. development of seed tubers, or wilting plants because of the sun).

Because both *suei* and *mwayen* are the discourse topic of the second paragraph, and because they are compared with each other, there are many instances of topicalisation in this section. *suei* 'short yam' is topicalised in line 21, followed by *mwayen* 'yam' in line 23. Probably to emphasise the fact that she is still talking about long yam, the speaker again topicalises *mwayen* in lines 24 and 25. In line 26, she switches to short yam once more, and so it is *suei* that is again topicalised.

12.4.2.2.2 Participants and props

Because this is a procedural text meant to instruct, there is only one participant or group of participants: the planter or planters, referred to by the second singular pronoun *wo* or, on one occasion, by the first plural inclusive pronoun *tap* (line 3). There are, however, many props in the story, for instance *suei* 'short yam', *mwayen* '(long) yam', *yêpin kanei* and *yêpin nanaop* 'leaves of particular plant species', *lik* 'carrying bag', *sapon kei* 'herbs', *nganngan* 'tuber (lit. 'food')', *nop* 'mound', *kap* 'bamboo stake' and *sin* 'sun'. A prop left implicit is *samel* 'knife', for instance to cut the herbs, but it is clear from the story that one would need such an instrument. This procedural text adheres very nicely to the observation made for this genre by Farr (1999), who mentions that scripts for procedural texts require them to be very precise in their description, whereas, for example, a description of planting in another type of narrative would receive much less detail.

Appendix I Recordings metadata

Code	Genre/subject	Speaker(s)	Length (hh:) mm:ss
052b*	Elicited conversation, plus some spontaneous conversation	2 male, 20–30	14:08
AK160411_1	Procedure (grass skirts)	Female, 40–50	01:39
AK160411_2	Procedure (coconut leaf baskets)	Female, 40–50	03:05
ANK020995*	Anecdote (meeting Ton)	Female, unknown	01:42
BK040311	Short speech about customs	Male, 50–60	05:59
CA120211_1	Procedure (frying sago)	Female, 40–50	05:01
CA120211_2	Procedure (coconut oil)	Female, 40–50	04:31
Game1_021012	Picture matching task	2 male, 50–60 and 60–70	42:36
Game2_021012	Picture matching task	2 male, 50–60 and 60–70	22:01
Game1_280812	Picture matching task	2 female, 40–30	16:14
Game2_280812	Picture matching task	2 female, 40–30	13:53
Game3_280812	Picture matching task	2 female, 40–30	25:08
Game4_280812	Picture matching task	2 female, 40–30	26:16
KM040311	Short speech about customs	Male, 50–60	03:48
KM050995*	Anecdote (walking trip)	Male, 30–40	02:38
KM060111	Procedure (making gardens)	Male, 50–60	10:12
KM190211	Procedure (bride price ceremony)	Male, 50–60	04:53
KS030611_1	Traditional legend (yams)	Male, 80–90	07:12
KW290311	Procedure (choosing bride)	Male, 40–50	03:25
KW290611	Traditional legend (snake and eagle)	Male, 40–50	07:42
LK100411	Traditional legend (Casta reef)	Male, 30–40	15:19
LK250111	Traditional legend (two stones)	Male, 30–40	08:59
LM190611	Traditional legend (man carried away by spirits)	Male, 40–50	06:33
LM240611	Family history	Male, 40–50	07:55
LM260511_1	Traditional legend (rain)	Male, 40–50	07:22
LM260511_2	Procedure (conch shell)	Male, 40–50	02:30
LL010711	Children's story (dog and cuscus)	Female, 30–40	11:41
LL030611	Anecdote (Alim)	Female, 30–40	13:22
LL300511_1	Children's story (dog's tails)	Female, 30–40	10:49
LL300511_2	Procedure (betel nut)	Female, 30–40	04:56
MK050311	Procedure (tree bark baskets)	Female, 40–50	03:56
MK060211	Anecdote (fishing trip)	Female, 40–50	04:51
MK220995*	Anecdote (gardening)	Female, 20–30	00:49
MS250311	Traditional legend (Koleyep)	Male, 50–60	09:41
NP210511_1	Traditional legend (coconut)	Male, 40–50	06:09
NP210511_2	Anecdote (hunting trip)	Male, 40–50	08:06

https://doi.org/10.1515/9783110675177-013

(continued)

Code	Genre/subject	Speaker(s)	Length (hh:)mm:ss
NS220511_1	Traditional legend (rain)	Male, 50–60	06:41
NS220511_2	Family history	Male, 50–60	04:10
NK290311_1	Procedure (planting yams)	Female, 40–50	03:30
NK290311_2	Procedure (ceremony for girl)	Female, 40–50	03:01
NP190511_2	Traditional legend (Mapou)	Female, 70–80	05:15
NP220611_1	Procedure (planting yams)	Female, 60–70	03:12
NP220611_2	Family history	Female, 60–70	02:06
NP260511	Procedure (food taboo)	Female, 60–70	02:34
OBK040311	Semi-structured conversation about customs	3 male, two 50–60, one 60–70	16:48
OL040311	Short speech about customs	Male, 60–70	05:31
OL200111	Family history	Male, 60–70	11:39
OL201210	Family history	Male, 60–70	16:31
PN100411	Traditional legend (fishing with net)	Male, 50–60	04:42
PK290411_1	Public speeches	All male, 40–70	03:59
PK290411_3	Public speeches	All male, 40–70	09:04
SY100411	Procedure (mourning ceremonies)	Male, 50–60	05:32
SP190311	Procedure (mourning ceremonies)	Male, 50–60	12:43
WL020611	Traditional legend (gardening)	Female, 30–40	05:45
WL020711	Children's story (two brothers)	Female, 30–40	12:44
YK290411_1	Public speeches	All male, 40–70	05:49
YK290411_2	Public speeches	All male, 40–70	04:47
Total			**07:53:45**

*Recording made by Ton Otto

Appendix II Texts

Text 1 Parulabei and Komou (8 min 59 sec)

This story was told on 25 January 2011 by Lalau Kanau (ca. 40 years old) of Loye village. His clan is Munukut. The story is a traditional legend about two chiefs, Parulabei and Komou, who are fighting because Komou broke a taboo: fishing at sunrise. It is unclear whether the two protagonists turn into stones because this taboo was broken or not, but they do. Parulabei eventually hits Komou and causes him to roll down into the crater of the extinct volcano, Malsu, where he will lie among the bushes and will get overgrown by weeds and moss. Komou is thus punished for breaking the taboo. Parulabei, however, stays put on top of the mountain so people can go and visit him.

(1) *Wong tepwo nirasip, ngayong Lalau Kanau.*
wong   te-pwo   panurasip   [ngaya-ng]$_S$ [Lalau Kanau]$_{NVPRED}$
1sg.FREE LIG-DEM.PROX first   name-1sg.PERT   Lalau Kanau
Me here first, my name is Lalau Kanau.

(2) *Pwapwa reo, i puron te rapkape rêk e?*
[pwapwae te-yo]$_S$ i   [puron [te   tap$_A$=ka-pe
story   LIG-DEM.INT   3sg   thing   REL   1pl.INCL=IRR.NS-PFV
têk=Ø$_O$]$_{RC}$]$_{NVPRED}$ e
tell=3sg.ZERO   TAG
The story, it's a thing we tell, right?

(3) *Mai re rapkasa rêk la mangsilan pwên.*
ma i te tap$_A$=ka-sa   têk=Ø$_O$ la   mangsilan   pwên
but   1pl.INCL=IRR.NS-MOD tell=3sg.ZERO go.to   meaningless NEG
But we cannot tell [it] just for nothing.

(4) *Te rapkaro yek salan la masayen palosi, a monokin a tapkape ro poposek.*
te   tap$_A$=ka-to   yek   [sale-n]$_O$ la   masayen palosi   a
SUB   1pl.INCL=IRR.NS-HAB hit   road-PERT   go.to   clear   past   and
monoki-n   ya   tap$_A$=ka-pe   to   pwapwasek=Ø$_O$
behind-PERT   then   1pl.INCL=IRR.NS-PFV HAB   speak.out=3sg.ZERO
For ususally we clarify its history [*lit.* 'road'] first, and afterwards we will tell [it].

(5) *Wong tepwo, wong Lalau Kanau.*
[wong te-pwo]₍ₜₒₚₛ₎ [wong]₍ₛ₎ [Lalau Kanau]₍ₙᵥₚᵣₑᴅ₎
1sg.FREE LIG-DEM.PROX 1sg.FREE Lalau Kanau
Me here, I am Lalau Kanau.

(6) *Tinang teo i Alup Nareng, i pari Parioi, Um Dan Sikai.*
[tina-ng te-yo]₍ₛ₎ i [Alup Nareng]₍ₙᵥₚᵣₑᴅ₎ i
mother-1sg.PERT LIG-DEM.INT 3sg Alup Nareng 3sg
[pari Parioi wumwa ta-n Sikai]₍ₙᵥₚᵣₑᴅ₎
belonging.to P. house CLF.POSS-PERT Sikai
My mother is Alup Nareng, she is from Parioi, the lineage of Sikai.

(7) *Tamong teo i Kanau, Kanau Kanawi.*
[tama-ng te-yo]₍ₛ₎ i [Kanau Kanau Kanawi]₍ₙᵥₚᵣₑᴅ₎
father-1sg.PERT LIG-DEM.INT 3sg Kanau Kanau Kanawi
My father is Kanau, Kanau Kanawi.

(8) *I pari Parilou, Munukut sêk paye.*
i [pari Parilou Munukut sêk paye]₍ₙᵥₚᵣₑᴅ₎
3sg belonging.to Parilou Munukut half.long down.below
He is from Parilou, the lower half of Munukut [located lower on the mountain].

(9) *Um turê reo, urê, urê mwat.*
wumwa ta-urê te-yo [wurê]₍ₛ₎ wurê
house CLF.POSS-1pc.EXCL.PERT LIG-DEM.INT 1pc.EXCL 1pc.EXCL
[mwat]₍ₙᵥₚᵣₑᴅ₎
bandicoot
[As for] our lineage, we are like the bandicoot.

(10) *Urê mwat, la are ipkape arei rurê la mwoyai.*
[wurê]₍ₛ₎ [mwat]₍ₙᵥₚᵣₑᴅ₎ la a=te [ip₍ₛ₎=ka-pe arei
1pc.EXCL bandicoot go.to OBL=SUB 3pl=IRR.NS-PFV say
[ta-urê]₍ₑ₎ la mwoyai]₍ₐ𝒹ᵥCl₎
CLF.POSS-1p.EXCL.PERT go.to peace
We are like the bandicoot, so they will call us peaceful.

(11) *Pwapwa reo ngapwa korêk tepwo...*
 [pwapwae te-yo]_TopO nga_A=pwa ko-têk=Ø te-pwo
 story LIG-DEM.INT 1sg=want.to IRR.1sg-tell=3sg.ZERO LIG-DEM.PROX
 The story I am going to tell now...

(12) *Pwapwa repwo i pari ai yoy re ngayan Parulabei, a yoy nisip te ngayan yoy Komou.*
 [pwapwae te-pwo]_S i [pari a=i yoy te
 story LIG-DEM.PROX 3sg belonging.to OBL=3sg stone REL
 ngaya-n Parulabei a yoy ni-sip te ngaya-n
 name-PERT Parulabei and stone other-one.INANIM REL name-PERT
 yoy Komou]_NVPRED
 stone Komou
 This story is about a stone called Parulabei and another stone called Komou.

(13) *A yoy reo, i mare kapwa sip pusungop te iro Paluai repwo re kapwa som yamasen te ipwak ai pwên.*
 a [yoy_i te-yo]_TopObl i ma=te kapwa sip pusungop_j
 and stone LIG-DEM.INT 3sg NEG_1=SUB if one.INANIM clan
 [te i_S=to Paluai]_RC te-pwo te kapwa som
 REL 3sg=be Paluai LIG-DEM.PROX SUB if one.ANIM
 yamase-n_j [te i_S=pwak a=i_i]_RC pwên
 person-PERT REL 3sg=meet OBL=3sg NEG_2
 These stones, it is supposedly not the case that there is a member of any clan on Baluan Island who encountered them.

(14) *Te yoy reo, yamat te i pari ai pusungop turê tepwo mwanen teo ipwak ai*
 te [yoy_i te-yo]_TopObl [yamat [te [i]_S [pari a=i
 SUB stone LIG-DEM.INT person REL 3sg belonging.to OBL=3sg
 pusungop ta-urê te-pwo mwanenen]_NVPRED]_RC
 clan CLF.POSS-1pc.EXCL.PERT LIG-DEM.PROX straight
 [te-yo]_S i=pwak a=i_i
 LIG-DEM.INT 3sg=meet OBL=3sg
 As for these stones, this person who is straight from our clan encountered them.

(15) *Yamat te ipwak ai yoy reo, i reo i apuong, apuong te ngayan Lalau Alipul.*
[yamat [te i₍ₛ₎=pwak a=i yoy te-yo]₍ᴿC₎]₍ₜₒₚₛ₎ [i te-yo]₍ₛ₎
person REL 3sg=meet OBL=3sg stone LIG-DEM.INT 3sg LIG-DEM.INT
i [apua-ng]₍ₙᵥₚᴿᴇᴅ₎ apua-ng [te
3sg great.grandparent-1sg.PERT great.grandparent-1sg.PERT REL
[ngaya-n]₍ₛ₎ [Lalau Alipul]₍ₙᵥₚᴿᴇᴅ₎]₍ᴿC₎
name-PERT Lalau Alipul
As for the person who encountered these stones, he is my great-grandfather; my great-grandfather who is named Lalau Alipul.

(16) *Lalau Alipul leo irou Ngat Kanawi.*
[Lalau Alipul te-yo]₍ₐ₎ i=tou [Ngat Kanawi]₍ₒ₎
Lalau Alipul LIG-DEM.INT 3sg=give.birth.to Ngat Kanawi
This Lalau Alipul begot Ngat Kanawi.

(17) *Ngat Kanawi irou Kanau Alipul liliu.*
[Ngat Kanawi]₍ₐ₎ i=tou [Kanau Alipul]₍ₒ₎ liliu
Ngat Kanawi 3sg=give.birth.to Kanau Alipul again
In turn Ngat Kanawi begot Kanau Alipul.

(18) *Kanau Alipul leo ipe rou Lalau Alipul le wong tepwo Lalau Kanau.*
[Kanau Alipul te-yo]₍ₐ₎ i=pe tou [Lalau Alipul le
Kanau Alipul LIG-DEM.INT 3sg=PFV give.birth.to Lalau Alipul or
wong te-pwo Lalau Kanau]₍ₒ₎
1sg.FREE LIG-DEM.PROX Lalau Kanau
Kanau Alipul begot Lalau Alipul or [that is to say] me here, Lalau Kanau.

(19) *Te wong tepwo rea, ngaro rêk pwapwa repwo.*
te [wong te-pwo te-ya]₍ꜰₒc:ₐ₎ nga₍ₐ₎=to têk [pwapwae
SUB 1sg.FREE LIG-DEM.PROX LIG-then 1sg=HAB tell story
te-pwo]₍ₒ₎
LIG-DEM.PROX
So it is *me* who usually tells this story.

(20) *A pwapwa reo ila remenan telo.*
a [pwapwae te-yo]₍ₛ₎ i=la temenin te-lo
and story LIG-DEM.INT 3sg=go.to like LIG-DEM.DIST
And the story goes like this.

(21) *Palosi reo ilak, a ip yamat nisopol leo, ipto pei si sopol suk tepwo ai peinan net.*
palosi te-yo   i=lak   ya   [ip yamat ni-sopol
past   LIG-DEM.INT 3sg=go.to   then   3pl person other-one.half.round
te-yo]$_{\text{TopS}}$ ip$_S$=to pei si   sopol   suk
LIG-DEM.INT   3pl=HAB   come come.down one.half.round shore
te-pwo   a=i   peinan   net
LIG-DEM.PROX   OBL=3sg   making.of   sea
A long time ago, the people who are dead now [*lit.* 'the people on the other side'], they used to come down to the seaside here, to do fishing.

(22) *Ma i re napun ip teo i pulêng, te ipkaro pe mayai net teo.*
ma i te [napu-n   ip   te-yo]$_S$ i   [pulêng]$_{\text{NVPRED}}$ te
but   taboo-PERT   3pl LIG-DEM.INT 3sg   daybreak   SUB
ip$_A$=ka-to   pe   mayai   [net   te-yo]$_O$
3pl=IRR.NS-HAB   make   be.quick   sea   LIG-DEM.INT
But it was forbidden for them to do it at daybreak [*lit.* 'their taboo was the daybreak'], so they used to do the fishing quickly.

(23) *La are kapwa net teo kipepwên, a ipkaliliu la lopwan ip panuan te pulêng teo kipe masai.*
la   a=te   kapwa   [net te-yo]$_S$ ki-pepwên   ya
go.to   OBL=SUB   if   sea   LIG-DEM.INT   IRR.3sg-be.finished   then
ip$_S$=ka-liliu   la   lopwa-n   ip   [panua-n   te   [pulêng
3pl=IRR.NS-return   go.to   place-PERT   3pl   front-3sg.PERT SUB   dawn
te-yo]$_S$ ki-pe   masai]$_{\text{AdvCl}}$
LIG-DEM.INT   IRR.3sg-PFV   be.clear
When the fishing would be finished, they would go back to their place before the day would break.

(24) *A ip aronan ip temenan teo, ipan pe ai pêng naringiai re ilak.*
a   [ip   arona-n   ip   temenin   te-yo]$_{\text{TopO}}$ ip$_A$=an
and   3pl   procedure-PERT   3pl   like   LIG-DEM.INT   3pl=PRF
pe=Ø   [a=i   pêngi   naringiai   te   i=lak]$_{\text{AdvCl}}$
do=3sg.ZERO   OBL=3sg   day   many   REL   3sg=go
These procedures of theirs like that, they have followed [them] for many days that went by.

(25) *Ma pêng sip teo, ippe pe kasiek.*
ma   pêngi sip   te-yo   ip$_A$=pe   pe=Ø   kasiek
but day   one.INANIM   LIG-DEM.INT 3pl=PFV   do=3sg.ZERO   not.well
But this one day, things went awry. [*lit.* 'they did [it] incorrectly']

(26) *Iplak te ipto pe ner a yamat te ngayan Parulabei reo, iro ning are panu reo iro masai.*
ip$_S$=lak   te   ip$_A$=to   pe   [net]$_O$ a   [yamat te   [ngaya-n]$_S$
3pl=go   SUB   3pl=CONT   make   sea   and   person REL   name-PERT
[Parulabei]$_{NVPRED}$ te-yo]$_{Tops}$ i$_S$=to   ning [a=te   [panua
Parulabei   LIG-DEM.INT   3sg=CONT   see   OBL=COMP   place
te-yo]$_S$ i=to   masai]$_{Compl:Obl}$
LIG-DEM.INT   3sg=CONT   be.clear
They went to do fishing and the person whose name was Parulabei, he saw the sky getting lighter.

(27) *A ipe pe pwaypwayei raip.*
a   i$_A$=pe   pe   [pwai.pwayei]$_O$ [ta-ip]$_E$
and   3sg=PFV   make   REDUP.fast   CLF.POSS-3pl.PERT
And he hurried them up.

(28) *Ipwa, "Rapkamaleu," ipul la ran yamar i re ngayan Komou reo.*
i$_A$=pwa   [tap$_S$=ka-maleu]$_{Compl:O}$ i=pul=Ø   la
3sg=say   1pl.INCL=IRR.NS-hurry   3sg=tell=3sg.ZERO go.to
[ta-n   yamat   i   [te   [ngaya-n]$_S$ [Komou]$_{NVPRED}$]$_{RC}$
CLF.POSS-PERT   person 3sg REL   name-3sg.PERT   Komou
te-yo]$_{AG}$
LIG-DEM.INT
He said, "We should hurry," he told the person who was named Komou.

(29) *Ipwa,"Rapkamaleu, re panu ro masai o."*
i$_A$=pwa   [tap$_S$=ka-maleu   te   [panu]$_S$ to   masai   yo]$_{Compl:O}$
3sg=say   1pl.INCL=IRR.NS-hurry   SUB   place   CONT   be.clear DEM.INT
He said, "We should hurry, because the sky is getting light."

(30) *ma Komou reo ipwa, "Pwên, pulêng teo imala siei sonean papwên."*
ma   [Komou te-yo]$_A$ i=pwa   [pwên   [pulêng   te-yo]$_S$
but   Komou   LIG-DEM.INT   3sg=say   NEG   daybreak   LIG-DEM.INT

```
        i=ma=la        siei   sonea-n    pa-pwên]Compl:O
        3sg=NEG₁=go.to tear   near-PERT  yet-NEG₂
```
But Komou said, "No, the day is not about to break yet."

(31) *I ya, Parulabei reo ino yauyau wot tai o.*
```
     i    ya   [Parulabei  te-yo]S    i=no     yauyau  wot
     3sg  then Parulabei   LIG-DEM.INT 3sg=IPFV bend    go.level
     [ta-i]AG     yo
     CLF.POSS-3sg DEM.INT
```
Well, Parulabei gave in to him.

(32) *A ipno ru pe net lak lak lak.*
```
     a    ipA=no   tu         pe   [net]O  lak lak lak
     and  3pl=IPFV STAT.CONT  make sea     go  go  go
```
And they continued fishing for a while.

(33) *Ma Komou reo ipwa kining nonot te nuanan te pulêng teo iro masai me a ipe pwa, "Kay, rap."*
```
     ma  Komou  te-yo]A      i=pwa        ki-ning     nonot      [te
     but Komou  LIG-DEM.INT  3sg=want.to  IRR.3sg-see understand COMP
     nuanan te   [pulêng   te-yo]S      i=to      masai     me]Compl:O
     truth  SUB  daybreak  LIG-DEM.INT  3sg=CONT  be.clear  come
     a    iA=pe    pwa  [kay  tap]Compl:O
     and  3sg=PFV  say  okay  1pl.INCL
```
But Komou, he was about to understand that indeed dawn was approaching, and he said, "Okay, let's go."

(34) *Ippe nêm net a iplêp kaip nik.*
```
     ipA=pe       nêm          [net]O  a   ipA=lêp   [ka-ip            nik]O
     3pl=make     be.finished  sea     and 3pl=take  CLF.food-3pl.PERT fish
```
They stopped fishing and took their fish.

(35) *Ipno ro rer a rer a rer a ret.*
```
     ipS=no    to    tet   a    tet   a    tet   a    tet
     3pl=IPFV  CONT  move  and  move  and  move  and  move
```
They walked and walked.

(36) *Ma no nansê te ipto pwa kawau worup la Malsu, are panu raip teo lalon Malsu reo...*
ma no nan=sê [te ip_S=to pwa ka-wau worup
but only PART=DIM SUB 3pl=CONT want.to IRR.NS-walk descend
la Malsu]_{AdvCl} [a=te [panu ta-ip te-yo]_S
go.to Malsu OBL=SUB village CLF.POSS-3pl.PERT LIG-DEM.INT
[lalo-n Malsu te-yo]_{NVPRED}]_{AdvCl}
inside-PERT Malsu LIG-DEM.INT
But when they were almost about to go down into Malsu (since their village was inside Malsu [the crater of the volcano])...

(37) *... Ipto pwa kaworup lak a panu reo ipe wot masai, a i o.*
ip_S=to pwa ka-worup lak ya [panu te-yo]_S
3pl=CONT want.to IRR.NS-descend go then place LIG-DEM.INT
i=pe wot masai a i yo
3sg=PFV go.level be.clear and 3sg DEM.INT
[When] they were about to go down into the village, the sun came up. There.

(38) *Iwot masai re onga nopok pe wot pei rau lapanan i pwo.*
i_S=wot masai te onga [nopok]_S pe wot pei
3sg=go.level be.clear SUB and.so argument PFV go.level come
[ta-u lapana-n]_E i pwo
CLF.POSS-3du.PERT chief-PERT 3sg DEM.PROX
The sun came up, and so an argument was developing between the two chiefs here.

(39) *Nopok te ipei reo, ipei moyengan Parulabei a Komou.*
[nopok [te i_S=pei te-yo]_{RC}]_{TopS} i_S=pei moyenga-n
argument REL 3sg=come LIG-DEM.INT 3sg-come middle-PERT
Parulabei a Komou
Parulabei and Komou
The argument that came up, it came up between Parulabei and Komou.

(40) *La are Parulabei reo ipul la rai a ipwa, "Pwên, a ngan pul nipêng..."*
la a=te Parulabei te-yo]_A i=pul=Ø la
go.to OBL=SUB Parulabei LIG-DEM.INT 3sg=tell=3sg.ZERO go.to
[ta-i]_{AG} a i_A=pwa [pwên a nga=an pul
CLF.POSS-3sg.PERT and 3sg=say NEG and 1sg=PRF tell

ni-pêngi]~Compl:O~
other-time
Because Parulabei spoke to him [Komou] and said, "No, I have said another time..."

(41) *"Le ngan pul nirasip, te ngapwa, 'Pulêng to masai o...'*
le nga$_A$=an pul=Ø panurasip [te nga$_A$=pwa [pulêng
or 1sg=PRF tell=3sg.ZERO first SUB 1sg=say daybreak
to masai]~Compl:O~ yo]
CONT be.clear DEM.INT
"I have told you before, I said, "The day is breaking...""

(42) *"Womaro yong pwên."*
wo$_A$=ma=to yong=Ø$_O$ pwên
2sg=NEG$_1$=CONT hear=3sg.ZERO NEG$_2$
"[But] you weren't listening [to what I said]."

(43) *Onga upe pe kulkulu.*
onga u$_A$=pe pe [kul.kulua]$_O$
and.so 3du=PFV make REDUP.quarrel
And so they were quarreling.

(44) *Kulkulu re upe reo, kulkulu ipei lak lak lak a ipei la nin.*
[kul.kulua [te u=pe=Ø]$_{RC}$ te-yo]~TopS~ [kul.kulua]$_S$
REDUP.quarrel REL 3du=make=3sg.ZERO LIG-DEM.INT REDUP.quarrel
i$_S$=pei lak lak lak a i=pei la nine
3sg=appear go go go and 3sg=appear go.to fight
The quarrel that they had, it went on and on and it became a fight.

(45) *Te upe nin te paran koroan.*
te u$_A$=pe [nine te paran koroan]$_O$
SUB 3du=make fight REL really violent
They had a fight that was really violent.

(46) *A u yamat teo, u re u tinawayen yoy taywêp, u yoy taywêp te upe nin.*
a [u yamat te-yo]~TopS~ u te [u]$_S$ [tinawayen yoy
and 3du person LIG-DEM.INT 3du REL 3du huge stone
ta=yuêp]~NVPRED~ [u]$_S$ [yoy ta=yuêp [te
SPEC.COLL=two.INANIM 3du stone SPEC.COLL=two.INANIM REL

u_A=pe   [nine]_O]_RC]_NVPRED
3du=make   fight
And those two persons, they were two huge stones, two stones that fought.

(47) *A kila nel le iro areo, murin.*
    a   ki-la   neli   [te   i_S=to   a=te-yo]_RC   murin
    and IRR.3sg-go.to rope REL 3sg=be OBL=LIG-DEM.INT broken
    The vines that were there would turn out to be snapped.

(48) *Kila kei re iro areo onga, pôrin.*
    ki-la   kei   [te   i_S=to   a=te-yo]_RC   onga   pôrin
    IRR.3sg-go.to tree REL 3sg=be OBL=LIG-DEM.INT and.so broken
    The trees that were there would turn out to be broken.

(49) *Uno pe lak lak lak a mamaroun ip palosi...*
    u_A=no   pe=Ø_O   lak lak lak a   mamarou-n ip   palosi
    3du=IPFV make=3sg.ZERO go go go and way-PERT 3pl past
    They were doing [this] for a long long time, and [according to] the ways of those before...

(50) *Yamat te ngayan Lalau Alipul le irou Ngat Kanawi a Ngat Kanawi ipe rou tamong teo...*
    yamat [te [ngaya-n]_S [Lalau Alipul]_NVPRED] [te i_S=tou
    person REL name-PERT Lalau Alipul   REL 3sg=give.birth.to
    [Ngat Kanawi]_O]_RC a   [Ngat Kanawi]_S i=pe   tou
    Ngat Kanawi   and Ngat Kanawi 3sg=PFV give.birth.to
    [tama-ng   te-yo]_O
    father-1sg.PERT LIG-DEM.INT
    [like] this man whose name was Lalau Alipul, who gave birth to Ngat Kanawi and Ngat Kanawi gave birth to my father...

(51) *Pasin taip palosi reo konan ip ngan kein mangat to kanen ip o.*
    pasin ta-ip   palosi te-yo   konan   ip=ngan
    way   CLF.POSS-3pl.PERT past   LIG-DEM.INT such.as 3pl=eat
    kei-n   mangat to kane-n   ip   yo
    magic.herbs-PERT work   be body-PERT 3pl DEM.INT
    The ways of those before were such that they ate stimulating herbs to help them work.

(52) *Ma kilot wot patpat, wosave iro mari la pian te pwên.*
ma ki$_S$-lot   wot   patpat wo=save   i$_S$=to   mari   la
and IRR.3sg-fall go.level bed   2sg=know 3sg=HAB be.asleep go.to
pian te pwên
good SUB neg
And when he [Lalau Alipul] would go lie down in bed, you know he did not sleep too well.

(53) *A in sok samel lai are ipe mari o.*
a   i$_A$=an   sok   [samel ta-i]$_O$ a=te   i$_S$=pe
and 3sg=PRF sharpen knife   CLF.POSS-3sg OBL=SUB 3sg=PFV
mari   yo
be.asleep DEM.INT
And he had already sharpened his knife when he went to sleep.

(54) *A in sarek lak to pwoyon patpat.*
a   i$_A$=an   satek=Ø$_O$ lak to pwoyo-n   patpat
and 3sg=PRF put.into=3sg.ZERO go   be under-PERT bed
And had put [it] under the bed.

(55) *Ma pwempwem som to reng a isirak.*
ma   [pwempwem som]$_S$ to   teng a   i$_S$=sirak
EMP bird.sp   one.ANIM CONT cry and 3sg=get.up
A *pwempwem* bird was crying, and he woke up.

(56) *Isirak a iro au wot sal ken teo.*
i$_S$=sirak   a   i$_S$=to   wau   wot   sale ke-n   te-yo
3sg=get.up and 3sg=CONT walk go.level road leg-PERT LIG-DEM.INT
He got up and he was walking along this little track.

(57) *iro au wor a ipe yong tukuan u lapan taymou re uro pe nin.*
i$_S$=to   wau   wot   a   i$_A$=pe   yong [tungua-n   u
3sg=CONT walk go.level and 3sg=PFV hear tremor-PERT 3du
lapan ta=yumou   [te   u$_A$=to   pe   [nine]$_O$]$_{RC}$]$_O$
chief SPEC.COLL=two.ANIM REL 3du=CONT make fight
He was walking and he heard a tremor of the two chiefs who were fighting.

(58) *Iyong are uro ret mut nel, uret pông kopup a kei iro pôpôr.*
i$_S$=yong [a=te u$_A$=to tet mut [neli]$_O$ u$_A$=tet
3sg=hear OBL=COMP 3du=CONT walk break rope 3du=walk
pông [kopup]$_O$]$_{Compl:Obl}$ a [kei]$_S$ i=to pô.pôt
popping.sound bamboo and tree 3sg=CONT REDUP.break
He heard them noisily breaking vines and bamboo, and trees were breaking.

(59) *A i a, imaret lai lopwan mangat tai reo pwên.*
a i ya i$_S$=ma=tet la a=i [lopwa-n mangat
and 3sg then 3sg=NEG$_1$=walk go.to obl=3sg place-PERT work
ta-i te-yo]$_{LG}$ pwên
CLF.POSS-3sg.PERT LIG-DEM.INT NEG$_2$
And well, he did not go to the place of his work.

(60) *Ma ipe nengtou.*
ma i$_S$=pe neng.tou
but 3sg=PFV stop.walking
But he stopped walking.

(61) *Tareo, nengnengtou reo, ino pe.*
[ta=te-yo]$_{TopO}$ [neng.neng.tou te-yo]$_{TopO}$ i$_A$=no
SPEC.COLL=LIG-DEM.INT REDUP.stop.walking LIG-DEM.INT 3sg=IPFV
pe=Ø$_O$
do=3sg.ZERO
He stopped walking there. [*lit.* 'That, stopping, he did.']

(62) *Ma i re imaru mapwai la are kapwa u lapan teo, uro pe pun kuluan sa, le upe pun ninen sa pwên.*
ma i te i$_S$=ma=tu mapwai la [a=te kapwa [u
but 3sg=NEG$_1$=STAT.CONT know go.to OBL=COMP if 3du
lapana te-yo]$_{TopA}$ u$_A$=to pe pun [kulua-n sa]$_O$ le
chief LIG-DEM.INT 3du=CONT make INTF quarrel-PERT what or
u$_A$=pe pun [nine-n sa]$_O$]$_{Compl:Obl}$ pwên
3du=make INTF fight-PERT what NEG$_2$
But he did not know what exactly the two chiefs were apparently arguing or quarreling about.

(63) *Ma aronan te ula pe nêm nin o, a ipe yong lai poposekan kamou o:*
ma arona-n [te u$_A$=la pe nêm [nine]$_O$ yo]$_{RC}$
but procedure-PERT REL 3d=go.to make be.finished fight DEM.INT
a i$_A$=pe yong [la a=i pwapwasek-an kamou]$_O$ yo
and 3sg=PFV hear go.to OBL=3sg speak.out-NOM speech DEM.INT
But (from) the way in which they concluded their fight, and [what] he heard from their talking:

(64) *Komou reo isok sau Parulabei sopol.*
[Komou te-yo]$_A$ i=sok sau [Parulabei sopol]$_O$
Komou LIG-DEM.INT 3sg=strike cut Parulabei one.half.round
Komou struck off one half of Parulabei.

(65) *Minak tepwo, kapwa kalak...*
minak te-pwo kapwa ka$_S$-lak
present.time LIG-DEM.PROX if IRR.NS-go.to
Nowadays, if one goes there...

(66) *Asilon sopol le isok sau re ila yen pwapwan teo, Parulabei reo ipe apek tan Komou reo.*
[asilo-n sopol [te i=sok sau=Ø [te i=la
side-PERT one.half.round REL 3sg=hit cut=3sg.ZERO REL 3sg=go.to
yen pwapwa-n]$_{RC}$]$_{RC}$ te-yo]$_{TopO}$ [Parulabei te-yo]$_A$ i=pe
lie side-PERT LIG-DEM.INT Parulabei LIG-DEM.INT 3sg=PFV
apek=Ø [ta-n Komou te-yo]$_E$
hit=3sg.ZERO CLF.POSS-PERT Komou LIG-DEM.INT
This half part of him [Parulabei] that he [Komou] struck off, which came to lie at his [Parulabei's] side, Parulabei whacked [it] into Komou.

(67) *Iruk antek takaui re ila lot te ila yen lalon Malsu relo.*
i$_A$=tuk antek takau=i$_O$ te i$_S$=la lot te
3sg=beat put.away altogether=3sg SUB 3sg=go.to fall SUB
i$_S$=la yen lalo-n Malsu te-lo
3sg=go.to lie inside-PERT Malsu LIG-DEM.DIST
He knocked him away altogether, so that he [Komou] fell and came to lie down inside Malsu.

(68) *Ma ila lot lak a Parulabei reo ipe arei la suwot tai la remenan telo:*
ma   i$_S$=la   lot lak a   [Parulabei te-yo]$_A$ i=pe   arei
EMP 3sg=go.to fall go and   Parulabei LIG-DEM.INT 3sg=PFV say
la   suwot   [ta-i]$_{AG}$ la   temenin te-lo
go.to go.down   CLF.POSS-3sg.PERT   go.to   like   LIG-DEM.DIST
He fell down, and Parulabei spoke to him as follows:

(69) *"On suwot, a wope la lopwan te lumlum kipe nengo."*
wo$_S$=an suwot   a   wo$_S$=pe la   lopwa-n   [te
2sg=PRF   go.downwards   and   2sg=PFV   go.to place-PERT   REL
[lumlum]$_A$ ki-pe   neng=o$_O$]$_{RC}$
moss   IRR.3sg-PFV climb=2sg
"You have gone down, and you will go to a place where moss will grow on you."

(70) *"Leut kipe ret pwono, a ip lau kasa ro ning muyom pwên."*
[leut]$_A$ ki-pe   tet   pwon=o$_O$ a   [ip laue]$_A$
weed   IRR.3sg-PFV grow   be.covered=2sg   and   3pl people
ka-sa   to   ning [muya-m]$_O$ pwên
IRR.NS-MOD HAB   see   skin-2sg.PERT   NEG
"Weeds will cover you, and the people will not be able to see your skin."

(71) *"Ma wong, wong kope ro wat."*
ma   [wong]$_{TopS}$ [wong]$_S$ ko-pe   to   wat
but   1sg.FREE   1sg.FREE   IRR.1sg-PFV   be   up.high
"But I, I will be up high."

(72) *A ip laulau karo au kasông; ipkaro rou malalông, lapan a sayo.*
a   [ip lau.laue]$_S$ ka-to   wau   kaso-ng
and   3pl REDUP.people   IRR.NS-HAB   walk   near-1sg.PERT
ip$_A$=ka-to   tou   [malala-ng]$_O$ lapana   a   sayo
3pl=IRR.NS-HAB   put   clearness-1sg.PERT   chief   and   ordinary.man
"And all the people will come near to me; they will clear me [of weeds etc.], chiefs and ordinary people."

(73) *"Kiru au ma kikunawayut patung. Monokin a kisirak liliu a kipe ret lai rare ipwa kila ai."*
ki$_S$-tu   wau   ma   ki$_S$-kunawayut   pata-ng
IRR.3sg-STAT.CONT   walk   and   IRR.3sg-take.rest on.top-1sg.PERT

monoki-n   ya   ki$_S$-sirak   liliu   a   ki$_S$-pe   tet   la
behind-PERT then IRR.3sg-get.up again and IRR.3sg-PFV move go.to
a=i   ta=te-yo   i$_S$=pwa   ki-la   a=i
OBL=3sg SPEC.COLL=LIG-DEM.INT 3sg=want.to IRR.3sg-go.to OBL=3sg
"[Someone] will come and take a rest on top of me. Afterwards, he will get up again and he will go wherever he wants to go."

(74) *Lalau Alipul leo iyong nêm kamou rau, nopok tau reo, a ining nêm nin tau*
[Lalau Alipul te-yo]$_A$ i=yong   nêm   [kamou
Lalau Alipul LIG-DEM.INT 3sg=hear be.finished speech
ta-u   nopok   ta-u   te-yo]$_O$ a
CLF.POSS-3du.PERT argument CLF.POSS-3du.PERT LIG-DEM.INT and
i$_A$=ning nêm   [nine ta-u]$_O$
3sg=see be.finished fight CLF.POSS-3du.PERT
Lalau Alipul finished hearing their speech, their argument, and he watched their fight until it was over.

(75) *A imaret la lopwan mangat tai reo pwên.*
a   i$_S$=ma=tet   la   [lopwa-n   mangat ta-i
and 3sg=NEG$_1$=walk go.to place-PERT work CLF.POSS-3sg.PERT
te-yo]$_{LG}$ pwên
LIG-DEM.INT NEG$_2$
And he did not go to his work place.

(76) *Pari ai re puron te ining teo i puron paran poun, te puron puleau sip.*
pari   a=i   te   [puron [te   i$_A$=ning=Ø$_O$]$_{RC}$ te-yo]$_S$
belonging.to OBL=3sg SUB thing REL 3sg=see=3sg.ZERO LIG-DEM.INT
i   [puron paran poun te   puron puleau sip]$_{NVPRED}$
3sg thing really new SUB thing unusual one.INANIM
Because (of) this thing that he had seen, it was a really new thing, an unusual thing.

(77) *Ma in, sê yamat naringiai ining yoy re in pe nin, a iyong sê ngolan yoy, le pwên.*
ma in   [sê   yamat naringiai]$_A$ i=ning   [yoy   te   [i$_A$=an
don't.know who person many   3sg=see stone REL 3sg=PRF
pe   [nine]$_O$]$_{RC}$]$_O$ a   i$_A$=yong sê   [ngola-n   yoy]$_O$ le pwên
make fight   and 3sg=hear DIM language-PERT stone or NEG
I don't know whether a lot of people have seen stones that had a fight and heard a little of the stones' language, or not.

(78) *Ma i re apuong te narun Alipul leo, ining pun te i mare kapwa i pwapwa mangsilan pwên, a imala kanum tai pwên o.*
ma i te [apua-ng [te naru-n Alipul te-yo]_{RC}]_{TopS}
but great.grandparent-1sg.PERT REL son-PERT Alipul LIG-DEM.INT
i_S=ning pun te [i ma=te kapwa [i]_S [pwapwae
3sg=see INTF COMP 3sg NEG₁=SUB if 3sg story
mangsilan]_{NVPRED} pwên]_{Compl:O} a i_S=ma=la [kanum
meaningless NEG₂ and 3sg=NEG₁=go.to garden
ta-i]_{LG} pwên yo
CLF.POSS-3sg.PERT NEG₂ DEM.INT
But my great-grandfather, who was the son of Alipul, he clearly saw that it was not as if this was a meaningless story, and [so] he did not go to his garden.

(79) *A ipe liliu a ililiu la um tairê.*
a i_S=pe liliu a i_S=liliu la [wumwa ta-irê]_{LG}
and 3sg=PFV return and 3sg=return go.to house CLF.POSS-3pc.PERT
And he went back to their house.

(80) *Ililiu la um tairê a ipe la rêk pwapwaen la raip pusulop tai reo.*
i_S=liliu la [wumwa ta-irê]_{LG} a i_A=pe la
3sg=return go.to house CLF.POSS-3pc.PERT and 3sg=PFV go.to
têk [pwapwae-n]_O la [ta-ip pusungop ta-i
tell story-PERT go.to CLF.POSS-3pl.PERT clan CLF.POSS-3sg
te-yo]_{AG}
LIG-DEM.INT
He went back to their house and he told the story to his family.

(81) *Minak tepwo, pwapwa reo ipe ro lalon Paluai, are Parulabei a Komou reo upe nin palosi re ilak.*
minak te-pwo [pwapwae te-yo]_S i=pe to
present.time LIG-DEM.PROX story LIG-DEM.INT 3sg=PFV be
lalo-n Paluai [a=te [Parulabei a Komou te-yo]_A
inside-PERT Paluai OBL=SUB Parulabei and Komou LIG-DEM.INT
u=pe [nine]_O palosi te i=lak]_{AdvCl}
3du=make fight past REL 3sg=go.to
Nowadays, this story has been known on Baluan, about how Parulabei and Komou fought a long time ago.

(82) *Kamou reo Parulabei iarei suwot tan Komou reo, minak tepwo i nuanan.*
[[kamou te-yo]_TopO   [Parulabei]_A   i=arei=Ø   suwot
speech LIG-DEM.INT   Parulabei   3sg=say=3sg.ZERO   go.down
ta-n   [Komou te-yo]_AG]_TopS   minak   te-pwo   i
CLF.POSS-PERT   Komou   LIG-DEM.INT   present.time   LIG-DEM.PROX   3sg
[nuanan]_NVPRED
truth
The speech that Parulabei made to Komou, nowadays it is true.

(83) *Komou reo, no pwapwaen teo re rapto yong, ma pari ai re tapmaro ning muyan pwên.*
[Komou te-yo]_TopPoss   [no   pwapwae-n   te-yo]_TopO   te
Komou   LIG-DEM.INT   only story-PERT LIG-DEM.INT   REL
tap_A=to   yong=Ø_O   ma   pari   a=i   te
1pl.INCL=HAB   hear=3sg.ZERO EMP belonging.to OBL=3sg   SUB
tap_A=ma=to   ning   [muya-n]_O   pwên
1pl.INCL=NEG_1=HAB   see   skin-PERT   NEG_2
Regarding Komou, it is only his story that we usually hear as to why we do not see his skin.

(84) *Parulabei iro wat a rapto au ai a rapto ning muyan.*
Parulabei   i_S=to   wat   a   tap_S=to   wau   a=i   a
Parulabei   3sg=be   high   and   1pl.INCL=HAB   walk   OBL=3sg   and
tap_A=to   ning   [muya-n]_O
1pl.INCL=HAB   see   skin-PERT
Parulabei is high up and we go to him and see his skin.

(85) *A rapto kunawayut patan a rappe ro ret lai rare rappa kala ai.*
a   tap_S=to   kunawayut   pata-n   a   tap_S=pe   to
and   1pl.INCL=HAB   take.rest   on.top-PERT   and   1pl.INCL=PFV   HAB
tet   la   a=i   ta=te-yo   tap_S=pwa
move   go.to   OBL=3sg   SPEC.COLL=LIG-DEM.INT   1pl.INCL=want.to
ka-la   a=i
IRR.NS-go.to   OBL=3sg
And we take a rest on top of him and we go to whatever we intend to go to.

(86) *Pwapwa reo iyen Paluai la remenan de ino rok a ino rok a ino rok a ino rok au me me me me me me.*
[pwapwae te-yo]$_S$ i=yen Paluai la temenin te i$_S$=no
story LIG-DEM.INT 3sg=lie Paluai go.to like SUB 3sg=IPFV
tok a i$_S$=no tok a i$_S$=no tok a i$_S$=no
stay and 3sg=IPFV stay and 3sg=IPFV stay and 3sg=IPFV
tok wau me me me me me me
stay move come come come come come come
This story has been on Baluan so that it stayed and stayed and stayed for a long long time...

(87) *A monok pun tepwo a Lapanin Lapanin Solok ipe me sukeleyeki la polpolot.*
a monoki pun te-pwo ya [Lapanin Solok]$_A$ i=pe
and behind INTF LIG-DEM.PROX then Lapanin Solok 3sg=PFV
me sukeleyek=i$_O$ la polpoloti
come turn.around=3sg go.to chant.type
Until recently, Lapanin Solok turned it into a *polpolot* [traditional chant].

(88) *Ipe me pei la polpolor, onga...*
i$_A$=pe me pe=i$_O$ la polpoloti onga
3sg=PFV come make=3sg go.to chant.type and.so
He has made it into a *polpolot*, and so...

(89) *Polpolot teo, u re ukape pe polpolot teo, won yoy re iptaneki la polpolot teo...*
[polpolot te-yo]$_{TopO}$ [u]$_{TopA}$ te u$_A$=ka-pe pe [polpoloti
chant.type LIG-DEM.INT 3du REL 3du=IRR.NS-PFV do chant.type
te-yo won yoy [te ip$_A$=tanek=i$_O$ la polpoloti
LIG-DEM.INT history stone REL 3pl=create=3sg go.to chant.type
te-yo]$_{RC}$]$_O$
LIG-DEM.INT
(As for) this *polpolot*, the two that will chant [*lit.* 'do'] the *polpolot*, the history of the stones which was turned into a polpolot...

(90) *Kisawin, Alup Kisawin Pwaril, te i pên tan Ngat Kanawi re taman ining yoy reo, a Alup Pilan Pokut, te ukape pe polpolorin yoy reo.*
[Kisawin Alup Kisawin Pwaril [te [i]$_S$ [pên ta-n
Kisawin Alup Kisawin Pwaril REL 3sg daughter CLF.POSS-PERT

Ngat Kanawi]~NVPRED~ [te   [tama-n]~S~   i=ning   [yoy]~O~]~RC~]~RC~ te-yo]~A~
Ngat Kanawi   [REL father-PERT   3sg=see   stone   LIG-DEM.INT
a   [Alup Pilan Pokut]~A~   te   u~A~=ka-pe   pe [polpoloti-n
and   Alup Pilan Pokut   SUB   3du=IRR.NS-PFV   do   chant.type-PERT
yoy te-yo]~O~
yoy te-yo]~O~
[It is] Kisawin, Alup Kisawin Pwaril, who is the daughter of Ngat Kanawi whose father has seen these stones, and Alup Pilan Pokut, who are the ones who will sing the *polpolot* of the two stones.

(91) *Inêm.*
      i~S~=nêm
      3sg=be.finished
      It's finished.

Text 2 Planting yams (3 min 30 sec)

This story was told on 29 March 2011 by Ninou (45 years old), daughter of Ngat Selan of Lipan village and married to Kireng Wari of Pumbanin. It is a short procedural text about planting yams. Making a garden involves a lot of communal effort. Before planting can commence, the men go and clear a patch of forest for a garden (*apaput*). They remove the large trees and set fire to the remaining vegetation after it is sufficiently dry. Then the smaller branches are removed (*antek yut*). After that the women come in to remove remaining leaves and small branches (*arariyek*) and sweep the garden patch clear of debris (*sisiyek*). Finally, the men come in again to make mounds for planting the tubers (*kururupa*).

Yams are considered a prized food. Planting may only be done by women and is surrounded by many taboos and procedures. There is a special verb (*lolomêek*) to refer to planting yams, as opposed to a general verb (*pat*), used to refer to planting everything else. This text skips the garden preparation phase and goes straight to the planting phase, explaining the different procedures for planting *mwayen* 'yam' and *suei* 'short yam'. Often, the owner of the garden will ask several kinswomen to assist. They will receive cooked food (*nauwenan*) in return.

(1)   *Wong Ninou. Ninou parian Kireng teo.*
      [wong]~S~   [Ninou Ninou paria-n   Kireng te-yo]~NVPRED~
      1sg.FREE   Ninou   Ninou wife-PERT   Kireng LIG-DEM.INT
      I am Ninou. Ninou, the wife of Kireng.

(2) *Ngapwa kopoposek aronan lolomêek te rapto lolomêek.*
nga$_A$=pwa   ko-pwapwasek   [arona-n   lo.lomêek   [te
1sg=want.to   IRR.1sg-speak.out   procedure-PERT   REDUP.plant   REL
tap$_S$=to   lo.lomêek]$_{RC}$]$_O$
1pl.INCL=HAB   REDUP.plant
I am going to tell about the planting procedure according to which we plant yams.

(3) *A taim te rappwa kalolomêek o, woriu pit palsi.*
a   taim   [te   tap$_S$=pwa   ka-lo.lomêek   yo]$_{RC}$
and   time   REL   1pl.INCL=want.to   IRR.NS-REDUP.plant   DEM.INT
wo$_A$=tiu   pit=Ø$_O$ palosi
2sg=collect   be.close=3sg.ZERO   past
When we are going to plant yams, you will first put [them] together.

(4) *Suei kila ai napukun sip, mwayen kila ai napukun sip.*
[suei]$_S$ ki-la   a=i   [napukun sip]$_{LOC}$ [mwayen]$_S$
short.yam   IRR.3sg-go.to   OBL=3sg   heap   one.INANIM   yam
ki-la   a=i   [napukun sip]$_{LOC}$
IRR.3sg-go.to   OBL=3sg   heap   one.INANIM
The short yams will go onto one heap, the long yams will go onto one heap.

(5) *A woruptup.*
a   wo$_S$=tup.tup
and   2sg=REDUP.cover
And you cover [them].

(6) *Worou pokpok yêbin kanei, kila paye.*
wo$_A$=tou   [pokpok   yêpin kanei]$_O$ ki$_S$-la   [paye]$_{LOC}$
2sg=put   fern.species   leaf   chestnut   IRR.3sg-go.to   down.below
You will put a *kanei* leaf; [it] will go on the ground.

(7) *Monokin, worou suei suwot kôn a wope lêp yêpin nanaop a worou lak to kôn.*
monoki-n   wo$_A$=tou   [suei]$_O$ suwot   kô-n   a
behind-PERT   2sg=put   short.yam   go.downwards   on.top-PERT   and
wo$_A$=pe   lêp   [yêpin nanaop]$_O$ a   wo$_A$=tou=Ø$_O$ lak   to
2sg=PFV   take   leaf   tree.species   and   2sg=put=3sg.ZERO   go   be
[kô-n]$_{LOC}$
on.top-PERT

Afterwards, you put the short yams down on top and you will take a *nanaop* leaf and put [it] on top [of the yams].

(8) *I reo pari ai re kipe pe... mwayen teo, le suei reo, kipe arei pal kumun, a wope lomêek.*
    i   te-yo   pari   a=i   te   ki-pe   pe
    3sg LIG-DEM.INT belonging.to OBL=3sg SUB IRR.3sg-PFV make
    [mwayen te-yo   le suei   te-yo]$_A$ ki-pe
    yam   LIG-DEM.INT or short.yam LIG-DEM.INT IRR.3sg-PFV
    arei pal   [kumu-n]$_O$ ya   wo$_A$=pe lomêek=Ø$_O$
    bite break sprout-PERT then 2sg=PFV plant=3sg.ZERO
    This is because the long or short yams, when they sprout, you will plant [them].

(9) *Ila ai mwayen pulek temenin teo, wolêp pulek yêpin kanei, yêpin pokpok, yêpin kanei menengan teo.*
    i=la   a=i   mwayen pulêek temenin te-yo   wo$_A$=lêp
    3sg=go.to OBL=3sg yam   too   like   LIG-DEM.INT 2sg=take
    pulêek [yêpin kanei   yêpin pokpok   yêpin kanei
    too   leaf   chestnut leaf   fern.species leaf chestnut
    menengan te-yo]$_O$
    big   LIG-DEM.INT
    For long yams, it is the same. You also take this big *kanei* leaf.

(10) *Woangek palsi suwot paye; wope rou mwayen lak to kôn.*
    wo$_A$=angek=Ø$_O$ palosi suwot   [paye]$_{LOC}$
    2sg=spread.out=3sg.ZERO past   go.downwards down.below
    wo$_A$=pe tou [mwayen]$_O$ lak   to [kô-n]$_{LOC}$
    2sg=PFV put yam   go.to be on.top-PERT
    You spread [it] out on the ground; you will put the yams on top of it.

(11) *A wope lêp yêpin nanaop; worou la patan.*
    a   wo$_A$=pe lêp   [yêpin nanaop]$_O$ wo$_A$=tou=Ø   la
    and 2sg=PFV take leaf   tree.species   2sg=put=3sg.ZERO go.to
    [pata-n]$_{LOC}$
    on.top-PERT
    And you will take a *nanaop* leaf; you put [it] on top [of the yams].

(12) *Woruptup, ma pêng tarulêp tu wot nêm.*
wo_S=tup.tup   ma   [pêngi ta=tulêp]_S   tu
2sg=REDUP.cover and   day   SPEC.COLL=three.INANIM   STAT.CONT
wot   nêm
go.horizontally be.finished
You cover [them], and three days [have to] go by.

(13) *Wope lêp, onga wope lêp lik.*
wo_A=pe lêp=Ø_O   onga   wo_A=pe lêp   [lik]_O
2sg=PFV take=3sg.ZERO   and.so 2sg=PFV take carrying.bag
You will take [them], and take a carrying bag.

(14) *Wolêp lik a wolêp sapon kei nangin pari lalon kanum, wolêp yêpin môkei.*
wo_A=lêp [lik]_O   a   wo_A=lêp [sapo-n   kei   nangin
2sg=take carrying.bag and 2sg=take top-PERT   tree   smell
pari   lalo-n   kanum]_O wo_A=lêp [yêpin môkei]_O
belonging.to   inside-PERT   garden   2sg=take leaf   plant.sp
You take a carrying bag and you collect fragrant herbs for inside the garden, you take *môkei* leaves.

(15) *Yêpin ip môkei, a pwalingan kowei poron nik, kowei pang.*
yêpin ip   môkei   a   pwalinga-n kowei   poro-n   nik
leaf 3pl plant.sp and   with-PERT   plant.sp skeleton-PERT fish
kowei   pang
plant.sp rain
Leaves of *môkei*, together with *kowei poron nik* and *kowei pang*.

(16) *Wolêp nêmnêmti, putpuron teo nêmnêmti, onga wope lêp; wope san lalon lik tao.*
wo_A=lêp [nêmnêmti put.puron   te-yo   nêmnêmti]_O   onga
2sg=take all   REDUP.thing   LIG-DEM.INT   all   and.so
wo_A=pe lêp=Ø_O   wo_A=pe san=Ø_O   [lalo-n   lik
2sg=PFV take=3sg.ZERO   2sg=PFV cut=3sg.ZERO   inside-PERT   carrying.bag
ta-o]_LOC
CLF.POSS-2sg.PERT
You take all the herbs; you take [them] and you chop [them] inside your carrying bag.

(17) *Wolêp lik onga wope lak onga wope lêp suei; wolêp yokosek.*
wo$_A$=lêp [lik]$_O$ onga   wo$_S$=pe lak onga wo$_A$=pe lêp
2sg=take carrying.bag and.so 2sg=PFV go and.so 2sg=PFV take
[suei]$_O$ wo$_A$=lêp yokosek=Ø$_O$
short.yam 2sg=take mix=3sg.ZERO
You take the bag and you will go and take the short yams; you take and mix [them] [with the herbs].

(18) *Suei a mwayen a kila lalon lik teo.*
[suei   a   mwayen]$_S$ ya   ki-la   lalo-n
short.yam and yam   then IRR.3sg-go.to inside-PERT
lik   te-yo
carrying.bag LIG-DEM.INT
The long and short yams will go into the bag.

(19) *Wope yen antek, woantek la ai nop onga wopwa wololomêek onga*
wo$_A$=pe yen   antek=Ø$_O$ wo=antek=Ø$_O$ la
2sg=PFV PROG put.away=3sg.ZERO 2sg=put.away=3sg.ZERO go.to
a=i   nop   onga   wo$_S$=pwa   wo=lo.lomêek   onga
OBL=3sg mound and.so 2sg=want.to 2sg=REDUP.plant and.so
You will be distributing [them], you distribute [them] to the mounds. You are about to go planting, and...

(20) *Nganngan te kapwa wolomêek teo kiro pung nangin sapon kei reo, nangin paran mwalolon, te onga kipe sa pei la paran pian.*
[nganngan [te   kapwa   wo$_A$=lomêek=Ø$_O$]$_{RC}$ te-yo]$_A$ ki-to
food   REL if   2sg=plant=3sg=ZERO   LIG-DEM.INT   IRR.3sg-HAB
pung   [nangin   sapo-n   kei   te-yo   nangin   paran mwalolon]$_O$
smell   scent.of   top-PERT herb LIG-DEM.INT   scent.of really fragrant
[te   onga   ki$_S$-pe   sa   pei   la   paran pian]$_{AdvCl}$
SUB   and.so IRR.3sg-PFV come.up   appear   go.to   really good
The food that you will plant, it will smell the scent of the plant tops, the very fragrant scent, so that it will come up very well.

(21) *Aronan lolomêek teo, suei reo sôkôm a worou maran teo, kimwangmwang sa wat.*
arona-n   lo.lomêek   te-yo   [suei$_i$ te-yo]$_{TopPoss}$ sôkôm
procedure-PERT REDUP.plant   LIG-DEM.INT   short.yam   LIG-DEM.INT   some

ya wo_A=tou [mata-n_i te-yo]_O ki_S-mwang.mwang sa
then 2sg=put eye-PERT LIG-DEM.INT IRR.3sg-REDUP.watch come.up
[wat]_LOC
up.high
As for the way of planting: the short yams, some you will put [into the ground] with their eyes looking up. [*lit.* 'you put its eye, it will look come up']

(22) *Sôkôm a worou maran teo; kila luek paye.*
sôkôm ya wo_A=tou [mata-n te-yo]_O ki_S-la luek
some then 2sg=put eye-PERT LIG-DEM.INT IRR.3sg-go.to come.out
[paye]_LOC
down.below
Some, you will put [into the ground] with their eyes facing down. [*lit.* 'you put its eye, it will go down']

(23) *Ma ila ai i ramwayen, mwayen teo woro lomêeki la ai re koropun nganngan teo iro au sa wat.*
ma i=la a=i i ta=mwayen [mwayen te-yo]_TopO
but 3sg=go.to OBL=3sg 3sg SPEC.COLL=yam yam LIG-DEM.INT
wo_A=to lomêek=i_O la a=i te [koropu-n nganngan
2sg=HAB plant=3sg go.to OBL=3sg SUB base-PERT food
te-yo]_S i=to wau sa [wat]_LOC
LIG-DEM.INT 3sg=HAB walk come.up up.high
But when it comes to (long) yam, you will plant it in such a way that the bottom of the tuber is facing upwards.

(24) *A i reo. Mwayen teo iro ret la paye.*
a i te-yo [mwayen te-yo]_S i=to tet la
and 3sg LIG-DEM.INT yam LIG-DEM.INT 3sg=HAB grow go.to
[paye]_LOC
down.below
That's it. Long yam usually grows down.

(25) *A mwayen teo kumun kisa pei, a kanen kipe ret suwot paye.*
a [mwayen_i te-yo]_TopPoss [kumu-n_i]_S ki-sa pei
and yam LIG-DEM.INT sprout-PERT IRR.3sg-come.up appear
ya [kane-n_i]_S ki-pe tet suwot [paye]_LOC
then meat-PERT IRR.3sg-PFV grow go.down down.below
So as for the long yams, when their sprout will appear, their meat will grow downwards.

(26) *Ma suei reo, worou maran la paye, are kanen teo iro pei sa wat.*
ma   [suei$_i$ te-yo]$_{TopPoss}$ wo$_A$=tou [mata-n$_i$]$_O$ la   [paye]$_{LOC}$
but short.yam   LIG-DEM.INT   2sg=put   eye-PERT   go.to down.below
a=te   [kane-n$_i$ te-yo]$_S$ i=to   pei   sa   [wat]$_{LOC}$
OBL=SUB   meat-PERT   LIG-DEM.INT   3sg=HAB appear   come.up up.high
But as for short yams, you put their eyes downwards, as their meat usually grows upwards.

(27) *Ma worou suei suwot paye, a kanen kipe pei la menton.*
ma   wo$_A$=tou [suei]$_O$ suwot   [paye]$_{LOC}$ ya   [kane-n]$_S$
EMP   2sg=put   short.yam   go.down   down.below   then meat-PERT
ki-pe   pei   [la   menton]$_{Adv}$
IRR.3sg-PFV   appear   go.to many
So when you plant short yams facing downwards, then their meat [i.e. tubers] will become plenty.

(28) *Ma kapwa worou maran sa wat, a kanen teo kipe sa pei wat a kipe pei lano tupun.*
ma   kapwa   wo$_A$=tou [mata-n]$_O$ sa   [wat]$_{LOC}$ ya   [kane-n
but   if   2sg=put   eye-PERT   come.up   up.high   then meat-PERT
te-yo]$_S$ ki-pe   sa   pei   [wat]$_{LOC}$ a   ki$_S$-pe
LIG-DEM.INT   IRR.3sg-PFV   come.up   appear   up.high   and IRR.3sg-PFV
pei   la   [no   tupu-n]$_{Adv}$
appear   go.to only grandparent-PERT
But if you plant them facing upwards, then their meat will grow upwards and will only turn into seed tubers [*lit.* 'grandparents'].

(29) *Ma aronan pun teo worou luek, maran nganngan teo kisuwot liliu luek lalon lem paye.*
ma   arona-n   pun   te-yo   wo$_A$=tou=Ø   luek
but   procedure-PERT   INTF   LIG-DEM.INT   2sg=put=3sg.ZERO   come.out
[mata-n   nganngan   te-yo]$_S$ ki-suwot   liliu   luek
eye-PERT food   LIG-DEM.INT   IRR.3sg-go.down   again   come.out
[lalo-n   lem   paye]$_{LOC}$
inside-PERT   hole   down.below
But [according to] the right way of how you put [it], the 'face' of the tuber has to go down away from you into the hole.

(30) *Are kanen teo kipe pei la pian.*
a=te   [kane-n   te-yo]$_S$ ki-pe   pei   [la   pian]$_{Adv}$
OBL=SUB meat-PERT LIG-DEM.INT IRR.3sg-PFV appear go.to good
So that its meat will come up well.

(31) *Wolomêek nêm onga, raim te wosôpui onga kumun teo kipe sa wau la wat.*
wo$_A$=lomêek=Ø$_O$ nêm   onga   taim te   wo$_S$=sôpui
2sg=plant=3sg.ZERO be.finished and.so time REL 2sg=cover.hole
onga [kumu-n   te-yo]$_S$ ki-pe   sa   wau   la
and.so sprout-PERT LIG-DEM.INT IRR.3sg-PFV come.up move go.to
[wat]$_{LOC}$
up.high
You finish planting [it], and when you cover the hole its sprout will come up and grow high.

(32) *Taim te kisa pei onga wope lêp kap.*
taim te   ki$_S$-sa   pei   onga   wo$_A$=pe lêp   [kap]$_O$
time REL IRR.3sg-come.up appear and.so 2sg=PFV take stalk
When it comes up, you will take bamboo stalks.

(33) *Wola san kap onga wope yiniek la ai onga wope pang nou rea.*
wo$_A$=la   san   [kap]$_O$ onga   wo$_A$=pe yiniek=Ø$_O$ la
2sg=go.to cut stake and.so 2sg=PFV erect=3sg.ZERO go.to
[a=i]$_{LOC}$ onga   wo$_S$=pe pang nou   te-ya
OBL=3sg and.so 2sg=PFV feed central.stick LIG-then
You go and cut stalks and you will erect [them] at [the places where the yams are growing] and you will lead the yam vines onto them.

(34) *Wo pang nou a kumun teo kisa yop lêp ip kupwen kap teo.*
wo$_S$=pang nou   a   [kumu-n   te-yo]$_A$
2sg=feed central.stick and sprout-PERT LIG-DEM.INT
ki-sa   yop lêp [ip kupwe-n   kap te-yo]$_O$
IRR.3sg-come.up jump take 3pl branch-PERT stalk LIG-DEM.INT
You lead [them] and the vines will grow along the stalks.

(35) *Wope weiek lak to paran kap, are kiro wat.*
wo$_A$=pe weiek=Ø$_O$ lak to [pata-n   kap]$_{LOC}$ a=te
2sg=PFV coil=3sg.ZERO go be stick-PERT stalk   OBL=SUB

```
       kiS-to          [wat]LOC
       IRR.3sg-be      up.high
```
You will coil [them] to the stems of the stalks so that they are high up.

(36) *Kiro ret wat, pari ai re kisa suwot lalon ponat a sin teo kipe rap nêm ai ponat a ponat kipe kôk...*
```
       kiS-to          tet       [wat]LOC    pari            a=i     te      kiS-sa
       IRR.3sg-HAB     grow      up.high     belonging.to    OBL=3sg SUB     IRR.3sg-MOD
       suwot           [lalo-n             ponat]LOC   a    [sin te-yo]S       ki-pe
       go.down         inside-PERT         soil        and  sun LIG-DEM.INT    IRR.3sg-PFV
       tap             nêm            a=i        [ponat]LOC a    [ponat]S
       shine.strong    be.finished    OBL=3sg    soil       and  soil
       ki-pe           kôk
       IRR.3SG-PFV     heat.up
```
They should grow up high, because if they would stay down on the ground and the sun would shine strongly onto the soil and the soil would heat up...

(37) *A kipe run maran nganngan tao, a nganngan tao isa ro ret la pian pwên.*
```
       a      kiA-pe         tun      [mata-n     nganngan ta-o]O             a
       and    IRR.3sg-PFV    boil     eye-PERT    food     CLF.POSS-2sg.PERT  and
       [nganngan ta-o]S                i=sa         to      tet    [la     pian]Adv
       food     CLF.POSS-2sg.PERT      3sg=MOD      CONT    grow    go.to   good
       pwên
       NEG
```
And it would burn the buds of the yam plants, and then your tubers are not able to grow well.

(38) *Ma woroui ro wat a poyon teo kiro masai, wope ro rou malalan la pian, a kipe ret la no pian.*
```
       ma     woA=tou=iO       to      [wat]LOC    a     [pwoyo-n       te-yo]S
       but    2sg=put=3sg      be      up.high     and   under-PERT     LIG-DEM.INT
       ki-to              masai        woA=pe     to       tou     [malala-n]O    [la
       IRR.3sg-CONT       be.clear     2sg=PFV    CONT     give    clearness-PERT go.to
       pian]Adv    ya     kiS-pe tet         la     [no  pian]Adv
       good        then   IRR.3sg-PFV grow    go.to  only  good
```
But you put it up, and if the underground is clear, you will clear it nicely, then it will just grow very well.

(39) *A monokin a kila lom, a wope ngan tea.*
    a   monoki-n   ya   ki$_S$-la   lom   a   wo$_A$-pe
    and behind-PERT then IRR.3sg-go.to be.ready and 2sg-PFV
    ngan=Ø$_O$ te-ya
    eat=3sg.ZERO LIG-then
    And afterwards it will become ready for harvesting, and then you will eat [it].

(40) *Inêm.*
    i$_S$=nêm
    3sg=be.finished
    It's finished.

References

Aikhenvald, Alexandra Y. 2003. *Classifiers: a typology of noun categorization devices*. Oxford: Oxford University Press.
Aikhenvald, Alexandra Y. 2006. Serial Verb Constructions in typological perspective. In Alexandra Y. Aikhenvald & R. M. W. Dixon (eds.), *Serial Verb Constructions: A Cross-Linguistic Typology*, 1–68. Oxford: Oxford University Press.
Aikhenvald, Alexandra Y., & R. M. W. Dixon. 1998. Dependencies between grammatical systems. *Language* 74. 56–80.
Aikhenvald, Alexandra Y., & R. M. W. Dixon. 2013. *Possession and ownership: a cross-linguistic typology*. Oxford: Oxford University Press.
Ambrose, Wallace. 2002. From very old to new; obsidian artefacts in the Admiralty Islands. In Christian Kaufmann, Christin Kocher Schmid, & Sylvia Ohnemus (eds.), *Admiralty Islands: Art from the South Seas*, 67–72. Zurich: Museum Rietberg.
Ameka, Felix. 1992. Interjections: The universal yet neglected part of speech. *Journal of Pragmatics* 18. 101–118.
Arms, David G. 1973. Whence the Fijian transitive endings? *Oceanic Linguistics* 12(1/2). 503–558.
Baker, Mark, & Lisa DeMena Travis. 1998. Events, times, and Mohawk verbal inflection. *Canadian Journal of Linguistics* 43(2). 149–203.
Blust, Robert A. 1981. Some remarks on labiovelar correspondences in Oceanic languages. In J. Hollyman & Andrew Pawley (eds.), *Studies in Pacific languages in honour of Bruce Biggs*, 229–253. Auckland: Linguistic Society of New Zealand.
Blust, Robert A. 1998. Seimat vowel nasality: a typological anomaly. *Oceanic Linguistics* 37. 298–322.
Blust, Robert A. 2007. The prenasalised trills of Manus. In Jeff Siegel, John Lynch, & Diana Eades (eds.), *Language description, history and development. Linguistic indulgence in memory of Terry Crowley*, 297–312. Amsterdam: John Benjamins.
Blust, Robert A. 2008. A reanalysis of Wuvulu phonology. *Oceanic Linguistics* 47. 275–293.
Blust, Robert A. 2013. *The Austronesian languages* (2nd edn.). Canberra: Pacific Linguistics. https://openresearch-repository.anu.edu.au/bitstream/1885/10191/6/Blust-2013-AustronesianLanguages.pdf (accessed 8 July 2019).
Blust, Robert A., & Stephen Trussel. 2019. The Austronesian Comparative Dictionary: web edition. http://www.trussel2.com/ACD/ (last accessed 27 September 2017).
Boersma, Paul, & David Weenink. 2018. Praat: doing phonetics by computer (Version 6.0.43). Retrieved from http://www.praat.org/.
Boettger, Juliane. 2015. *Topics in the grammar of Lele: a language of Manus Island, Papua New Guinea*. Cairns: James Cook University dissertation.
Bowern, Claire. 2011. *Sivisa Titan: sketch grammar, texts, vocabulary based on material collected by P. Josef Meier and Po Minis*. Honolulu: University of Hawai'i Press.
Brownie, John, & Marjo Brownie. 2007. *Mussau grammar essentials*. Ukarumpa: SIL-PNG Academic Publications.
Budd, Peter. 2014. Partitives in Oceanic languages. In Silvia Luraghi & Tuomas Huumo (eds.), *Partitive cases and related categories*, 523–562. Berlin: De Gruyter Mouton.
Bühler, Alfred. 1935. Versuch einer Bevölkerungs- und Kulturanalyse auf den Admiralitätsinseln. *Zeitschrift für Ethnologie* 67(1). 1–32.

Burridge, Kate. 2002. Changes within Pennsylvania German grammar as enactments of anabaptist world view. In Nick J. Enfield (ed.), *Ethnosyntax*, 207–230. Oxford: Oxford University Press.

Bybee, Joan L., Revere D. Perkins, & William Pagliuca. 1994. *The evolution of grammar: tense, aspect, and modality in the languages of the world*. Chicago: University of Chicago Press.

Carrier, James G., & Achsah H. Carrier. 1989. *Wage, trade, and exchange in Melanesia: a Manus society in the modern state*. Berkeley: University of California Press.

Chafe, Wallace. 1976. Givenness, contrastiveness, definiteness, subjects and topics. In Charles N. Li (ed.), *Subject and topic*, 25–56. New York: Academic Press.

Chappell, Hilary, & William McGregor. 1989. Alienability, inalienability and nominal classification. *Proceedings of the Berkeley Linguistic Society*, 24–36.

Chappell, Hilary, & William McGregor. 1996. *The grammar of inalienability: A typological perspective on body part terms and the part-whole relation*. Berlin & New York: Walter de Gruyter.

Cleary-Kemp, Jessica. 2015. *Serial Verb Constructions revisited: A case study from Koro*. Berkeley: University of California dissertation.

Comrie, Bernard. 1976. *Aspect; an introduction to the study of verbal aspect and related problems*. New York: Cambridge University Press.

Comrie, Bernard. 2005. Endangered numeral systems. In J. Wohlgemuth & T. Dirksmeyer (eds.), *Bedrohte Vielfalt: aspects of language death*, 203–230. Berlin: Weißensee Verlag.

Crowley, Terry. 2002. *Serial verbs in Oceanic: a descriptive typology*. Oxford: Oxford University Press.

Cruse, Alan. 2006. *A glossary of semantics and pragmatics*. Edinburgh: Edinburgh University Press.

Cutler, Anne. 2012. *Native listening: Language experience and the recognition of spoken words*. Cambridge, MA: MIT Press.

Dalsgaard, Steffen. 2009. Claiming culture: new definitions and ownership of cultural practices in Manus Province, Papua New Guinea. *The Asia Pacific Journal of Anthropology* 10. 20–32.

Dalsgaard, Steffen, & Ton Otto. 2011. From kastam to kulsa? Leadership, cultural heritage and modernization in Manus Province, Papua New Guinea. In Edvard Hviding & Knut M. Rio (eds.), *Made in Oceania: social movements, cultural heritage and the state in the Pacific*, 141–160. Wantage: Sean Kingston Publishing.

Dik, Simon C. 1997. *The theory of functional grammar* (second revised edn.). Berlin: Mouton de Gruyter.

Dixon, R. M. W. 1988. *A grammar of Boumaa Fijian*. Chicago: University of Chicago Press.

Dixon, R. M. W. 2009. The semantics of clause linking in typological perspective. In R. M. W. Dixon & Alexandra Y. Aikhenvald (eds.), *The semantics of clause linking: a cross-linguistic typology*, 1–55. Oxford: Oxford University Press.

Dixon, R. M. W. 2010a. *Basic linguistic theory: Methodology* (Vol. 1). Oxford: Oxford University Press.

Dixon, R. M. W. 2010b. *Basic linguistic theory: Grammatical topics* (Vol. 2). Oxford: Oxford University Press.

Dixon, R. M. W. 2012. *Basic linguistic theory: Further grammatical topics* (Vol. 3). Oxford: Oxford University Press.

Durie, Mark. 1988. Verb serialization and "verbal-prepositions" in Oceanic languages. *Oceanic Linguistics* 27. 1–23.

Elliot, Jennifer R. 2000. Realis and irrealis: Forms and concepts of the grammaticalisation of reality. *Linguistic Typology* 4. 55–90.

Erteschik-Shir, Nomi. 2007. *Information structure: The syntax-discourse interface*. Oxford: Oxford University Press.

Evans, Bethwyn. 2003. *A study of valency-changing devices in Proto Oceanic*. Canberra: Pacific Linguistics.

Farr, Cynthia J. M. 1999. *The interface between syntax and discourse in Korafe: A Papuan language of Papua New Guinea*. Canberra: Pacific Linguistics.

Fillmore, Charles J. 1976. Pragmatics and the description of discourse. In S. Schmidt (ed.), *Pragmatik II*. Munich: Wilhelm Fink Verlag.

Fortune, Reo F. 1965. *Manus religion: an ethnological study of the Manus natives of the Admiralty Islands*. Lincoln: University of Nebraska Press.

Gabelentz, Georg von der. 1869. Ideen zu einer vergleichenden Syntax. Wort und Satzstellung. *Zeitschrift für Völkerpsychologie und Sprachwissenschaft* 6. 376–384.

Hafford, James A. 1999. *Elements of Wuvulu grammar*. Arlington: University of Texas MA thesis.

Hafford, James A. 2014. *Wuvulu grammar and vocabulary*. Honululu: University of Hawai'i dissertation.

Hamel, Patricia J. 1994. *A grammar and lexicon of Loniu, Papua New Guinea*. Canberra: Pacific Linguistics.

Heine, Bernd, & Tania Kuteva. 2002. *World lexicon of grammaticalization*. Cambridge: Cambridge University Press.

Hengeveld, Kees. 1992. *Non-verbal predication: Theory, typology, diachrony*. Berlin: Mouton de Gruyter.

Holt, J. 1943. Etudes d'aspect. *Acta Jutlandica* 15(2). 1–13.

Keenan, Edward L., & Bernard Comrie. 1977. Noun phrase accessibility and universal grammar. *Linguistic Inquiry* 8. 63–99.

Kiyomi, Setsuko. 1995. A new approach to reduplication: a semantic study of noun and verb reduplication in the Malayo-Polynesian languages. *Linguistics* 33. 1145–1167.

Krifka, Manfred. 2008. Basic notions of information structure. *Acta Linguistica Hungarica* 55(3–4). 243–276.

Ladefoged, Peter, & Ian Maddieson. 1996. *The sounds of the world's languages*. Oxford: Blackwell Publishers.

Lambrecht, Knud. 1994. *Information structure and sentence form: Topic, focus, and the mental representations of discourse referents*. Cambridge: Cambridge University Press.

Levinson, Stephen C. 2003. *Space in language and cognition: Explorations in cognitive diversity*. New York: Cambridge University Press.

Levinson, Stephen C., P. Brown, E. Danziger, L. De León, J.B. Haviland, E. Pederson, & Gunter Senft. 1992. Man and tree & space games. In Stephen C. Levinson (ed.), *Space stimuli kit 1.2: November 1992*. Nijmegen: Max Planck Institute for Psycholinguistics.

Levinson, Stephen C., & David Wilkins. 2006. *Grammars of space: Explorations in cognitive diversity*. Cambridge: Cambridge University Press.

Lévy-Bruhl, Lucien. 1914. L'expression de la possession dans les langues mélanésiennes. *Mémoires de la Société de Linguistique de Paris* 19(2). 96–104.

Lichtenberk, Frantisek. 1991. Semantic change and heterosemy in grammaticalization. *Language* 67(3). 475–509.

Lichtenberk, Frantisek. 1995. Apprehensional epistemics. In Joan L. Bybee & Suzanne Fleischman (eds.), *Modality in grammar and discourse*, 293–327. Amsterdam: John Benjamins.
Lichtenberk, Frantisek. 2000. Inclusory pronominals. *Oceanic Linguistics* 39. 1–32.
Lichtenberk, Frantisek. 2006. Associative and possessive constructions in Oceanic: The links and the differences. In H.Y. Cheng, L.M. Huang, & D. Ho (eds.), *Streams converging into an ocean: Festschrift in honor of Professor Paul Jen-Kuei Li on his 70th birthday*, 19–47. Taipei: Academia Sinica.
Lichtenberk, Frantisek. 2009a. Attributive possessive constructions in Oceanic. In William McGregor (ed.), *The expression of possession*, 249–292. Berlin: Mouton de Gruyter.
Lichtenberk, Frantisek. 2009b. The semantics of clause linking in Toqabaqita. In R.M.W. Dixon & Alexandra Y. Aikhenvald (eds.), *The semantics of clause linking: a cross-linguistic typology*, 239–260. Oxford: Oxford University Press.
Lynch, John. 1998. *Pacific languages: An introduction*. Honolulu: The University of Hawai'i Press.
Lynch, John. 2002. The Proto-Oceanic labiovelars: some new observations. *Oceanic Linguistics* 41(2). 310–362.
Lynch, John, Malcolm Ross, & Terry Crowley. 2002. *The Oceanic languages*. Richmond, Surrey: Curzon.
Marantz, Alec. 1982. Re Reduplication. *Linguistic Inquiry* 13. 435–482.
Margetts, Anna, & Peter Austin. 2007. Three-participant events in the languages of the world: Towards a crosslinguistic typology. *Linguistics* 45(3). 393–451.
McCloy, Daniel R. 2016. phonR: tools for phoneticians and phonologists. R package version 1.0-7.
McGregor, William. 2009. *The expression of possession*. Berlin: Mouton de Gruyter.
Mead, Margaret. 1930. *Growing up in New Guinea: a comparative study of primitive education*. New York: William Morrow.
Mead, Margaret. 1934. *Kinship in the Admiralty Islands*. New York: The American Museum of Natural History.
Mead, Margaret. 1956. *New lives for old: cultural transformation – Manus, 1928–1953*. New York: William Morrow.
Meier, P. Josef. 1907–1912. Mythen und Sagen der Admiralitätsinsulaner. *Anthropos 2, 3, 4, 7*. 646–667, 933–941, 193–206, 651–671, 354–674, 501–502.
Mithun, Marianne. 1984. The evolution of noun incorporation. *Language* 60. 847–894.
Mithun, Marianne. 1999. *The languages of native North America*. Cambridge: Cambridge University Press.
Mithun, Marianne. 2001. Understanding and explaining applicatives. *Chicago Linguistic Society Papers* 37(2). 73–97.
National Statistical Office Papua New Guinea. 2014. *National population & housing census 2011*. Port Moresby: NSOPNG.
Nettle, Daniel. 1999. *Linguistic diversity*. New York: Oxford University Press.
Nevermann, H. 1934. *Ergebnisse der Südsee-Expedition 1908–1910 – Admiralitäts-inseln*. Friedrichsen: De Gruyter.
Nishida, Hiroko. 1999. A cognitive approach to intercultural communication based on schema theory. *International Journal of Intercultural Relations* 23(5). 753–777.
Ohnemus, Sylvia. 2002. *An ethnology of the Admiralty Islanders*. Bathurst: Crawford House.

Otto, Ton. 1991. *The politics of tradition in Baluan*. Canberra: The Australian National University dissertation.
Otto, Ton. 1992. The ways of kastam: tradition as category and practice in a Manus village. *Oceania* 62. 264–283.
Otto, Ton. 1997. Informed participation and participating informants. *Canberra Anthropology* 20. 96–108.
Otto, Ton. 2002. Manus: the historical and social context. In Christian Kaufmann, Christin Kocher Schmid, & Sylvia Ohnemus (eds.), *Admiralty Islands: Art from the South Seas*, 29–39. Zurich: Museum Rietberg.
Otto, Ton. 2008. Fieldwork in Manus, Papua New Guinea: on change, exchange and anthropological knowledge. In J. Kommers & E. Venbrux (eds.), *Cultural styles of knowledge transmission. Essays in honour of Ad Borsboom*. Amsterdam: Aksant.
Otto, Ton, & Poul Pedersen. 2005. *Tradition and agency: tracing cultural continuity and invention*. Aarhus: Aarhus University Press.
Pawley, Andrew. 1973. Some problems in Proto-Oceanic grammar. *Oceanic Linguistics* 12. 102–188.
Pawley, Andrew. 2003. Grammatical categories and grammaticisation in the Oceanic verb complex. In Anastasia Riehl & Thess Savella (eds.), *Cornell Working Papers in Linguistics*, 149–172. Ithaca, NY: Cornell University.
R Core Team. 2017. R: A Language and Environment for Statistical Computing. Vienna: R Foundation for Statistical Computing. Retrieved from https://www.R-project.org/.
Ross, Malcolm. 1988. *Proto Oceanic and the Austronesian languages of Western Melanesia*. Canberra: Pacific Linguistics.
Schokkin, Dineke. 2013. Directionals in Paluai: semantics, use, and grammaticalization paths. *Oceanic Linguistics* 52(1). 169–191.
Schokkin, Dineke. 2014. Discourse practices as an areal feature in the New Guinea region? Explorations in Paluai, an Austronesian language of the Admiralties. *Journal of Pragmatics* 62. 107–120.
Schokkin, Dineke. 2017. Contact-induced change in an Oceanic language: The Paluai – Tok Pisin case. *Journal of Language Contact* 10(1). 76–97.
Schokkin, Dineke. 2018. Language Contexts: Paluai, also known as Pam-Baluan (Papua New Guinea). In Peter Austin (ed.), *Language Documentation and Description* 15. 65–86.
Schokkin, Dineke. Forthcoming. Preverbal directionals as markers of associated motion in Paluai (Austronesian; Oceanic). In Antione Guillaume & Harold Koch (eds.), *Associated Motion*. Berlin: De Gruyter Mouton.
Schokkin, Dineke, & Ton Otto. 2017. Relatives and Relations in Paluai. *Oceanic Linguistics* 56(1). 226–246.
Schooling, Stephen, & Janice Schooling. 1980. A preliminary sociolinguistic and linguistic survey of Manus Province, Papua New Guinea. *Papers in New Guinea Linguistics* 26. 211–241.
Schwartz, Theodore. 1963. Systems of areal integration: some considerations based on the Admirality Islands of Northern Melanesia. *Anthropological Forum* 1. 56–97.
Sgall, Petr, Eva Hajičová, Jarmila Panevová, & Jacob Mey. 1986. *The meaning of the sentence in its semantic and pragmatic aspects*. Dordrecht: D. Reidel.
SIL International. 2017. Toolbox (Version 1.6.2). Retrieved from https://software.sil.org/toolbox/download/.

Simons, Gary F., & Charles D. Fennig. 2018. Ethnologue: Languages of the world, twentieth edition (online version). http://www.ethnologue.com (last accessed 20 December 2018).

Smith-Stark, Thomas C. 1974. The plurality split. *Papers from the 10th regional meeting of the Chicago Linguistic Society.* 657–671.

Stassen, Leon. 1997. *Intransitive Predication.* Oxford: Clarendon Press.

Stutzman, Verna. 1997. *A study of the Lou verb phrase.* Winnipeg: University of Manitoba MA thesis.

Tannen, Deborah, & Cynthia Wallat. 1993. Interactive frames and knowledge schemas in interaction: Examples from a medical examination/interview. In Deborah Tannen (ed.), *Framing in discourse,* 57–76. New York/Oxford: Oxford University Press.

Timberlake, Alan. 2007. Aspect, tense, mood. In Timothy Shopen (ed.), *Language typology and syntactic description,* 280–333. Cambridge: Cambridge University Press.

Trubetzkoy, Nikolai S. 1969. *Principles of phonology.* Berkeley: University of California Press.

Vries, Lourens de. 2006. Areal pragmatics of New Guinea: Thematization, distribution and recapitulative linkage in Papuan narratives. *Journal of Pragmatics* 38(6). 811–828.

Wanek, Alexander. 1996. *The state and its enemies in Papua New Guinea.* Richmond: Curzon Press.

Wilkins, David, & Deborah Hill. 1995. When "go" means "come": questioning the basicness of basic motion verbs. *Cognitive Linguistics* 6. 209–259.

Wozna, Beata, & Theresa Wilson. 2005. *Seimat grammar essentials.* Ukarumpa: SIL-PNG Academic Publications.

Subject Index

adjectives
- and stative verbs 86–88, 103
- as predicate head 227–228
- criteria to distinguish 116
- place in the NP 103, 168–169
- semantic classes of 102–103

Admiralties languages 1, 3, 4–5, 9, 14, 24, 32, 43, 47, 59, 60, 133, 204
ambitransitive verbs 84, 249–250
anaphora 124, 126, 256, 365, 370–376
animacy
- and argument marking 84, 86, 122, 129–133, 178–180, 238–250
- and number marking 58, 165–166
- and numeral classifiers 76–79, 133–137

applicatives 43, 83, 240–241, 243–246, 255–265, 269–271
articles, lack of 41, 58, 59
aspect 85, 91, 95–96, 193–196
- and reality status/modality 217–221
- core aspect 181–189
- secondary aspect 189–193

bound pronouns
- and indexing 176–180
- and parts of speech 116
- paradigms 121–122

cataphora 370–376
causatives 251–255
classifiers
- numeral classifiers 76–79, 133–136, 139–140
- possessive classifiers 79–80, 122–123, 170–171, 228–229

clitics 122, 125, 128, 129, 157–158, 162, 207, 337
- and stress assignment 45–47
- bound pronoun clitics. See bound pronouns

comitative 69, 120, 233, 278
comparative constructions 116, 129–130, 235–237
complement clauses 210–212, 307, 337–345
compounding
- and nouns 80–81
- and verbs 94
conditional clauses 213–214, 354–358
consonants 8–21, 33, 35, 37–39, 44, 54–56
- thematic consonants 261–265
constituent order 151, 238, 247, 302, 306, 377, 381
coordination 151–153, 173–174, 358–361
copulas 232–235
core arguments, marking of 238–246

definiteness 46, 125–126, 137, 140–142, 146, 147, 157–158, 335, 364, 367–369
deixis 112
- and demonstratives 124–129, 226, 372–376
- and directionals 88–90, 197, 281–289, 299

demonstratives
- as definiteness markers 124–126
- paradigms 129

derivational processes 43–45, 103, 105, 115, 117, 158–163
- and nouns 20, 80–83
- and verbs 41, 86–88, 93–101, 251–267, 269–270

diphthongs 22, 23, 26–28
directionals
- and argument marking 243–244, 284–289
- as preverbal particles 196–201
- in SVCs 281–289
- paradigm 88–90

discourse topics 178, 368, 375, 388–389, 391
ditransitive verbs 86, 243–246
dual number 120, 165

epenthesis 26, 28, 39
existential predicates 16–17, 317–318

focus 46, 52–54, 303, 381–386
fortition 20, 32
fricatives 10, 11–12, 14, 20–21, 30–31, 34, 37–39

gender marking, absence of 119, 124, 370
grammaticalisation 110, 115, 212, 234, 265
– and aspect markers 91, 181, 185, 189, 195–196
– and SVCs 298–300

imperative 308–313
imperfective aspect 180, 182–183, 190, 208, 218–219, 319
inclusory pronouns 173–174
intensity 40, 46–54
interrogative 146–150, 231–232, 302–308, 312–313, 319, 335, 340
intonation 46, 48–54, 310
intransitive verbs 84, 85, 86–90, 95–99, 100–101, 116, 197, 247, 250, 251, 260–262, 265–267, 276, 277, 281–284, 290–293
– extended intransitive verbs 240–243, 258, 270
irrealis. See reality status

labialisation 15–17, 18–19, 24, 30, 33–34
laterals 11, 13–14, 21, 35
left-dislocation. See topicalisation
lenition 14, 24, 34
lexicalisation 94, 259, 291–294, 298
light verbs 92–93, 252
loan words 37–39, 100, 110, 228, 257, 314, 317–318, 342–344, 347–348
local nouns 59, 246–248, 282, 285
Loniu language 5, 29, 32, 47, 76, 77, 90, 174, 189
Lou language 3, 5, 24

modality 113–115, 202, 209–221, 317, 324, 339, 341, 342–344, 351, 352–353
morphophonological processes 34–37

nasals 13, 14, 15, 17–19, 22, 30, 34, 35
negation 215–217, 222, 296, 307, 311–312, 313–323, 344, 361
nominalisation 44–45, 73, 88, 93, 94–97, 99–102, 105, 130, 252, 300–301
non-verbal predicates 86, 104, 138, 141, 160–161, 225–237, 303–304, 315, 344–345

noun phrase 58, 99, 104–105, 118, 139, 164–174, 321, 367–368
– structure of 164–165
numeral classifiers. See classifiers
numerals 76–78, 133–138, 140, 141, 145, 158, 161–162, 167–168, 228

partitive 143–144, 158–163
passive voice, lack of 85, 381
paucal number 119, 120, 165–166, 206
perfect aspect 185–189, 207, 208, 319
perfective aspect 180, 184–185, 187, 207–208, 217, 319
pitch 40, 46–54
plosives 10, 11–17, 30–31, 34, 37, 39
possessive classifiers. See classifiers
possessive constructions 59, 80–81, 96, 99–100, 131–133, 148–149, 228–231, 232–233, 315, 318, 333, 368–369
– direct possession 61–74, 169–170
– indirect possession 122–123, 169, 170–171
– semantics of 60–76, 79–80
prefixation 58, 201–202, 251–252, 258, 266–267
prepositions 59, 68, 129–133, 161, 171–172, 230–231, 298–300
Proto-Oceanic 43, 74, 80, 88, 120, 133, 134, 135, 157, 158, 251, 256, 258, 262, 264–265, 266, 269

reality status 176, 181, 185, 201–209, 222, 233, 310–312, 339, 340, 341, 342–345, 354–357
– and modality 209–210, 212–214, 217–221
– and negation 216–217, 317
– and past habitual 204–206
– formal expression of 202–203
reciprocal 258–259, 269–271
reduplication 13, 20, 41, 43–45, 104–105, 115, 116
– and nouns 80–81
– and verbs 94–98, 265–266, 300–301
reflexive 260, 268–269
relative clauses 106, 325, 356
– functions of the common argument 326–337

– position relative to the noun 164
rhotics 13–14

Seimat language 5, 76, 174
serialisation 94, 272–274
– adverbial 276–278
– and lexicalisation 289–294
– and wordhood 300–301
– cause-effect 274–276
– formal properties 294–297
– valency-increasing 278–279
– with a directional 196, 281–289, 297–298, 310
– with a posture verb 279–281
spatial nouns 68–70, 247
specificity 74–76, 81, 136–137, 142, 146, 157–158, 161, 163, 165–168, 202, 206–207, 325, 369
stative verbs 95, 109, 185, 186–18, 191, 194, 251–255, 266–267, 274–278
– and adjectives 86–88, 102–103, 235
– criteria for distinguishing 116
stops. *See* plosives
stress 24–25, 33, 40–48, 118–119, 126, 196, 303
suffixation 105, 122–123, 133–136, 170–171
– and (morpho)phonological processes 17, 18, 20, 24, 35–37, 43, 44
– suffixes on nouns 60–61, 74, 80–81, 83, 169–170

– suffixes on verbs 93, 99–100, 251, 255–265
syllabification 8–9, 20, 26–28, 56
– and vowel-glide alternation 24–26

tail–head linkage 360
tense, lack of 84, 180, 184, 203
Titan language 1, 3, 4, 5, 174
Tok Pisin language 3, 7, 37–39, 100, 129, 136, 138, 152, 195, 208, 228, 257, 314, 348, 351, 355, 384, 385
– grammatical borrowing from 85, 92, 315, 317–318, 342–344, 347–348
topicalisation 46, 177, 362, 365, 370, 376–381, 384, 391
transitive verbs 85–86, 93, 95, 96, 99–100, 195, 238–239, 242–243, 248–251, 254, 261, 268–269, 292–293, 320

valency 238–239, 248–249
– increasing valency 251–259, 278–279
– reducing valency 94–95, 265–271
verb complex 107, 112, 118, 121–122, 176, 202, 238, 247, 280, 297, 314–315
– structure of 221–224
vowels 8, 9, 15, 18, 20, 21–29, 31–32, 36–37, 39, 40, 41, 54, 60, 103, 123, 182

zero derivation 100–101, 106

Author Index

Aikhenvald, Alexandra 62, 79, 84, 272, 273, 289, 367
Ambrose, Wallace 4
Ameka, Felix 156, 157
Arms, David 262
Austin, Peter 284

Baker, Mark 206
Blust, Robert 4, 5, 9, 15, 18, 32, 58, 60, 135
Boersma, Paul 29
Boettger, Juliane 5, 112
Bowern, Claire 5, 174
Brownie, John and Marjo 5, 76
Budd, Peter 158
Bühler, Alfred 4
Burridge, Kate 217
Bybee, Joan 180, 187, 190

Carrier, James and Achsah 4
Chafe, Wallace 363
Chappell, Hilary 62
Cleary-Kemp, Jessica 5, 204
Comrie, Bernard 136, 180, 182, 184, 185, 189, 336
Crowley, Terry 1, 4, 5, 32, 41, 58, 59, 76, 79, 80, 120, 134, 158, 273, 273, 281
Cruse, Alan 124, 370, 372
Cutler, Anne 47

Dalsgaard, Steffen 4
de Vries, Lourens 164, 360
Dik, Simon 363, 365, 366
Dixon, R.M.W. 6, 62, 85, 102, 119, 175, 184, 189, 192, 238, 239, 241, 244, 255, 305, 325, 337, 345, 367
Durie, Mark 299

Elliot, Jennifer 209
Erteschik-Shir, Nomi 365
Evans, Bethwyn 43, 95, 251, 252, 253, 256, 258, 260, 264, 266

Farr, Cynthia 389, 390, 391
Fennig, Charles 1, 3
Fillmore, Charles 362
Fortune, Reo 4

Hafford, James 5
Hajičová, Eva 362
Hamel, Patricia 5, 29, 32, 47, 75, 76, 77, 90, 174, 189, 204
Heine, Bernd 299, 368
Hengeveld, Kees 225, 234
Hill, Deborah 88
Holt, J. 180

Keenan, Edward 336
Kiyomi, Setsuko 82
Krifka, Manfred 363, 367
Kuteva, Tania 299, 368

Ladefoged, Peter 24
Lambrecht, Knud 362, 363, 364, 365, 366, 375, 381, 384
Levinson, Stephen 6, 89, 90, 126, 221, 383
Lévy-Bruhl, Lucien 62
Lichtenberk, Frantisek 60, 62, 72, 74, 170, 173, 197, 212, 216, 270, 290, 359
Lynch, John 1, 4, 5, 15, 18, 32, 41, 58, 59, 76, 79, 80, 120, 134, 158

Maddieson, Ian 24
Marantz, Alec 43
Margetts, Anna 284
McGregor, William 62
Mead, Margaret 4
Meier, P. Josef 5
Mey, Jacob 362
Mithun, Marianne 93, 202, 255

Nettle, Daniel 1
Nevermann, H. 4
Nishida, Hiriko 365

Ohnemus, Sylvia 4
Otto, Ton 4, 5, 6, 63, 76, 79

Pagliuca, William 180
Panevová, Jarmila 362
Pawley, Andrew 176, 253
Pedersen, Poul 4, 79
Perkins, Revere 180

Ross, Malcolm 1, 3, 4, 5, 32, 41, 58, 59, 76, 79, 80, 120, 133, 134, 135, 158

Schokkin, Dineke 1, 3, 59, 63, 90, 164, 197, 358, 360
Schooling, Stephen and Janice 5
Schwartz, Theodore 4
Sgall, Petr 362
Simons, Gary 1, 3

Smith-Stark, Thomas 166
Stassen, Leon 84, 86, 88, 227, 232

Tannen, Deborah 364
Timberlake, Alan 186, 190, 209, 210, 215, 344, 354
Travis, Lisa DeMena 206
Trubetzkoy, Nikolai 35

von der Gabelentz, Georg 378

Wallat, Cynthia 364
Wanek, Alexander 4
Weenink, David 29
Wilkins, David 88, 90
Wilson, Theresa 5, 76, 174
Wozna, Beata 5, 76, 174